# NATURAL TURF FOR SPORT AND AMENITY
## Science and Practice

# Natural Turf for Sport and Amenity
# Science and Practice

W.A. Adams

*Institute of Biological Sciences*
*University of Wales*
*Aberystwyth*
*UK*

and

R.J. Gibbs

*New Zealand Turf Culture Institute*
*Fitzherbert Science Centre*
*Palmerston North*
*New Zealand*

with contributions from

T.R.O. Field, E.J.P. Marshall and J.R. Watson

CAB INTERNATIONAL

CAB INTERNATIONAL
Wallingford
Oxon OX10 8DE
UK

Tel: Wallingford (0491) 832111
Telex: 847964 (COMAGG G)
Telecom Gold/Dialcom: 84: CAU001
Fax: (0491) 833508

A catalogue entry for this book is available from the
British Library.

ISBN 0 85198 720 6

Typeset by Colset Pte Ltd, Singapore
Printed and bound in Great Britain at the University Press, Cambridge

# Contents

# FOREWORD

If you, the prospective purchaser or borrower, happen to read this foreword before making your decision, I say, 'Don't hesitate: you are getting a first-class work of reference by two first-class authors'. If, as is more likely, you come across these words in the course of reading or referring to this book, let me assure you, in more leisurely detail, that you have at your disposal a wide-ranging, up-to-date and really authoritative textbook, written primarily from a UK standpoint by two world-renowned authors. Bill Adams has been involved in turfgrass research and advisory work for 30 years, coming to it at first with the special expertise of soil science and developing a national reputation second to none, recognized by a special award from the National Turfgrass Council in 1990. Richard Gibbs also made his mark in turfgrass research and education before moving to New Zealand; he worked with Bill Adams at Aberystwyth and undertook some classic work to compare winter pitch constructions in terms of playing quality and cost benefit.

Over recent years there have been produced in the UK, a number of descriptive manuals and alphabetical reference works on practical turfgrass management and problems, but no scientifically grounded textbooks to stand comparison with those emanating from the USA. I believe that this book remedies this deficiency. Readers in the UK at every level – and especially students – will appreciate text and examples unquestionably applicable to their own conditions. But equally readers in the USA and many other countries will welcome and benefit from a clear presentation of the best current thinking from a country long renowned for its natural turf.

The care of intensively-managed sportsturf becomes ever more complex and demanding, as the numbers – and expectations – of players increase. The pressures on amenity grassland become ever more severe. Research is still needed in many sectors, but most of all is needed the direct translation of existing scientific knowledge into practical terms. This book comes just at the right time to fill this gap. New educational qualifications, and new awareness of proper professional standards among greenkeepers and groundstaff, demand new and reliable works of reference. This book will give everyone – from student to contractor and manager – an invaluable foundation of good basic information and well-proven practical guidance. I commend it to all those working with turfgrass and amenity grass.

JOHN SHILDRICK
*Chief Executive*
*National Turfgrass Council*
*UK*

# PREFACE

Turf culture whether for sport or amenity is still considered by many to be a craft rather than a science. This is not totally unjustified because a few head groundsmen/greenkeepers still try to maintain a mystique by being secretive about apparently successful cultural practices. Moreover most scientists give a low rating to any research devoted to improving recreational facilities.

Sportsturf research is usually considered as 'near market' because it is perceived that findings can be applied immediately and are of commercial benefit. To some extent this is true, but history has shown that, on occasions, research which had the greatest impact received negligible funding from industry because the opportunity for commercial exploitation was questionable. Probably the best example was in the 1970s when research and development on soil and drainage design for sportsturf, which received little funding, resulted in a major improvement in the quality of winter games pitches at all levels. We are fortunate in the UK that the Sports Council has not only helped support research which would otherwise have been held back, but has also helped to disseminate information and advice to providers of facilities for sport.

The culture of turf for sport of all kinds is now an advanced technology built on areas of research and development as diverse as plant physiology and applied soil physics. The use of modern cultivars of turfgrasses and the implementation of developments in construction and maintenance have greatly improved the quality and wear tolerance of sportsturf. Recently much attention has been paid to establishing criteria for playing quality not only in the interests of player enjoyment and safety but also in order to prescribe the standards to be achieved in the construction or maintenance of turfgrass facilities. The requirement for compulsory competitive tendering for Local Authority contracts has been a major spur in this area.

Natural turf fulfils valuable roles as a playing surface for sport and as an amenity for relaxation and enjoyment. Public appreciation of the quality of the environment favours natural over synthetic turf but there are legitimate concerns over, for example, the extravagant use of irrigation water and high inputs of fertilizers and biocides in some categories of sportsturf. Inputs, especially when their effect is purely cosmetic, will have to be decreased. Also the potential for golf courses in particular to enhance environmental quality through biodiversity should be exploited more generally.

The main aim of this book is to provide an understanding of the science upon which technical developments in sportsturf construction and maintenance have been built and to explain the character and basis for current practices and procedures. Research data on soils and grasses relevant to turf culture are presented. The authors have been mindful of the needs of teachers and students of turf culture for sport or amenity especially those requiring sound technical knowledge.

In addition to its use as a textbook for academic training up to degree level and for professional or vocational qualifications, the book provides information required by consultants, contractors, managers of sports facilities and practising groundstaff and greenkeepers.

The first five chapters explore the science background of soils and construction materials, turfgrasses, the theory and practice of sportsfield drainage and principles of turf establishment and maintenance. These are followed by four chapters dealing in detail with different games or sports: soccer and rugby, golf and bowls, cricket and tennis and horse racing. The penultimate chapter describes the amenity uses of turf and the final chapter introduces grass species and cultivars used in tropical areas and the characteristics which make them suitable for use in particular circumstances.

# Acknowledgements

The authors are grateful to Elaine Lowe who typed the manuscript and Rebecca Adams, Bill Corrie and Mohammed Afzal who assisted with the preparation of figures. Keith McAuliffe and John Shildrick are thanked, along with others who made valuable and constructive comments on the text.

Several organizations assisted in a variety of ways. Myerscough College allowed the use of facilities by Richard Gibbs. The Sports Turf Research Institute allowed us to reproduce some technical information. The New Zealand Turf Culture Institute assisted in reviewing some chapters and provided photographs. The United States Golf Association Green Section allowed us to reproduce the latest specifications on golf green design and the Sports Council permitted us to use unpublished research data.

Finally we thank three companies who through their financial assistance in the production of the book allowed us to include colour photographs without affecting the price of the publication. These are Boughton Loam Ltd, Rufford Top-dress Supplies Ltd and Surrey Loams Ltd.

# 1 SOIL CONSTITUENTS AND PROPERTIES IN TURFGRASS SYSTEMS

## INTRODUCTION

Soils support the world's natural vegetation. Where this has been altered or replaced by man they support arable crops, forests and grassland used for agriculture or recreation. Evocative terms such as 'Mother Earth' and 'the living soil' reflect the importance attached to this shallow covering on most of the world's land surface.

Most definitions of soil stress its function in supporting plant life, but life within the medium is equally important. The essential property of a soil is that it provides an environment for the biological cycling of carbon. There are two components to this cycling: (i) the fixation of carbon by photosynthetic organisms (usually plants); (ii) the decomposition of carbon compounds derived from these by heterotrophic organisms (mainly microorganisms) living within the medium. There are many other aspects to soil, but a recognition of its central importance as an environment facilitating the biological cycling of carbon clarifies the distinction between a soil and a material from which a soil may form. It is important in turfgrass science to restrict the term 'soil' to the *in situ* material supporting turf to avoid confusion with the wide range of materials which may be used to ameliorate existing soils or to produce media from which soils form.

Soils vary morphologically and in physical and chemical properties on a local and regional scale. Local differences are usually due to either variations in the soil parent material (the rock or deposit from which the soil formed), or topography, or the history of land use, or a combination of these. On a regional or world scale differences in climate and the age of soils, in addition to the other factors, can cause major variations. The interplay of the several

1

soil-forming factors results not only in major differences between soils on a local scale, but also in similar soils occurring in different geographical areas.

The fabric of soils comprises mineral particles and organic matter. Mineral particles usually have a dominant effect on the framework of soils, but organic matter begins to control physical behaviour at quite small contents by weight. The mineral and organic constituents of soils are not distributed randomly. The type and quality of organization (termed soil structure) varies and can be changed very quickly in sportsfields subjected to intensive use. The distribution of primary particle sizes and their association into aggregates in a soil determine both the amount and size distribution of the pore space. This space is available for air or water storage and pore size distribution determines the air–water balance in soils. Total porosity is increased when particles aggregate to produce crumbs which are porous and create large pores between them. Very small total porosities occur when there are no aggregates and small primary particles inter-pack with larger ones.

In addition to providing plants with water, oxygen and physical support, soils supply the 13 essential mineral nutrients required by plants. These are taken up from the soil solution as dissolved cations or anions and knowledge of the mechanisms by which soils replenish the nutrients in solution is vital to an understanding of nutrient availability. Mechanisms include physical, chemical and biological processes and not only are different processes involved with different nutrients, but also the relative importance of the processes differs between soils.

## CONSTITUENTS OF SOILS

## Mineral constituents

The inorganic or mineral constituents of soils can be classified according to size and mineralogy. Most physical properties of a soil are controlled or affected by the size distribution of its particles. Mineralogy, that is the chemical composition and structure of particles, not only affects physical properties, but also influences the retention and supply of plant nutrients.

### Size distribution of mineral particles

In most soils it is the particles smaller than 2 mm which affect behaviour. For this reason particle size classes are based on the size distribution of the fine earth, that is soil sieved through a 2 mm aperture. Sand, silt and clay are the names given to the particle size categories. The size limits for these categories differ somewhat between different systems, as do subdivisions between the categories. The system in general use in Britain is that of the Soil Survey of

**Table 1.1.** Particle size grade categories for sportsturf soils.

| Description | Very coarse sand | Coarse sand | Medium sand | Fine sand |
|---|---|---|---|---|
| Particle size range (diameter in $\mu$m) | 2000-1000 | 1000-500 | 500-250 | 250-125 |

| Description | Very fine sand | Coarse silt | Fine silt | Clay |
|---|---|---|---|---|
| Particle size range (diameter in $\mu$m) | 125-60 | 60-20 | 20-2 | less than 2 |

**Table 1.2.** Typical commercial grades for particles larger than 2 mm in diameter.

| Description | Fine gravel | Coarse gravel | Fine stone | Medium stone | Coarse stone |
|---|---|---|---|---|---|
| Particle size range (diameter in mm) | 2-5 | 5-10 | 10-20 | 20-50 | 50-100 |

England and Wales (Hodgson, 1974). Unfortunately, this provides inadequate subdivision of the sand category for sportsturf use. For this reason the American system for subdividing the sand categories is usually combined with the British grades for silt and clay. Recommended categories and subdivisions within categories are shown in Table 1.1.

In many soils the proportion of gravel and stones, that is particles larger than 2 mm, is too large to be ignored and even small proportions must be recorded in cricket square topdressings. Small amounts of gravel (less than 10% by weight) have a negligible effect on physical properties and are best quoted along with sand, silt and clay to give a total of 100%. When gravel and stone content exceeds 10% of the soil it is advisable to quote the particle size distribution of the fine earth as normal and report separately the gravel plus stone content of the total soil. Recommended size grades for gravel and stones are given in Table 1.2. Gravel and stone sized materials are used as permeable fill over drains and as drainage layers in constructed profiles. Particle size grades and ranges for these are normally described according to the mesh sizes used in their separation rather than by any standard system of size classification.

The 'Triangle of Texture' relates the field properties of a soil to particle size distribution and it has been traditional to plot particle size distribution on a triangle to obtain a shorthand soil description, such as 'clay loam'. The term 'texture' is becoming replaced by 'particle size class' and this is to be recommended because of the widespread confusion between soil texture and soil structure. Whilst particle size class categories give a general idea of the types of soils appropriate for different sportsturf uses, soils cannot be prescribed solely on particle size class. For example, the strongly binding soils used on cricket squares require a more precise description than 'clay loam' or 'clay', and the categories of 'loamy sand' and 'sandy loam' are too broad to either prescribe or describe adequately a golf green soil. The soils plotted in Fig. 1.1 illustrate the general spread of variation found in different sportsturf soils.

Because of the need for more precise descriptions of the particle size distribution of, in particular, sands, sandy soils and drainage materials, accumulation curves are used extensively. These enable both visual and rapid numerical assessment of key criteria in particle size distribution. Very sandy soils do not contain complex particle aggregates so their physical properties can be more closely predicted from particle size distribution than clayey soils. Various properties can be derived from accumulation curves which have been found useful in describing the particle size distribution of sands and sand-dominant sportsturf soils (Fig. 1.2). These properties are often used in specifying materials used in construction and maintenance.

**1.**   *D*-value. A $D_x$ value is the mesh size of sieve through which $x\%$ of the particles in a sample pass. Two *D*-values which are commonly used are $D_{50}$ and $D_{20}$. The $D_{50}$ value is termed the 'median' particle diameter and it gives a general indication of the coarseness of the sample. The $D_{20}$ value is useful because fine-grained sandy soils generally behave as though made up of particles of the $D_{20}$ value.

**2.**   Gradation Index. This index is used to describe the uniformity of the distribution of particle sizes and thus the potential for interpacking. It expresses the size ratio of larger to smaller particles and for sportsturf soils the index used in the UK is the $D_{90}/D_{10}$ (Adams *et al.*, 1971).

**3.**   Percentage of particles within a desirable size range. Various size ranges have been examined for their suitability for sportsturf use. The size range 100–600 μm was recommended following work at the University College of Wales (Adams *et al.*, 1971), but this has been superseded by the 125–500 μm size range because it uses accepted size grade categories.

*Structure and properties of soil minerals*

The different particle size categories of soil affect its behaviour partly because of their size, but also because they differ in shape and chemical structure.

The predominant minerals in soils are crystalline silicates. The basic building block of all these is the silica tetrahedron in which a silicon atom sits in the space created by four closely packed oxygen atoms. There are essentially four types of organization of this $SiO_4^{4-}$ unit which give rise to the main groups of crystalline silicates. In the first group tetrahedra are not chemically bonded but exist as separate ions in the crystal with the charge on the silicate anion neutralized by cations within the crystal. The olivine minerals found in basic igneous rock are of this type, but this group is of no quantitative importance in developed soils. All other crystalline silicates involve polymerization of the silicate anion. Polymerization occurs via the sharing of oxygen atoms

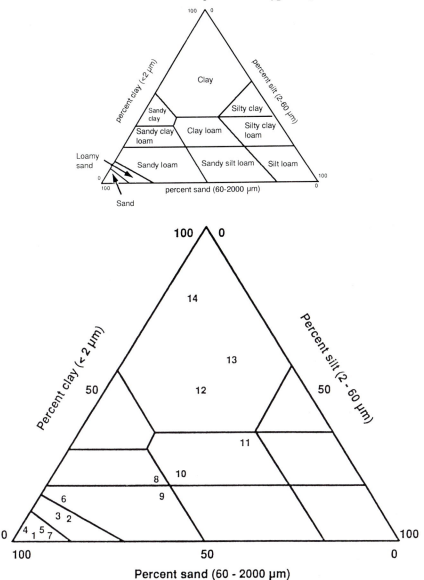

**Fig. 1.1.** Particle size classes of a range of sportsturf soils. 1) Wembley Stadium, 0-50 mm (1991); 2) Welsh National Stadium, 0-50 mm (1989); 3) Headingley Rugby League pitch, 0-50 mm (1989); 4) Llandrindod Wells bowling greens, 0-50 mm (1991); 5) St Andrews golf greens, 0-50 mm (1990); 6) Ascot horse racing track, 0-100 mm (1990); 7) Tokyo horse racing track, 0-50 mm (1989); 8) Wimbledon Centre Court, 0-50 mm (1990); 9) Wolverhampton tennis courts, 0-50 mm (1988); 10) Royal Berkshire Polo Club, 0-100 mm (1991); 11) Lord's cricket square, 0-50 mm (1991); 12) Sabina Park, Jamaica, cricket square, 0-50 mm (1967); 13) Asgiriya, Sri Lanka, cricket square, 0-50 mm (1984); 14) Perth, Western Australia, cricket square, 0-50 mm (1982);

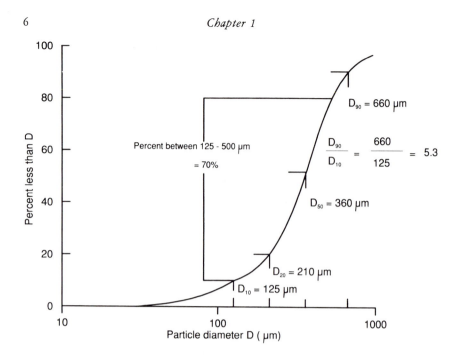

**Fig. 1.2.** Accumulation curve showing criteria important in the description of sands and sand-dominant sportsturf soils.

between adjacent tetrahedra. Thus the simplest type of polymerization is a linear development where two of the four oxygen atoms are shared between adjacent tetrahedra. This usually results in chains of indefinite length, but ring structures can be found. The repeating unit is $SiO_3^{2-}$ illustrated below:

$$
\begin{array}{cccc}
O & O & O & O \\
| & | & | & | \\
O-Si-O-Si-\boxed{O-Si-}O-Si \\
| & | & | & | \\
O & O & O & O
\end{array}
$$

The pyroxene minerals are of this form. These occur in basic igneous rocks along with the closely related amphiboles where a degree of cross-linkage between chains occurs. These somewhat linear crystal minerals are easily weathered and do not persist in soils.

The two groups of crystalline silicates of major importance in soils are those where either three or four of the tetrahedral oxygens are shared between adjacent silica tetrahedra. In the former case the sharing of three oxygens results in the creation of a sheet of connected tetrahedra. This silica sheet is a constituent of all clay minerals.

In addition to sheets of silica tetrahedra, clay minerals also include sheets of metal hydroxides. In most cases the two types of sheet are covalently bonded.

The hydroxide sheet may be predominantly aluminium or magnesium or iron hydroxide but in many cases more than one species of cation is present. The hydroxide sheet is different from the silica sheet in that six rather than four oxygen atoms surround the cation and this packing creates a larger 'hole'. Thus cations such as magnesium which cannot fit into the space created by four oxygen atoms can fit into the space created by six. Aluminium is the only major element in soils which can fit into either.

Only three structural organizations of component sheets are found in clay minerals. These are shown in Fig. 1.3 together with the names of common soil clay minerals.

All clay minerals are stable in two dimensions because of their sheet structure. Therefore particles are platy. Not only are they platy but also small so that in most soils clay minerals dominate the clay size fraction. Small size and platyness result in a very high specific surface area (surface per unit weight). An average specific surface area for a clay mineral in soil would be around $10\,\mathrm{m^2\,g^{-1}}$, which is about 1000 times greater than that of medium sand particles.

An important property of clay minerals is surface charge. The aluminium cation can fit into the space normally occupied by silicon in a silica tetrahedron. When aluminium replaces silicon, because of the difference in charge between the two cations this replacement results in a residual negative charge on the clay mineral. In micas around 25% of the silicon is commonly replaced by aluminium. Replacement, especially of silicon by aluminium and in general by cations of lower charge but of similar size to other cations, can result in a 'permanent' negative charge on clay minerals. This charge confers a 'cation exchange capacity' to clays and it is a property which differs between clay minerals. In addition to retaining cations this charge can cause clay particles to repel each other and disperse as individual microscopic units in suspension in water.

All clay minerals swell when wet because they are platy and their faces attract hydrated cations. In most clays this swelling is between particles but in vermiculite and especially smectites, swelling occurs both between particles and between layers within clay particles. Smectitic clays form under moist tropical weather conditions and are rather uncommon in British soils. The extreme stickiness of smectites when wet, and their substantial shrinkage and high mechanical strength when dry, explain the 'sticky dog' character when wet and the cracking and extremely fast pace of some cricket pitches in Australia, South Africa and the West Indies.

The sharing of all four oxygen atoms of silica tetrahedra with adjacent tetrahedra results in a three-dimensional structure. In soils, quartz ($SiO_2$) is the dominant member of the group although in rocks and some soils feldspars which have a similar structure may be important. Being a crystal form with three-dimensional stability, quartz is usually dominant in the sand fractions of soils. Also because of this, quartz particles are usually rounded or blocky. Whilst quartz is normally the dominant mineral in natural sands and sand fractions of soils, this is not always the case. Fresh rocks can be fragmented by

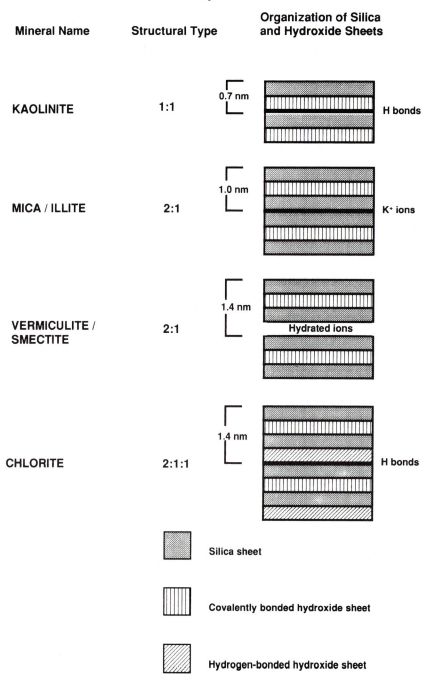

**Fig. 1.3.** Simplified structure of different clay minerals.

natural processes such as frost or glacial action and also by mechanical rock crushing. In these cases sand fractions may not only contain a wide range of weatherable minerals capable of releasing some plant nutrients, but also particles would not have been rounded by abrasion through movement by wind or water.

Quartz neither contains nor retains plant nutrients; indeed as a clean mineral its only contribution to soil–plant systems is to help control pore size distribution. However, quartz is rarely 'clean' but is usually coated by oxides of iron and aluminium together with some organic matter giving it a gold or brown colour. Surface coatings on quartz sand particles are important because they counteract its inertness.

Oxides of iron and aluminium occur as coatings on quartz and clay particles and as discrete particles of clay size and larger. Although not normally exceeding 5 % of the soil weight, oxides of iron and aluminium have important properties. They constitute a source of some plant nutrients because elements including manganese and zinc co-precipitate with them. Also they adsorb and retain several plant nutrients taken up as anions including phosphate, borate and molybdate. These anions are retained strongly but in the nutrient-poor conditions of sandy soils they act as important, slowly available reserves of nutrients.

For detailed information on the structure of minerals in soils consult Dixon and Weed (1989).

## Organic constituents

The organic fraction of soils comprises living organisms and dead organic matter. The dead organic matter exists in different physical and chemical forms and although some fractions are chemically stable the organic component of soils is more easily broken down than the mineral fraction.

### Soil organisms

Soil dwellers range in size from bacteria and fungi to earthworms and insect larvae. Most live on dead organic matter but some consume living plant tissue (e.g. leatherjackets which are an insect pest and some fungi which cause pathogenic diseases of turfgrasses).

Soil organisms can be divided into three categories important to the turfgrass scientist.

Firstly, there are the main decomposers of organic matter, the major group of heterotrophic bacteria and fungi. Secondly, there is a group of soil movers. This is a small group comprising, in the main, earthworms. Thirdly there are the damaging pathogens and pests of turf.

Bacteria and fungi are responsible for over 90% of the breakdown of organic matter in soils. There are many different species and the population as a whole can decompose not only natural residues but also materials ranging from fuel oil to exotic substances used as herbicides or pesticides. Bacteria are generally less tolerant of acidity than fungi so that in soils with a pH below 5 (see Soil pH section for an explanation of this term) organic matter breakdown tends to be slower because of a decrease in the activity and metabolic diversity of the microbial population. Thus acidification normally results in an increase in total soil organic matter. Fungi are able to grow through soil and a surface layer of organic matter or thatch is especially liable to be exploited by fungal mycelium. The waxy substances produced by fungi can aggravate the hydrophobic nature of organic matter when it becomes dry, contributing to the 'dry patch' phenomenon.

The soil movers are extremely important in affecting turfgrass soils. Turf-grass soils are essentially uncultivated and earthworms are key agents in preventing the surface accumulation of organic matter in these soils. They achieve this by two means; firstly by both consuming and carrying down dead plant debris into the soil and secondly by casting mineral soil onto the soil surface thereby diluting surface organic matter. The species primarily responsible are *Lumbricus terrestris*, *Aporrectodea longa* and *Aporrectodea caliginosa*. The non-casting species of earthworm are of minor significance. In different turf-grass situations the soil-moving activity of earthworms ranges from being greatly beneficial to extremely damaging. The range of activities and their effects are summarized in Table 1.3. The key visual consequence of an absence of earthworms is the accumulation of surface organic matter. This can be seen in many upland areas of grassland where earthworms cannot survive because of high soil acidity (Plate 1). The same phenomenon follows the chemical elimination of earthworms, so that the development of thatch on golf or bowling greens is not unexpected but is a natural consequence of eliminating worms.

There are a number of pests of turf and also some soil fungi which are turf pathogens. These pests, and strategy for their control, will be examined in later sections but one important point can be made here. All of the common fungal pathogens survive in soil for the most part by living on dead plant material. That is they are facultative pathogens. Fungal diseases occur on grasses in extensive amenity areas of parks and hills and also on agricultural grassland, but they rarely cause significant damage. It is on some intensively used and maintained turf that diseases are seriously damaging because the grass is stressed by wear and close mowing.

### Soil organic matter

The living organisms in soil account for only about 0.1% of the total organic matter. The remainder exists in forms ranging in complexity and stability.

**Table 1.3.** Activities and effects of casting earthworms in sportsturf.

| Activity | Beneficial effects | Adverse effects |
|---|---|---|
| Drawing down of grass debris on or near the soil surface and mixing with mineral soil | Prevention of thatch. Acceleration of the cycling of nitrogen and other nutrients | May bury weed seeds |
| Casting onto the surface of fine mineral soil mixed with variable amounts of organic matter | Equivalent to a normal soil topdressing, beneficial on low maintenance cricket squares but not necessarily elsewhere | Casts smother grass especially if compressed onto the soil and affect adversely both surface trueness and ball roll. Soil brought up from below increases silt and clay in sand-dominant rootzones |
| Creation of burrowing holes within soils and up to the soil surface | Creation of rapid drainage channels in compact soils. Increases macroporosity/aeration (see Fig. 4.15) | Produces a 'softer' soil more subsceptible to shear and surface damage |

Organic matter in turfgrass soils arises primarily from the leaf, stem and root residues of grasses. These residues may, in some instances, be supplemented by additions of peat, which is lacking in plant nutrients and is decomposed slowly. For soils in an unchanging type of land use, the amount of soil organic matter is a balance between input and decomposition. For a given land use the warmer the climate the lower is the soil organic matter content because of the more rapid breakdown of plant residues with higher temperatures. Grasslands return more plant residues than most other agronomic uses of land, so grassland soils have typically large organic matter contents.

The return of residues to turfgrass soils is affected substantially by the removal or non-removal of clippings. The clippings from a ryegrass-dominant sportsturf can amount to over $8000 \, \text{kg ha}^{-1} \text{y}^{-1}$ dry matter but fine turf, mown closely, returns around half this amount. Because of variations in turf type and maintenance, the return of plant residues to soils ranges from less than $4000 \, \text{kg ha}^{-1}$ to over $10{,}000 \, \text{kg ha}^{-1}$. When clippings are allowed to 'fly' they may provide more than half of the total returns of plant residues to the soil. However clippings are less lignified than stem bases and roots and decompose more quickly. Because of this the contribution of clippings to residual soil organic matter is less than that of stem bases and roots. When clippings are removed the annual returns from stem bases and roots are around $6000 \, \text{kg ha}^{-1}$ (Riem Vis, 1981) which, unlike that of clippings, varies little with changes in fertilizer input.

Grass residues returned to well-aerated soils break down rapidly at first

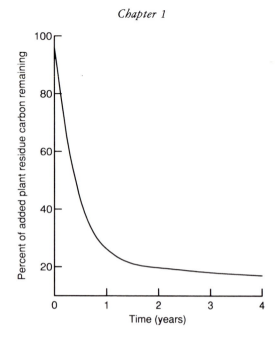

**Fig. 1.4.** The proportion of turfgrass residue carbon added to soil which remains in subsequent years.

and then progressively more slowly. Figure 1.4 shows the pattern of change in the rate of decomposition of plant residues. With time the residual material contributes to soil humus which is colloidal material (fine dispersible particles) with some properties in common with clay.

Soil organic matter is often spoken of as though it were relatively uniform. This is untrue because although the main component is humus, which is chemically quite stable to microbial attack, it also includes decomposing plant residues and dead microorganisms. These newly decomposing constituents have a disproportionately large effect on the release of plant nutrients. Decaying fragments of plants are capable of creating large continuous pores, such as those produced by old root channels. Humus varies chemically but consists of large molecules rather than pieces of tissue and exhibits surface active properties. In general, with regard to sportsturf soils, plant fragments create pores which aid drainage but humus, which improves water and nutrient retention, may also block pores. The fact that humus is a product of the decay of plant tissues highlights the danger in formulating rootzone media where fibrous organic matter such as peat, which is subject to chemical change, controls its physical properties. The blocking of macropores caused by colloidal organic matter in intensively used and compacted sportsturf soils contrasts with most natural, horticultural or agricultural soils where almost without exception its effects are beneficial.

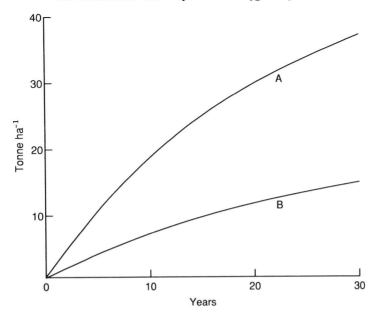

**Fig. 1.5.** Increase in soil organic matter with time with two inputs of turfgrass residues. ($A = 10\,t\,ha^{-1}\,y^{-1}$; $B = 4\,t\,ha^{-1}\,y^{-1}$.)

The proportion of the carbon returned as plant residue which remains at the end of a year is known as the humification coefficient. For residues from turfgrasses this coefficient ranges from 0.25 to 0.3. The rapid loss of plant residue carbon in the first year and relatively slower losses in subsequent years which is illustrated in Fig. 1.4 make it possible to produce a model describing changes in soil organic matter content with time. The net increase in soil organic matter is illustrated in Fig. 1.5 for two contrasting turfgrass residue inputs. Two points should be noted. Firstly, the rate of increase in soil organic matter content as a percentage of soil is quite slow. For example, the increase in soil organic matter with an annual plant residue input of $10\,t\,ha^{-1}$ would only be from 0% to 3.0% in 20 years when expressed as a percentage of the top 10 cm of soil. However, looked at in another way, this accumulated amount of organic matter would give a thatch thickness of about 15 mm were it to have accumulated on the soil surface.

Humus, which is the main constituent of soil organic matter, is comprised of molecules which are very large and variable in size. They should be considered to be dispersed in suspension rather than being in true solution. Its colloidal molecules are rounded or spherical in shape and spongy in character. Humus is a macromolecular acid with carboxyl ($-COOH$) as the main functional group as it is in other simple organic acids. This is the only functional group which dissociates to create a residual negative charge at normal soil pHs. However, because of its complexity dissociation takes place over a wide range of pHs. Thus the negative charge on humus colloids and hence its cation exchange capacity increase about four-fold from pH 4 to pH 7.

In addition to its physical properties, soil organic matter contains essential plant nutrients. The main nutrients are nitrogen, phosphorus and sulphur. Their ratios by weight in relation to carbon are approximately: C,100: N,10: P,1: S,0.5.

Trace elements, especially copper, zinc and boron, are also retained in organic matter. The plant nutrient with the closest dependency on organic matter is nitrogen. This is the only major plant nutrient whose reserves in soil depend entirely on its occurrence in organic matter. Allison (1973) provides useful reference material on the role of organic matter in soils.


## PHYSICAL AND CHEMICAL PROPERTIES OF SOILS

## Physical properties

Sport played on natural turf has requirements in terms of soil physical properties which are far more demanding than other soil uses and it is for this reason that it is common for soils to be designed and created for particular sports. Requirements vary from guaranteed drainage potential in intensive use for golf and winter games to high binding strength in cricket square soils. Whereas over recent decades increases in fertilizer input have increased the perceived value of agricultural soils, major upgrading of sportsturf soils has been gained by better control over their soil physical characteristics.

Poor drainage has been and is still the main soil problem on winter games pitches and fine turf areas. Two factors are important. Firstly, players wish to play in all weather conditions which they are prepared to tolerate. Secondly, many sportsturf areas are subjected to high intensities of use. The pressure of use, especially in wet conditions, causes severe deterioration in the drainage characteristics of many soils. Thus a soil which behaved as a well-drained agricultural soil may turn rapidly into a quagmire when used as a soccer pitch. In order to understand and remedy the problems of sportsfield drainage it is first necessary to examine some basic soil–water relationships.


### Soil–water relationships

Water is retained in soils by surface tension at air–water interfaces within the pore systems. The surfaces of soil constituents are generally hydrophilic (water loving) and the behaviour of water in soil pores can be modelled on the well-known phenomenon of capillary rise. When a glass capillary tube is dipped

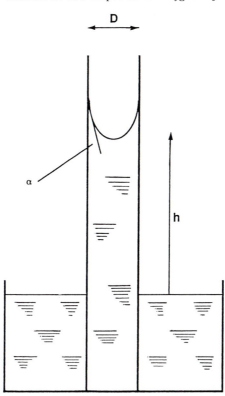

**Fig. 1.6.** Illustration of the capillary rise phenomenon.

into water, a column of water rises up the tube until the upward lift due to surface tension acting around the internal circumference of the tube balances the downward weight of water within the tube (Fig. 1.6). The phenomenon is described by the relationship:

$$D = \frac{4T \cos \alpha}{h d g}$$

where $D$ is the diameter of the tube; $T$ is surface tension; $h$ is the height of capillary rise; $d$ is the density of water; $g$ is the acceleration due to gravity and $\alpha$ is the angle of contact between water and surface. $T$, $d$ and $g$ are constant or approximately so and $\cos \alpha$ can be taken as 1, so that $D$ is inversely proportional to $h$ and the equation simplifies to:

$$D(\text{mm}) = \frac{30}{h(\text{mm})}$$

If soil pores are considered as a system of small tubes, this enables the calculation of the diameter of the largest water-filled pore at a position *h* millimetres above the watertable. All pores smaller than this will be water-filled and all larger pores drained.

If the average depth to the watertable in soils is known, it is possible to calculate the maximum size of water-filled pore when drainage has virtually ceased. In sportsfield soils this depth is taken to be 400 mm because constructions are seldom deeper (500 mm is used generally for agricultural soils). This depth of 400 mm represents 400 mm of water tension and is equivalent to a pressure of $-40$ mbar or $-4$ kPa (suction would be positive). A sportsturf soil would be expected to reach its wettest stable state in equilibrium with a tension of 400 mm about one day after saturation. In this 'field capacity' condition, the largest water-filled pore would have a diameter of 0.075 mm or 75 $\mu$m. Since it is particles which create pores it should be recognized that rounded particles in close packing create pores about 0.4 of their own diameter. The interpretation of this is that for a soil to contain any air-filled porosity at all, when the soil is in equilibrium with a watertable 400 mm deep, some pores have to be present which are created by either primary particles or aggregates of particles larger than about 200 $\mu$m in diameter. A very small proportion of natural soils in Britain or elsewhere in the world have a sufficiently dominant sand content for primary particles to create air-filled pore space at field capacity. Thus, in the vast majority of soils, air-filled porosity is created by aggregates of clay and other small particles, rather than by sand size primary particles.

*Air and water storage*

When a soil is saturated and then allowed to drain, water is lost from the large pore spaces under gravity. The water removed is called 'gravitational water' and the soil is at field capacity once drainage is complete. As a soil loses water by evaporation from the surface and transpiration by plants, water becomes restricted to progressively smaller pores and greater suction must be applied to remove it. In effect the energy state of the water falls and in order to maintain a transpiration flow into plants, the energy state of water within roots must be lower than the soil water. Plants achieve this by increasing the osmotic potential of cells in the roots. As soils dry further, plants reach a point where they cannot sustain the evaporative demand at the leaves and close their stomata to increase the resistance to water loss. Plants may wilt temporarily in these conditions but they will recover if evaporative demand falls.

Temporary wilting or loss of turgidity occurs in circumstances when plants struggle to maintain the rate of water flow into them, but a point of soil dryness is recognized where plants cannot recover. This situation, called 'permanent wilting point', is taken to be $1.5 \times 10^{3}$ mm of tension or 15 bar, or 1500 kPa. The amount of water held between field capacity and permanent

**Table 1.4.** Relationships between terms used in the description of the water retention categories of soils and the size of pores and the particles creating them.

| Description | Gravitational water | Easily available water | Difficult available water | Unavailable water |
|---|---|---|---|---|
| Equivalent tension (bar) | 0 | ←———— 0.04 ———— 2 ———— 15 ————→ | | more than 15 |
| Pore size ($\mu$m) | greater than 75 | ←———— 75 ———— 1.5 ———— 0.2 ————→ | | less than 0.2 |
| Particle size creating pore ($\mu$m) | greater than 200 | ←———— 200 ———— 4 ———— 0.5 ————→ | | less than 0.5 |
|  | | field capacity | permanent wilting point | |

wilting point is called the 'available water capacity'. This may be expressed as a percentage of soil depth or as the total depth of available water (in mm) in the depth of soil exploitable by roots. The term 'soil moisture deficit' is used to describe the amount of water (in mm) required to raise the soil moisture content back to field capacity.

The available water within a soil is not equally available to plants. This is because as soils become drier both the energy state of water falls and the resistance to water movement within soils increases greatly. Because of these factors the growth rate of many plants falls well before the permanent wilting point is reached. For this reason the term 'easily available water' is used, with a range from field capacity to 2 bar (200 kPa) to quantify the available water reserve over a range where supply is non-limiting to plant growth. Virtually all the available water retained by sandy soils is easily available water. Table 1.4 summarizes the relationships between water storage and soil particle and pore sizes.

### Air-filled porosity

Large soil pores are required for aeration, easy penetration of grass roots and rapid drainage. Oxygen diffuses $10^4$ times quicker through air than water so a minimum continuity of air-filled porosity is required to allow roots and microorganisms to respire aerobically and to prevent anaerobic processes which reduce nitrogen availability and result in the formation of phytotoxic products.

There is no precise value for the minimum air-filled porosity because the critical level depends upon plant species and oxygen demand within the soil. Nevertheless the value of 10% (air-filled pore space in a soil volume) is often taken as a minimum. Grass roots range in diameter from about 60 $\mu$m to over 250 $\mu$m. Thus a good volume of air-filled porosity is required for unrestricted root development. Compacted fine sandy soils which create pores less than 60 $\mu$m in diameter are liable to restrict root development and remain waterlogged when drainage ceases.

Since rapid drainage normally requires the presence of large pores within soils, good drainage and good aeration are often considered to be synonymous. This is not justified, especially in consideration of sportsturf soils, if drainage is taken to mean the ability to transmit water from soil surface to subsoil. The best way to illustrate this is to imagine a compacted single size very fine sand of particle diameter 100 $\mu$m. At a tension of 400 mm this material would remain completely saturated when drainage ceased. Nevertheless it would be able to transmit water to the watertable at a rate of around 130 mm h$^{-1}$, which is much greater than would be required to prevent surface ponding in the heaviest rain.

### Hydraulic conductivity and critical tension–air-entry pressure

Hydraulic conductivity is the term used to describe the ability of a soil to transmit water and is usually measured in mm h$^{-1}$. It is the inverse of the resistance to water flow through the soil. In a saturated soil the hydraulic conductivity remains constant because all pores are available for flow. In an unsaturated soil, hydraulic conductivity decreases as water content decreases because the pathways for flow become constrained to smaller and smaller water-filled pores offering a greater resistance and providing a more tortuous route. During drainage, when a pressure head is created by rain falling on the surface, one is concerned mainly with saturated or near-saturated flow. Unsaturated flow occurs in response to gradients in water tension created by water uptake by roots and evaporation from the soil surface.

Materials used in the construction and modification of sportsturf soils and as permeable fill materials have quite large hydraulic conductivities. Figure 1.7 shows that over the range of interest there is an approximately linear relationship between $\log_{10}$ saturated hydraulic conductivity and $\log_{10}$ particle diameter. The gradient of nearly 2 indicates that hydraulic conductivity increases in relation to the square of particle diameter. Table 1.5 draws together some of the more important properties of single size sands. The hydraulic conductivity of the finest sand fractions is large, but such materials would remain saturated unless the watertable could be maintained at a depth much greater than could be contemplated for a construction.

The air-entry pressure or critical tension is the depth of pore water con-

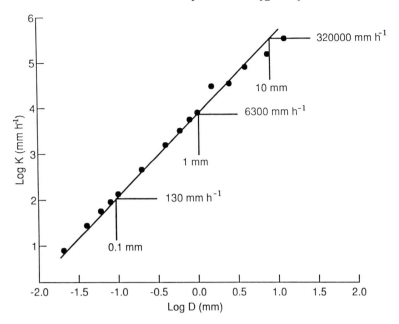

**Fig. 1.7.** Relationship between saturated hydraulic conductivity (*K*) and particle diameter (*D*).

**Table 1.5.** Some properties of single sized sands.

| Description and particle size range (mm) | Particle diameter (mm) | Critical tension (mm) | Saturated hydraulic conductivity $(mm\,h^{-1})$ |
|---|---|---|---|
| Very fine sand (0.06-0.125) | 0.100 | 900 | 130 |
| Fine sand (0.125-0.25) | 0.200 | 400 | 450 |
| Medium sand (0.25-0.5) | 0.400 | 220 | 1500 |
| Coarse sand (0.5-1.0) | 0.800 | 120 | 5500 |

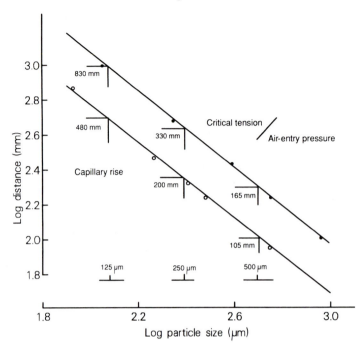

**Fig. 1.8.** Relationship between both critical tension and capillary rise and the particle size of single size sands.

tinuity required over a watertable to cause the maximum size of pore to empty and thus permit air entry. With respect to a constructed profile it is the depth to a capillary break layer of stone or gravel or the depth to the maintained watertable (Figs 3.9 and 3.11). It is impracticable and unnecessary for this depth to be greater than 400 mm. This means that pores created by particles greater than 200 µm in diameter must be present to ensure air entry even though the hydraulic conductivity of the profile may be greater than is necessary.

### Capillary rise versus air-entry pressure

A column or profile of sand will become completely saturated if the watertable is raised to the surface. If the watertable is lowered, water within pores will be subjected to a tension determined by the depth of the watertable below the surface and the size of pores drained will be related to the depth of the watertable. Conversely water will rise into a dry sand medium from a watertable introduced at its base. The wetted front will rise until equilibrium is reached. The equilibrium distance of the wetted zone above the watertable differs depending upon whether the material was drained from a saturated condition or was wetted by capillary rise from below. Figure 1.8 shows the relationship between critical tension and capillary rise equilibria for sands of different particle size. The appreciable difference between these values has implications for subsoil irrigation because a fixed watertable depth does not allow standard moisture conditions to be maintained. The height of the saturated zone above a watertable is about 30% greater if the profile was saturated initially than if wetted from a watertable below by capillary rise.

## Uniformity of particle size distribution

With rare exceptions the macropore space in natural and agricultural soils, that is the pores which are air-filled at field capacity, is created by aggregates of particles rather than primary particles themselves. In sportsturf subjected to intensive use in wet weather, these fragile structures cannot persist and primary particles of an appropriate size and distribution must be used to control both total porosity and pore size distribution. There is indeed a problem over and above that of being unable to rely on particle aggregation. This is that compaction in wet conditions pushes small particles into pores created by larger particles. In such conditions the macropore space produced by large particles can be occluded by no more than about 20% by weight of small particles, affecting both total porosity and pore size distribution.

A uniformity coefficient $D_{60}/D_{10}$ is used extensively in sedimentology and soil mechanics to describe the extent of particle sorting by natural processes (Terzaghi and Peck, 1967). It is now accepted that Gradation Index ($D_{90}/D_{10}$) provides a better criterion to assess the potential for interpacking not only of natural materials but also of artificial mixes of sands used for sportsturf. Figure 1.9 shows a close linear relationship between total porosity and the Gradation Index of a range of compacted sands. The theoretical total porosity of single size spheres in close packing is 40% by volume and sands with a Gradation Index of around three or less have total porosities of this order. Up to a quarter of this porosity can be lost in sands containing a broad range of particle sizes.

Sands with a wide range of particle sizes have the potential to create a wide range of pore sizes when uncompacted and when the distribution of particles is reasonably random. This is why such sands are favoured for horticultural use. Unfortunately, compaction causes interpacking and the less random distribution of particles eliminates macropores and reduces total porosity. The optimum sands for use on sportsturf are therefore those with a minimum opportunity for interpacking, even though this results in a small range in pore sizes.

## Particle shape

Consideration has been given to the relative benefits of rounded versus angular sands. When sands are compressed, rounded sands more readily adopt their

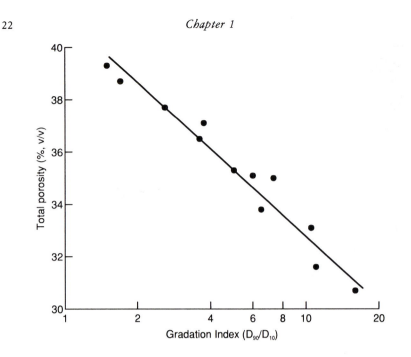

**Fig. 1.9.** Relationship between total porosity and Gradation Index.

closest packing whereas angular sands interlock and maintain a more open framework. The interlocking of angular materials can be easily demonstrated by comparing the stability to traffic of 6 mm crushed stone with 'Lytag'[1] or rounded shore gravel of the same particle size. Despite the potential benefits of angular sands such a specification is somewhat academic since the natural mechanisms of wind or water transport involved in the sorting of particle sizes cause abrasion resulting in the rounding of particles. As a consequence angular sands are rarely available. The description of the particle shape of sands is based on their angularity and sphericity and these can be assessed with reference to standard charts (Fig. 1.10).

Interpacking of sands with a large Gradation Index results in a reduced total pore space and reduced hydraulic conductivity. Thus a sand with 25 % in the 100–200 $\mu$m size range, but ranging up to particles larger than 2 mm, will usually when compacted have a hydraulic conductivity less than a sand with over 80 % in the 100–200 $\mu$m size range. The saturated hydraulic conductivity of a compacted sand is reduced dramatically by small quantities of particles of silt or clay size. Figure 1.11 illustrates the situation which in simple terms shows a logarithmic decrease in saturated hydraulic conductivity with increasing

---

[1]Lytag is a pelleted power station fly ash.

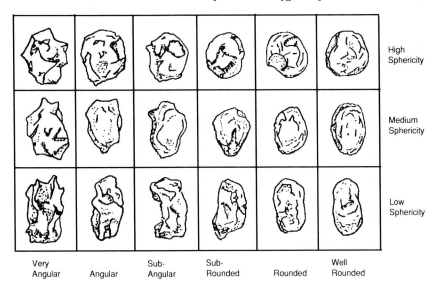

| | | | | | | |
|---|---|---|---|---|---|---|
| | | | | | | High Sphericity |
| | | | | | | Medium Sphericity |
| | | | | | | Low Sphericity |
| Very Angular | Angular | Sub-Angular | Sub-Rounded | Rounded | Well Rounded | |

**Fig. 1.10.** Diagram illustrating shape variation in grains and their description.

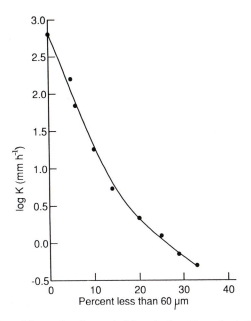

**Fig. 1.11.** The effect of increasing fine material content on the saturated hydraulic conductivity ($K$) of compacted sand/soil mixes.

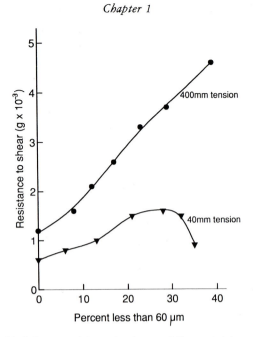

**Fig. 1.12.** Relationship between resistance to shear and fine material content for compacted sandy soils equilibrated at two water tensions.

fine material content. Less than about 12% silt plus clay is needed to ensure a hydraulic conductivity greater than $10 \, \text{mm} \, \text{h}^{-1}$.

However, small amounts of silt and clay in a 'clean' sand have the advantage of increasing its physical stability when moist. This benefit in stability and resistance to shear which is illustrated in Fig. 1.12 has to be balanced against the disadvantage of reduced hydraulic conductivity. The relative merits of pure sand versus sand/soil mixes with bypass drainage for rootzones for winter games pitches centre largely on this compromise (see Chapter 3).

The requirements for high hydraulic conductivity, surface stability, large total porosity and a sensible construction depth place tight limits on the particle size distribution of sand and sand-dominant mixes for sportsturf rootzones. For sands the constraints demand materials with not less than about 80% in the 125–500 $\mu$m size range. Even this quite narrow range can be refined because winter games pitches have a greater requirement for stability than golf and bowling greens. For intensively used winter games pitches, sands with a $D_{50}$ of 230 $\pm$ 30 $\mu$m and a Gradation Index of 3.3 or less are ideal, whereas for golf and bowling greens, sands with a $D_{50}$ of 330 $\pm$ 30 $\mu$m and a Gradation Index of 3.3 or less are most appropriate (Fig. 1.13). When sand-dominant soils are used for rootzones or topdressings the proportion less than 125 $\mu$m should not exceed 20% by weight.

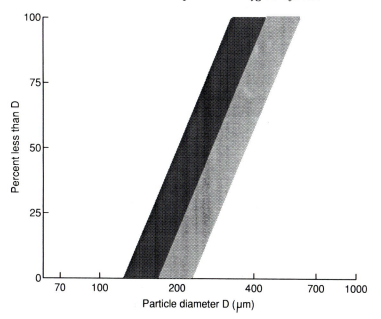

**Fig. 1.13.** Ideal particle size distributions for sands used in the construction of winter games pitches (dark shading) and golf and bowling greens (light shading).

It is important to recognize that sports such as cricket and tennis require soils with good binding characteristics and these requirements take precedence over free drainage. This aspect will be dealt with in Chapter 8.

### Dispersion and flocculation of clay and organic colloids

Clay minerals and organic colloids are negatively charged particles which repel each other. This repulsion can lead to particles becoming dispersed in suspension by their behaving as free individual particles. As separate particles they remain in suspension for many hours and when they settle out and dry they create a compact block impervious to water. A soil containing clay which is completely dispersed (deflocculated) is unmanageable because it has no structure and will not drain. Because of the importance of the condition of clay in soil quality, the extent of clay dispersion (clay dispersion coefficient) can be use as a criterion for soil quality.

Fortunately, it is rare for the clay and organic colloids to be dispersed. They are usually flocculated so that the particles are bound together into small aggregates (or domains) of silt size. These domains form the building blocks of much larger aggregates which incorporate silt and even sand particles. The forces which bind clay particles together are the molecular forces of attraction,

**Table 1.6.** Approximate minimum concentrations of chloride salts of basic cations required to flocculate soil clays.

| Salt | Concentration of cation (mM $l^{-1}$) |
|------|----------------------------------------|
| $CaCl_2$ | 1.0 |
| $MgCl_2$ | 2.0 |
| KCl | 30 |
| NaCl | 50 |

van der Waals forces. All molecules have this inherent attraction which is small for small molecules but is large for colloidal size molecules. Van der Waals forces act over very short distances of a few tens of nanometres, much shorter than the distance over which electrostatic repulsion acts.

In order for clay particles to flocculate their surface charge has to be neutralized to allow them to approach each other close enough for van der Waals forces to dominate. All cations are capable of achieving this neutralization of charge and therefore causing flocculation but there are large differences between cations in the concentration required. Thus calcium in near-neutral soils and aluminium in very acid soils will cause flocculation at very low concentrations. In contrast, sodium is a very poor 'flocculating agent' and high concentrations are needed (Table 1.6).

The main danger of soil clays becoming deflocculated is when sodium is a major exchangeable cation in the soil. Soil clays are liable to deflocculate when exchangeable sodium constitutes more than about 15% of the exchange capacity of the soil. This can occur through the use of very poor quality irrigation water or where salt is allowed to accumulate in the soil in arid and semi-arid areas or following flooding of soils by seawater. The crux of the problem is that such high concentrations of sodium are required to flocculate clay that slight leaching by rainwater will reduce the concentrations to a level where dispersion will occur. Soil management must ensure that exchangeable calcium remains dominant. This can be achieved following sea flooding for example by an application of gypsum ($CaSO_4$) which is sufficiently soluble to ensure that exchangeable sodium from the seawater is displaced by calcium as leaching proceeds.

## Chemical properties

The chemical properties of greatest importance are those which affect directly or indirectly the supply of essential plant nutrients which grasses obtain from soil.

**Table 1.7.** Typical cation exchange capacities of different clay minerals.

| Clay mineral | Cation exchange capacity at pH 7 ($mM_c kg^{-1}$) |
|---|---|
| Kaolinite | 40 |
| Chlorite | 200 |
| Illite | 400 |
| Vermiculite | 1400 |
| Smectite | 1000 |

Before examining the transformations in soils of specific plant nutrients it is necessary to explain soil properties which influence the behaviour of many nutrients.

### Cation exchange capacity

Clay minerals and organic colloids are negatively charged. The charge on clay minerals is predominantly permanent, that is, it is not affected by pH. As was explained earlier in this chapter the charge is derived from the isomorphous replacement of structural cations of higher charge (valency) by cations of smaller charge. Most common is the replacement of silicon by aluminium in the silica sheets of clays. Since different clays have different degrees of isomorphous replacement, their charge or cation exchange capacity varies (Table 1.7). Organic colloids differ from clay colloids in the source of the charge and also because the charge varies with pH. The charge on organic colloids originates through the dissociation of carboxyl groups (COOH) which are the colloids' predominant functional group. This process is akin to the dissociation of simple organic acids such as acetic acid ($CH_3COOH$) which dissociates into $CH_3COO^-$ and $H^+$. However, the main differences between organic colloids and simple organic acids are that organic colloids in soil are insoluble and each molecule contains many acid groups which dissociate over a wide pH range. The cation exchange capacity of humus colloids can be as high as $4500 \, mM_c \, kg^{-1}$ (see below for an explanation of the units) but it is usually much less than this especially in soils with a pH below 5.5. The cation exchange capacity of a soil is important because cations of all kinds are attracted to the surface of clay and organic colloids. The retention of cations reduces leaching losses and since exchangeable cations can exchange readily with the soil solution they remain as a readily available reserve for plants. The cation exchange capacity of the sand-dominant soils used for most sportsturf is small and depends almost entirely on the organic matter they contain. Nevertheless a 5% content of organic matter by weight confers a cation exchange capacity comparable to around a 30% content of an illitic clay.

Different cations have different affinities for the exchange sites on soil colloids. As a general rule the smaller the ion and the greater its valency the greater is its affinity. Thus, for example, calcium has a much greater affinity than either potassium or sodium. The cation exchange sites of soils will be occupied by a range of different cations although calcium dominates in soils above pH 5–5.5. Since cations with different valencies are involved, the quantity of exchangeable cations must be expressed as the amount of charge they comprise, taking into account their valency. Thus the amount of an exchangeable cation is given in millimoles of charge per kilogram of soil ($mM_c\,kg^{-1}$).

## Soil pH

The pH of a solution is $-\log_{10}$ of the hydrogen ion concentration (or strictly activity). Since it is a negative logarithm, pH decreases as the hydrogen ion concentration increases. Water ($H_2O$) dissociates into equal concentrations of $H^+$ and $OH^-$ ions and in pure water each has a concentration of $10^{-7}$ M. This concentration corresponds to a pH of 7 which is considered neutral. If a little hydrochloric acid is added to water it will dissociate into $H^+$ and $Cl^-$ ions, the $H^+$ ion concentration will increase and the pH will fall. Since the product of the concentration of $H^+$ and $OH^-$ ion concentration is constant ($10^{-14}$ M) the $OH^-$ ion concentration in turn will decrease.

It is not practicable to extract the soil solution from a soil to measure its pH so soil pHs are measured on a suspension of soil in water or in a neutral salt solution (e.g. M KCl). Both soil:solution ratio and the nature of the solution used affect the pH reading. Traditionally a ratio of 1 soil:2.5 solution (w/v) has been used but it is now more usual to use a ratio of 1:1 (v/v). The pH measured in salt solution is always lower than that measured in water. This is because the cation in the neutral salt displaces virtually all of the $H^+$ ions from the soil colloids into solution. The pH reading in salt solution is more stable than that read in water because when the particles in suspension in water settle out the $H^+$ ion concentration in the immediate vicinity of the electrode falls. Apart from this, there is little to commend measurement of pH in a salt solution.

Since clay and organic colloids are negatively charged, they behave as weak immobile acids. For a soil to be neutral therefore, these negatively charged sites must be satisfied by basic cations (e.g. calcium). When calcium is depleted through leaching it is replaced by hydrogen (and aluminium) ions and the soil becomes acid. Over the pH range 4–7 soil acidity reflects the degree of saturation of exchange sites with basic cations. Soil pH can be raised by increasing base saturation through the application of lime. Acidification occurs when $H^+$ ions displace exchangeable $Ca^{2+}$ and other basic cations. Exchangeable calcium is leached out naturally in drainage water as calcium bicarbonate, but

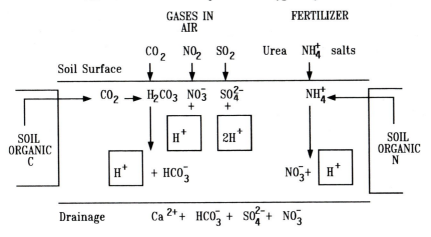

**Fig. 1.14.** Processes which lead to soil acidification in soils of the UK.

loss of $Ca^{2+}$ and thus acidification can be accelerated by other processes which are illustrated in Fig. 1.14.

## Soil aeration and oxidizing/reducing conditions

With few exceptions heterotrophic organisms, that is organisms which require preformed carbon compounds (e.g. carbohydrates) for energy, use oxygen as the terminal oxidant. In this process oxygen is reduced to water. Oxygen diffuses into turfgrass soils when its concentration therein is depleted by the metabolic activities of soil organisms (mainly microorganisms) and grass roots. Provided the air-filled porosity of the soil exceeds about 10% of the soil volume, sufficient oxygen can diffuse into the soil. If a soil becomes saturated, diffusion of oxygen virtually ceases and within around 24 hours its concentration falls to levels which cannot support aerobic metabolism. The soil becomes anaerobic (without oxygen). Once this situation is reached organisms which can use alternatives to oxygen come into their own. The first groups of bacteria to have a major effect are those capable of using nitrate ($NO_3^-$) and thus the process of denitrification proceeds. When the nitrate is all consumed, other oxidized constituents of soil will be used provided available organic matter is present. In consequence the soil environment will become chemically reduced not just anaerobic. Since organic matter is required to drive the reducing process the most severe consequences of waterlogging will occur at shallow soil depths where organic matter is concentrated.

The oxidation/reduction potential (redox potential) is a measure of the severity of reducing conditions and is defined by the Nernst equation:

**Table 1.8.** Sequence of important redox reactions in soils in anaerobic conditions.

|  | $E_h{}'$ (approx.) | Redox couple | Implication |
|---|---|---|---|
| Increasing severity of reducing conditions | +0.8V | $O_2/H_2O$ | Aerobic respiration |
|  | +0.4V | $NO_3{}^-/NO_2{}^-$ | Denitrification |
|  | +0.4V | $Mn^{4+}/Mn^{2+}$ | Increased Mn solubility and mobility |
|  | +0.2V | $Fe^{3+}/Fe^{2+}$ | Dissolution of Fe oxides, increase in $Fe^{2+}$ |
|  | −0.2V | $SO_4^{2-}/H_2S$ | Production of $H_2S$ and black metal sulphides |

$$E_h{}' = E_o{}' + \frac{RT}{zF} \log_e \frac{[\text{oxid}]}{[\text{red}]}$$

where $E_h{}'$ = redox potential at pH 7; $E_o{}'$ = standard electrode potential of the redox couple at pH 7; $R$ = gas constant; $T$ = absolute temperature; $z$ = electrons involved; $F$ = Faraday constant; and [oxid]/[red] = concentrations of oxidized and reduced forms (e.g. $NO_3{}^-$ /$NO_2{}^-$).

Redox potential is measured as the electrical potential of a platinum electrode which responds to the system's tendency to remove or donate electrons. In the equation given above it is evident that when the concentration of oxidized and reduced forms is equal, $E_h{}'$ will equal $E_o{}'$ since the logarithm of one is zero. The standard electrode potential differs between different redox systems and decreasing values for $E_o{}'$ of the different redox systems define the sequence of predominant reduction reactions as the severity of reducing conditions increases (Table 1.8). Thus, for example, denitrification proceeds as soon as oxygen is depleted but sulphide is not produced in significant quantities until most of the oxidized reserves of manganese and iron have been reduced. Reducing conditions in soils have important implications for the nutrition of turf and occurrence of toxicities. Clearly nitrate availability falls through denitrification, but also levels of soluble manganese and iron rise because the reduced forms are more soluble than the oxidized forms. Manganese may become toxic. Hydrogen sulphide is toxic to roots but its toxicity can

**Table 1.9.** Essential plant nutrients and typical ranges in concentration in young leaves of turfgrasses.

| Element | % in dry tissue |
| --- | --- |
| Nitrogen (N) | 2.0-4.5 |
| Phosphorus (P) | 0.2-0.5 |
| Potassium (K) | 2.0-4.0 |
| Chlorine (Cl) | 0.5-2.0 |
| Calcium (Ca) | 0.5-2.0 |
| Magnesium (Mg) | 0.1-0.5 |
| Sulphur (S) | 0.2-1.0 |
| Element | $\mu g\ g^{-1}$ in dry tissue |
| Iron (Fe) | 100-500 |
| Manganese (Mn) | 30-100 |
| Zinc (Zn) | 40-100 |
| Copper (Cu) | 5-50 |
| Boron (B) | 5-50 |
| Molybdenum (Mo) | 1-4 |

be reduced if it is precipitated as black ferrous sulphide by reaction with dissolved ferrous iron or residual ferric oxide.

$$Fe^{2+} + H_2S \rightarrow FeS + 2H^+$$

$$2FeOOH + 3H_2S \rightarrow 2FeS + + S + 4H_2O$$

This black precipitate causes the black layer phenomenon in turfgrass soils. Problems of toxicity to turfgrasses can also arise through the production of acetic acid and ethylene in anaerobic soils (Lynch, 1983).

## Processes in soil affecting nutrient availability

Grasses require 13 essential nutrients which they obtain from soil. Whilst grasses cannot survive in the absence of any of these, the amounts required differ widely so that the typical concentration of nitrogen in turfgrass leaves is around $10^4$ greater than that of molybdenum (Table 1.9).

Plants take up nutrients which are dissolved in the soil solution and uptake is usually related to the concentration in solution. The concentration of nutrients in the soil solution is affected by both external and internal factors. External factors include input in fertilizers, nutrients in irrigation water and deposition from the air. Internal factors are the amounts and chemical forms of nutrients and soil processes which affect the relationships between solid and solution. Some of these processes tend to buffer the concentration of a nutrient

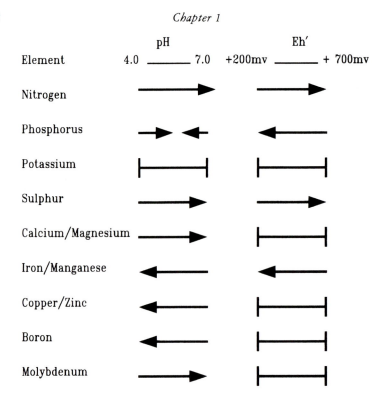

**Fig. 1.15.** Effect of soil pH and aeration conditions on the availability of essential plant nutrients.

in the soil solution so that a depletion in solution by plant uptake results in a compensating release from solid forms. Situations where cation exchange equilibria or the solubility of sparingly soluble precipitates control concentration in the soil solution fall into this category. Other processes are unidirectional, such as the decomposition of organic matter releasing available nitrogen and other nutrients and mineral weathering resulting in the release of nutrients. Conversely denitrification in waterlogged soils results in the depletion of nitrate from the soil solution. Processes of particular relevance to sportsturf soils will be covered here. For further details, reference may be made to Wild (1988).

*Trace elements*

The quantity of plant nutrients in soils is usually tens or hundreds of times greater than the amount available to turf because, in the main, they occur in chemically stable forms which do not release nutrients into solution. Often, however, their chemical stability varies with changes in soil pH and aeration so that deficiencies are a consequence of soil physical conditions rather than absolute lack of the nutrient. The availability of most essential nutrients required in small quantities (trace elements) falls into this category (Fig. 1.15).

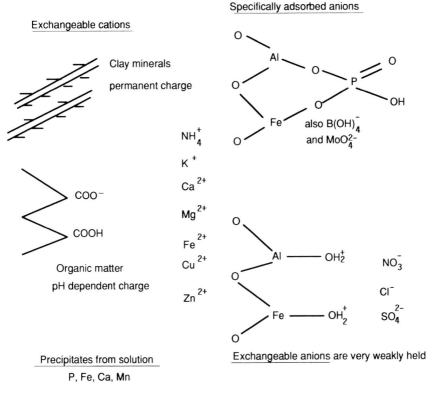

**Fig. 1.16.** Chemical forms of essential nutrients found in soils.

The figure summarizes a wide range of chemical processes but some key aspects can be highlighted. The availability of both iron and manganese is increased by acidity and poor aeration because the reduced forms of these elements are relatively soluble. Both zinc and copper are complexed by organic constituents of soils but also they both precipitate on iron oxides and calcium carbonates. The latter occur in soils of high pH. Boron and molybdenum are the two trace elements taken up as anions, $B(OH)_4^-$ and $MoO_4^{2-}$. Perhaps more accurately, boron is taken up as the undissociated acid, $B(OH)_3$. The effect of pH on the availability of these two nutrients is related to the pH dependency of their adsorption by oxides of iron and aluminium. This phenomenon of specific anion adsorption, illustrated for phosphate in Fig. 1.16, involves the anions of weak acids and is most effective over the range where the acid dissociates. Molybdic acid dissociates around pH 4 whereas boric acid dissociates around pH 9. The consequence of these different processes affecting trace elements is that the availability of all except molybdenum increases with decrease in pH.

Despite the occurrence of soil conditions which reduce the availability of trace elements, deficiencies are very rare in turf, even in sandy soils. The trace elements most likely to be deficient in these rare instances are either boron or copper. Copper is often very low in sandy soils and boron is easily leached out.

## Major nutrients

The nutrients most likely to restrict plant growth are the major nutrients nitrogen, phosphorus and potassium and it is for this reason that they are included in fertilizers. Sulphur is also a major nutrient but input from the air in industrial countries currently satisfies plant demand. Calcium and magnesium are required in intermediate amounts. A plant deficiency of the former is extremely rare even in acid soils where calcium is relatively low. Magnesium may occasionally be limiting in very sandy soils.

## Nitrogen

Nitrogen only accumulates in soils as a constituent of organic compounds. In this it differs from all other essential nutrients for, whilst in some soils a substantial proportion of the phosphorus, sulphur and trace elements such as boron may exist in organic compounds, inorganic mineral forms may be dominant.

Mineral forms of nitrogen which occur in soils are the plant-available forms of ammonium ($NH_4^+$) and nitrate ($NO_3^-$). These usually constitute less than 5% of the total nitrogen in soils but are vital components of the nitrogen cycle. The latter is summarized as a flow of nitrogen in turfgrass soils in Fig. 1.17.

Since the nitrogen content of soil organic matter is reasonably constant at around 4–5%, changes in soil nitrogen content reflect changes in soil organic matter. The mineralization of organic nitrogen to release ammonium is often illustrated as a one-way process. In fact the production of ammonium in soils is the net consequence of two opposing microbiological processes one of mineralization and one of immobilization. In essence soil microorganisms using nitrogen-rich materials such as young grass clippings release some of the nitrogen as ammonium ions but microorganisms utilizing low nitrogen materials such as peat require more nitrogen than is present in the material and will use ammonium present in the soil in competition with plants growing in the soil. Nitrogen fixed biologically by symbiotic or free-living microorganisms enters the soil system effectively as a high nitrogen organic residue.

There is little leakage of nitrogen from lightly grazed upland grassland and from non-intensively used turf where clippings are allowed to fly. Quite small

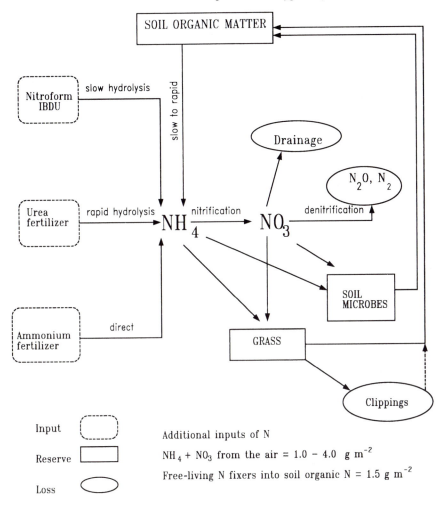

**Fig. 1.17.** The flow of nitrogen in turfgrass soils.

inputs of nitrogen from the air and biological fixation are sufficient to compensate for the minor losses such systems incur. In such circumstances nitrogen fertilizer may increase growth rates but it is not vital to a healthy sward.

The removal of turfgrass clippings constitutes a major drain on nitrogen cycling in the system. Nitrogen removal depends upon the nature of the turf and ranges from around $80 \, \text{kg ha}^{-1} \text{y}^{-1}$ in fine turf with a low input of fertilizer to around $300 \, \text{kg ha}^{-1} \text{y}^{-1}$ or more in heavily fertilized perennial ryegrass turf.

Since soil nitrogen reserve and nitrogen mineralization are generally related to soil organic matter content, nitrogen supply to turf is inadequate in soils of low organic matter content. These circumstances apply to the establishment of turf on either sand, subsoils or other low organic matter materials such as mining spoils and wastes or to the recovery of intensively used sportsturf where organic matter levels have been depleted by wear.

Ammonium has a pivotal position in nitrogen transformations. It is the form released from organic nitrogen, it is utilized by plants and microorganisms and is subject to oxidation to nitrate. Also it is directly or indirectly the main mineral form in nitrogenous fertilizers. Being a cation ammonium is retained in common with the other cationic plant nutrients on the cation exchange sites of clay minerals and soil organic matter (Fig. 1.16). In most loamy or clayey soils clay minerals provide the main source of exchange sites, but in very sandy sportsturf soils organic matter provides the only significant source.

Cations on exchange sites are in rapid equilibrium with cations in solution and so buffer soil solution concentrations. Thus, exchangeable cations, even though on the soil's solid phase, can be considered to be available to plants. In addition to acting as an available reserve of ammonium and other cations exchangeable cations are protected from leaching from soil.

Ammonium is oxidized to nitrate by groups of bacteria which use the energy derived from this oxidation for growth. These bacteria, the *Nitrosomonas* group which oxidize ammonium to nitrite and the *Nitrobacter* group which oxidize nitrite to nitrate belong to an interesting section of microorganisms called chemolithotrophs which oxidize the reduced forms of inorganic elements such as iron, manganese and sulphur as a source of energy. The overall process of nitrification is rapid in near-neutral soils but is reduced to less than one tenth of the rate in strongly acidic soils. In consequence turfgrasses in acidic soils obtain most of their nitrogen as ammonium. Indeed even in near-neutral soils this is probably the case because of the efficient root exploitation of soil by grasses and the relatively slow oxidation of ammonium.

At most soil pHs, oxides of iron and aluminium carry a positive charge on their surfaces. These attract anions electrostatically and thus can participate in anion exchange analogous to cation exchange on clay and organic colloids. Anions of strong acids such as nitrate, sulphate and chloride exchange on these positively charged surfaces but retention is very weak compared with cation retention. This contrasts with the much stronger covalent adsorption of anions of weak acids such as phosphate (Fig. 1.16). Anion exchange is the only mechanism by which nitrate can be held on the solid phase of soils and although the process may have some significance in oxide-rich soils of tropical areas, nitrate in soils can be considered to be in the soil solution and therefore highly susceptible to loss through leaching. Thus, despite the fact that ammonium and nitrate are taken up readily by plants, the concentration of nitrate in water draining from soils is usually greater than ammonium by ten to a hundred-fold.

Nitrification requires molecular oxygen and therefore aerated soil conditions. In contrast denitrification, the non-assimilatory reduction of nitrate to nitrous oxide and nitrogen gases occurs in poorly aerated and waterlogged soils. The compacted, organic-matter-rich surfaces of many sportsturf soils provide highly favourable conditions for denitrification for two reasons; firstly through an increased tendency for excessive wetness and secondly because denitrifying bacteria are most active when readily decomposible organic matter is present. Injudicious irrigation giving rise to alternating periods of good soil aeration and waterlogging results in substantial losses of nitrate through denitrification.

*Phosphorus*

Phosphorus is taken up by plants as an anion. The ionic species in the soil solution vary with pH. At a pH of 5, 90% is as $H_2PO_4^-$ but as the pH increases to pH 7 this dissociates to $HPO_4^{2-}$ and at this pH the amounts in the two forms are similar. The concentration of phosphorus in the soil solution is low because of its tendency to precipitate in solid forms and even in well-fertilized soils its concentration is unlikely to be greater than $1\ \mu g\ ml^{-1}$. Despite this, the high rooting density of grasses and infection by endotrophic mycorrhiza whose hyphae ramify within the soil and contribute to root exploitation, ensure that grasses are efficient scavengers for phosphate.

Phosphate reserves in soils are in organic and inorganic forms. In natural and semi-natural grassland areas receiving little or no fertilizer phosphate, organic forms predominate often accounting for 75% or more. The release of available phosphate from soil organic matter depends upon microbial activity so factors including temperature, pH and aeration which affect the mineralization of nitrogen also affect the mineralization of phosphorus from organic matter.

Inorganic phosphorus compounds predominate in well-fertilized soils and a range of forms occur depending on soil pH. Very sparingly soluble forms of iron, aluminium and calcium phosphates are involved and there are essentially two mechanisms by which these can form. The first mechanism is precipitation from solution. Thus if the phosphate concentration in the soil solution is increased by fertilizer application and the soil solution contains calcium, insoluble or sparingly soluble calcium phosphate will precipitate. Compounds formed in this way in addition to calcium phosphates are ferrous phosphates produced in poorly aerated soils and aluminium phosphates in strongly acidic soils. The other mechanism is where phosphate becomes adsorbed or precipitated onto an existing solid particle. Phosphate in solution can become precipitated onto particles of calcium carbonate in soils or can be adsorbed specifically onto the surfaces of iron and aluminium oxides. The calcium concentration in the soil solution increases with pH and free calcium carbonate is found in soils with pHs above about 6.5. Thus calcium phosphate forms predominate in

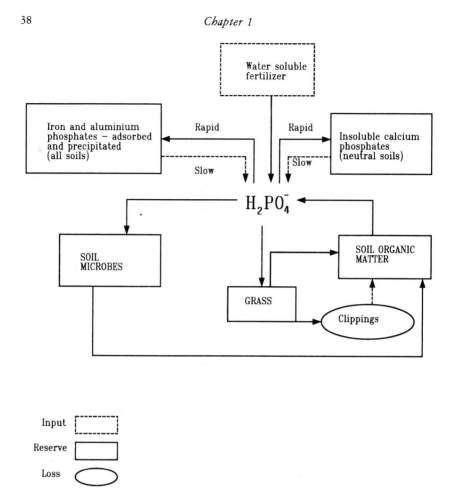

**Fig. 1.18.** The flow of phosphorus in turfgrass soils.

near-neutral and alkaline soils. The process of phosphate adsorption onto oxides of iron and aluminium is less pH dependent but is somewhat more significant in acidic soils. The mechanism of anion adsorption also occurs with other plant nutrients which are taken up as anions, notably boron and molybdenum (Fig. 1.16).

The transformations of soluble phosphorus into sparingly soluble forms – so-called immobilization has benefits and disadvantages. The main disadvantage is that turfgrasses·can be expected to utilize a small proportion of fertilizer phosphate in the year of application. The advantage is that immobilized phosphate is not completely unavailable and will benefit the turf for several years. The transformations of phosphorus in soil are summarized in Fig. 1.18.

## Potassium

Potassium does not occur in organic compounds in soils or plants. It is a major constituent of some soil minerals, notably micaceous clays and potash feldspar. These minerals release potassium very slowly but the rate of release is often adequate in loamy or clayey soils to supply the needs of lightly grazed grassland and amenity grassland when clippings are not removed. Very sandy soils are usually dominated by quartz and contain insignificant amounts of micaceous minerals. Some natural sand sized deposits derived from unweathered rock exist and can be produced by rock crushing and mechanical sorting. These may contain some micaceous minerals. Even when potassium containing minerals are present in sands, the rate of release of potassium is too slow to satisfy the requirements of sportsturf.

Potassium is retained on the cation exchange sites of clays and organic colloids and exchangeable potassium can constitute a substantial reserve in soils with a moderate cation exchange capacity. Exchangeable potassium can be considered to be available to plants but potassium can be 'fixed' more stably by some clay minerals. This property is related to the ionic size of potassium which enables it to sit comfortably in cavities between silica sheets in adjacent clay mineral layers. 'Fixed' potassium can exchange slowly with the soil solution and forms a reserve in soils rich in illite and/or vermiculite. In sportsturf soils fixed potassium has relevance only to soils on cricket squares.

The potassium ion is almost identical in size and behaviour to the ammonium ion. Both of these monovalent cations complete equally for exchange sites.

Because of their sandy nature the cation exchange capacity of sportsturf soils is often very small. They have little potential to retain potassium on exchange sites and potassium is readily leached from such soils in drainage water. Because of this, the ease of leaching of potassium approaches that of nitrate in very sandy soils.

The differences in susceptibility to leaching between nitrate, potassium and phosphorus from a loamy sand and a sand are illustrated in Fig. 1.19. The leaching circumstances illustrated are extreme with one bed volume being equivalent to 80 mm of rain on a 200 mm deep rootzone. It is clear that although nitrate leaches readily through sands and sandy soils, the presence of small amounts of clay causes a marked decrease in the leaching of potassium. The relative immobility of phosphate is very evident compared with nitrate and potassium and this is why its availability tends to build up even in very sandy sportsturf soils.

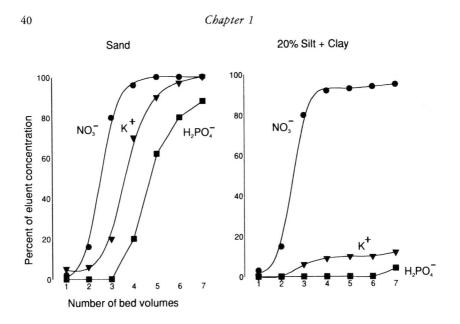

**Fig. 1.19.** The relative ease of leaching of nitrate, potassium and phosphate in fine sand and sandy soil containing 20% silt plus clay.

## References

Adams, W.A., Stewart, V.I. and Thornton, D.J. (1971) The assessment of sands suitable for use in sportsfields. *Journal of the Sports Turf Research Institute*, 47, 77–85.

Allison, F.E. (1973) *Soil Organic Matter and its Role in Crop Production*. Elsevier, Amsterdam.

Dixon, J.B. and Weed, S.B. (eds) (1989) *Minerals in Soil Environments*, 2nd edn. Soil Science Society of America, Madison, Wisconsin.

Hodgson, J.M. (1974) *Soil Survey Field Handbook*. Soil Survey Technical Monograph No. 5, Harpenden.

Lynch, J.W. (1983) *Soil Biotechnology: Microbiological Factors in Crop Production*. Blackwell Scientific, Oxford.

Riem Vis, F. (1981) Accumulation and decomposition of organic matter under sports turf. In: Sheard, E.W. (ed.) *Proceedings of the 4th International Turfgrass Research Conference*. Guelph, Canada, pp. 201–207.

Terzaghi, K. and Peck, R.B. (1967) *Soil Mechanics in Engineering Practice*, 2nd edn. John Wiley and Sons, New York.

Wild, A. (ed.) (1988) *Russell's Soil Conditions and Plant Growth*, 11th edn. Longmans, London.

# 2

## Turfgrasses for Sport and Amenity Use

---

### Introduction

Grasses occur naturally in most parts of the world. Perennial pasture grasses together with annual species which have been developed as cereals are the most important agricultural plants on earth. In nature, grasses are most successful in competition with other species in continental regions with rather low and seasonal rainfall. These climatic areas are now the main cereal-growing regions of the world typified by the prairies of north America and parts of continental Europe and Asia.

Perennial pasture grasses are best suited to amenity and sportsturf and within the British Isles these are found most extensively in the wetter west of the country where the climate is less suitable for arable farming. Grasses are not the natural climax vegetation in the British Isles. Their dominance is maintained in uncultivated areas by grazing animals preventing the establishment of shrubs and trees and by these animals and regular soil cultivation in other areas. The success and persistence of grasses in grazed or clipped conditions is because during vegetative growth the regenerative meristematic zones are close to the soil surface and remain undamaged by grazing animals and mowing machines.

Annual grasses are of little use in turf because they have evolved to fill a niche producing seed to survive situations of stress. The commonest stress is drought. This survival characteristic involving death of the vegetative plant is of no value in turf where increasingly there is a demand for round-the-year use. The classic alleged annual grass species, annual meadowgrass (*Poa annua*), behaves as a vegetative perennial in moist temperate climates, but retains the added advantage of profuse seed setting even at close mowing heights (Peel, 1982).

41

Although annual grasses are not used in turf, the seasonal growth of perennial grasses is exploited in the cool season/warm season system of some tropical and subtropical areas. In these situations temperate grasses cannot tolerate the high summer temperatures but tropical grasses become dormant in the cool season. In order to provide actively growing turf throughout the year the dormant period of tropical grasses is filled by temperate grasses overseeded into the turf prior to the dormant period. This complex turfgrass management, perfected on golf courses in moist subtropical regions of the USA, highlights an expertise practised by golf greenkeepers which is outside the experience of grassland farmers. This chapter and subsequent ones focus on temperate turfgrasses. Chapter 11 is devoted to tropical turfgrasses.

## CHARACTERISTICS OF TURFGRASSES

The term turfgrass is used loosely to describe species which when mown (or grazed) are capable of forming a more or less tightly knit and uniform vegetated surface. Some perennial grasses are tussock formers and tend to form a patchwork of crowns even when mown. Tufted hairgrass (*Deschampsia caespitosa*) and cocksfoot (*Dactylis glomerata*) which do not qualify as turfgrasses are of this type despite occurring in grazed pastures. Others found widely in permanent pasture on infertile soils, such as Yorkshire fog (*Holcus lanatus*), sweet vernal grass (*Anthoxanthum odoratum*) and crested dogstail (*Cynosurus cristatus*) may have a place in low maintenance amenity turf, but are either too patchy or have too short a season of active vegetative growth to warrant consideration as sown species in ornamental lawns or sportsturf.

Some grasses are intolerant of regular mowing or grazing. Typically these are found on wasteland or as weeds of arable land where circumstances do not expose them to regular defoliation. Couch grass (*Agropyron repens*) is an example of a species of this type which cannot be considered a turfgrass.

Grasses may change in character when mown regularly. One example, which is extremely unlikely to be used as a turfgrass species because of its deciduous character, is purple moor grass (*Molinia caerulea*). This species forms large tussocks in its natural habitat of highly acidic peaty soils. However, when mown regularly in the growing season, it forms a tight even turf not unlike that of tall fescue (*Festuca arundinacea*). The range of grasses found in sportsturf, amenity grassland and grazed pastures in the UK is listed in Table 2.1. Most of the species in the table have merits which justify their use in some types of maintained turf. However, several are only useful for low maintenance sites such as roadside verges, airfields and parkland. The temperate turfgrass species useful for summer or winter games pitches, golf and bowling greens and ornamental lawns are few and the following section concentrates on these species.

**Table 2.1.** The turf and pasture grasses in the UK and the situations under which they are most likely to occur.

| Common name | Botanical name | Occurrence[a] |
|---|---|---|
| Bentgrasses | | |
| Brown bent | *Agrostis canina* subsp. *montana* | AB--E |
| Common or browntop bent | *Agrostis capillaris/castellana* | AB--E |
| Creeping bent | *Agrostis stolonifera* | -B--E |
| Velvet bent | *Agrostis canina* subsp. *canina* | AB--E |
| Cocksfoot | *Dactylis glomerata* | -BC-- |
| Creeping softgrass | *Holcus mollis* | -B--- |
| Crested dogstail | *Cynosurus cristatus* | -B--- |
| Fescues | | |
| Chewings fescue | *Festuca rubra* subsp. *commutata* | ----E |
| Fine-leaved sheep's fescue | *Festuca tenuifolia* | -B--- |
| Hard fescue | *Festuca longifolia* | ----E |
| Meadow fescue | *Festuca pratensis* | -BC-- |
| Slender creeping red fescue | *Festuca rubra* subsp. *litoralis* | A---E |
| Sheep's fescue | *Festuca ovina* | A---E |
| Strong creeping red fescue | *Festuca rubra* subsp. *rubra* | AB-DE |
| Tall fescue | *Festuca arundinacea* | ---D- |
| Heath grass | *Danthonia decumbens* | AB--- |
| Meadow foxtail | *Alopecurus pratensis* | -B--- |
| Meadow grasses | | |
| Annual meadowgrass | *Poa annua* | --CDE |
| Rough stalked meadowgrass | *Poa trivialis* | -BC-- |
| Smooth stalked meadowgrass | *Poa pratensis* | -B-D- |
| Wood meadowgrass | *Poa nemoralis* | -B--- |
| Moor mat grass | *Nardus stricta* | A---- |
| Purple moor grass | *Molinia caerulea* | A---- |
| Ryegrass | | |
| Italian ryegrass | *Lolium multiflorum* | --C-- |
| Perennial ryegrass | *Lolium perenne* | -BCD- |
| Soft brome | *Bromus mollis* | -B--- |
| Sweet vernal grass | *Anthoxanthum odoratum* | -B--- |
| Timothy | | |
| Large-leaved timothy | *Phleum pratense* | -BCD- |
| Small-leaved timothy | *Phleum bertolonii* | -B--- |
| Tufted hairgrass | *Deschampsia caespitosa* | -B--- |
| Wavy hairgrass | *Deschampsia flexuosa* | A---- |
| Yorkshire fog | *Holcus lanatus* | -B--E |

[a]*Occurrence*: A, very acidic grassland; B, old pastures, ley meadows; C, intensively managed pastures; D, winter games pitches, horse racing tracks; E, fine sportsturf and ornamental lawns.

# Spreading ability

In general, grasses which are able to spread vegetatively by stolons (creeping above-ground stems) or rhizomes (underground stems) produce the most tightly knit turf. Examples include bents (*Agrostis* spp.), smooth stalked meadowgrass (*Poa pratensis*) and strong creeping red fescue (*Festuca rubra* subsp. *rubra*). However, this creeping behaviour is also shown by other species in regularly mown turf not normally recognized as having this ability. The process is called aerial tillering and involves the production of new tillers away from the parent plant on shoots extending above ground within or slightly above the turf. These tillers form roots in the protected moist environment. Annual meadowgrass has this ability as do some cultivars of perennial ryegrass (*Lolium perenne*). The tightly knit and even turf produced by species which spread by stolons or rhizomes is normally advantageous, but this characteristic is difficult to manage on a cricket square. It can also be disadvantageous in sportsturf subjected to severe tearing-type wear (e.g. horse racing).

# Verdure

Verdure is a qualitative assessment of the fresh green colour of turf and good verdure throughout the year is deemed desirable for many turf uses. The shade of green may vary but in essence greenness reflects active growth. A rare exception to this rule is a cultivar of meadow fescue (*Festuca pratensis*) selected at the Welsh Plant Breeding Station at Aberystwyth whose senescent leaves remain green. Many naturally selected strains of turfgrass species in the UK (especially those in the uplands) have a short growing season and are visually unattractive or at least non-green for around half of the year.

# Wear tolerance

Tolerance of wear and regular mowing are requirements of most turf but especially sportsturf. The games of golf and bowls demand a closeness of mowing on greens which can be tolerated by very few species. Only species of bent, fine-leaved fescues and annual meadowgrass of the temperate species can produce a tightly knit and reasonably vigorous turf when mown at heights of 3–8 mm. Other sport and amenity uses of turf do not demand such ultra-close mowing, and turfgrass cultivars of broadleaved species, including perennial ryegrass, timothy and smooth stalked meadowgrass grow well provided the cutting height is not much below 20 mm.

There are two components of wear. One is compression through treading to which all used turf is subjected to a greater or lesser extent. The other is tearing-type wear which is especially important in the winter games of soccer and rugby. Wear tolerance also has two components, one is the inherent resistance of grass to compression, scrubbing or tearing, which is especially important during weather conditions when growth is slow (i.e. durability). The second component, which is a major asset when growing conditions are good, is the ability to recover rapidly from the damage inflicted. All turfgrasses have a high tolerance of compression and the key differences in wear tolerance arise from the relative sensitivity to scrubbing or tearing and the ability to recover from damage. The tight mat-forming grasses typified by bents and smooth stalked meadowgrass are very tolerant of scrubbing wear, but the tightly knit nature of the turf means that studded boots tend to tear out sections of turf (divots). In contrast, studded boots tend to slide through perennial ryegrass turf without tearing out rooted tillers because of its open growth habit. Despite the poor ability of perennial ryegrass to spread vegetatively, its vigour and lower susceptibility to tearing-type wear means that in long-term practical situations it is more successful in winter games turf than any other sown species. The occurrence, features and uses of the important turfgrass species in the UK are summarized in Table 2.2.

## GROWTH AND DEVELOPMENT OF TURFGRASSES

## Establishment

Virtually all amenity turf and sportsturf in Britain is propagated initially from seed, notwithstanding the fact that much turf established from seed is transplanted to 'turf' virgin sites. This is not the case in tropical and subtropical areas where turfgrasses are frequently propagated vegetatively. Bermudagrass (*Cynodon dactylon*) is the most extensively used turfgrass in tropical areas and it is usually established from stolon transplants. Vegetative propagation is slower to give a tight turf cover but has the advantage of enabling a genetically uniform turf to be established of species which are natural outbreeders.

Successful establishment from seed requires a moist but adequately aerated soil and a soil temperature above 10°C. Germination is slow at temperatures below 10°C and this slowness results in a higher seedling mortality through attack by pathogenic fungi. Appropriate fungicides have been shown to improve establishment in such conditions (Lewis, 1988). Waterlogged anaerobic conditions can usually be avoided, but desiccation can occur on sandy soils. Rapid drying out of sand-dominant soils is a major hazard in the post-winter-game recovery period. Daylength is long in early May and the energy input of the sun is approaching its maximum. Establishment from seed at this time of year on sand-dominant rootzones is unreliable if irrigation is not available. The

**Table 2.2.** Uses and characteristics of the main turfgrasses in the UK.

| Name | Use/occurrence | Habit | Vegetative spreading | Cutting tolerance | Leaf fineness | Sensitivity tolerance |
|---|---|---|---|---|---|---|
| *Agrostis canina* | FT/A/OL | D,P | S | G | F | T-heat, cold |
| *Agrostis capillaris/castellana* | FT/A/OL | L-D,P | R/S | G | F | S-low P, T-acidity, cold |
| *Agrostis stolonifera* | FT/A/OL | L-D | S | M-G | M-F | T-wet soils |
| *Festuca arundinacea* | ST/A | L,E | T | P | C | T-drought |
| *Festuca longifolia* | FT/A/OL | D,E | T | G | F | T-drought |
| *Festuca rubra* subsp. *commutata* | FT/OL | D,E | T | G | F | S-wear, T-acidity |
| *Festuca rubra* subsp. *rubra* | FT/A/OL | L-D,E | R | M-G | F | S-wear, T-heavy metals drought, acidity, salt |
| *Lolium perenne* | ST/A/OL | L-D,E | T/AT | M-G | M-F | S-acidity, cold, T-wear |
| *Poa annua* | FT/ST | L-D, E-P | AT | M-G | M-F | S-low N, P, drought, T-acidity |
| *Phleum bertolonii* | ST/A | L-D | T | M | M | S-drought |
| *Poa pratensis* | ST/A | L-D,E | R | M | C-M | S-mild moist winters, T-wear |

Key:
FT, fine turf
ST, winter games/horse racing
A, amenity
OL, ornamental lawns

L, loose
D, dense
E, erect
P, prostrate

R, rhizomes
S, stolons
T, tillers
AT, aerial tillers

P, poor
M, moderate
G, good

C, coarse
M, moderate
F, fine

S, senstitive
T, tolerant

**Table 2.3.** Approximate length of time for emergence and establishment of some turfgrass species, soil temperature 10-13°C. (Adapted from Adams and Bryan, 1974.)

| Species | Days to emerge | Days to reach 50 mm height |
|---|---|---|
| Agrostis castellana | 9 | 32 |
| Festuca rubra | 9 | 27 |
| Lolium perenne | 9 | 18 |
| Poa annua | 9 | 27 |
| Poa pratensis | 12 | 33 |
| Phleum bertolonii | 8 | 26 |

greatest danger is that chitted seed or very young seedlings may become desiccated. Early autumn is a much more reliable period for grass establishment when this choice is possible. Decreasing daylength in September and the occurrence of dews on clear nights mean that good establishment can be achieved even in quite dry weather. Furthermore, competition from broadleaved weeds is reduced.

The speed of germination and establishment differs markedly between turfgrass species. These differences along with other characteristics influence the choice of species for specific uses. Table 2.3 illustrates the differences in germination and establishment between some common turfgrasses. There are three important points:

1. Perennial ryegrass germinates and establishes very rapidly.
2. Bentgrass, despite having very small seeds, germinates quite quickly but is slower to reach a height of 50 mm than its common companion in seed mixtures chewings fescue (*Festuca rubra* subsp. *commutata*).
3. Smooth stalked meadowgrass is slow to germinate and establish.

It is clear that smooth stalked meadowgrass would be unlikely to establish successfully when sown in a seed mixture with perennial ryegrass (Adams and Bryan, 1974). Also, because of the slow establishment of the former, it cannot be recommended as an overseeding species in the end of season renovation of winter games pitches (Canaway *et al.*, 1986). Chewings fescue, which rarely persists other than as a minor constituent of intensively used sportsturf, often acts as a nurse grass for bent in golf and bowling greens because of its more rapid establishment.

Differences in the rate of establishment of the different species within a seed mixture affect the composition of the young sward because rapidly germinating species shade the slower ones. In addition, the seeding rate of a specific mixture will affect competition amongst seedlings when the mixture contains grasses differing in establishment rate. Thus even a small proportion of perennial ryegrass in a seed mixture can result in it dominating the seedling turf if very high sowing rates are used. These two aspects, the relative speed of establishment and the influence of sowing rate, are as important as percentage content in a seed mixture in influencing the composition of the sward. In summary, slow-establishing species in a mixture are favoured by an increase in percentage content and decrease in sowing rate. Very rapidly establishing grasses will be favoured relative to others if very high seeding rates are used even though the percentage content in the mixture is small. In this context a high seeding rate is over $30 \, g \, m^{-2}$ and a low seeding rate less than $15 \, g \, m^{-2}$.

## Vegetative growth

Once established, unmown grasses grow vegetatively until flowering is initiated. After this a high proportion of the carbohydrate fixed by the leaves is diverted to flowering and seed production. This results in a marked decrease in the growth rate of leaves and roots of perennial grasses in mid-season. If grasses are mown regularly and flowering is prevented there is a much more even pattern of growth throughout the summer. Although there is still a surge of growth in late spring, variations in leaf growth rate throughout the summer are influenced mainly by nutrient and water supply (Shildrick, 1986).

Grasses grow by cell division and elongation at their base. Thus the oldest part of a leaf is its tip and the youngest its base. Mowing removes the mature parts of leaves which, through photosynthesis, make the main contribution to the energy balance of the plant. Severe mowing may not damage meristematic tissues but it removes much of the active photosythetic tissues, leaving stem bases and immature leaves which may not be energy self-sufficient. Recovery requires the use of the plant's carbohydrate reserves which are small in rapidly growing turf (Adams *et al.*, 1974). Mowing should be sufficiently frequent to minimize stress which is primarily determined by the proportion of leaf tissue removed rather than height of cut. Ideally the uncut height should not be more than 50% greater than the cut height (Adams, 1975). This demands a frequency of mowing which is not practicable in some situations. Nevertheless, uncut height should not be allowed to exceed double the cut height.

## Tillering

Buds in the axils of leaves on the initial grass shoot develop into new shoots. These are called tillers and the number and vigour of tillers controls the density of leaves in the sward. Tillering is affected by several environmental factors including temperature, light intensity and mineral nutrition. The optimum

**Table 2.4.** Tiller density (no. dm$^{-2}$) in three-month-old *Poa pratensis* cv. 'Fylking' maintained at three heights of cut and at three levels of available nitrogen. (Adapted from Adams *et al.* 1974.)

| Nitrogen level (mM) | Height of cut (mm) | | |
|---|---|---|---|
| | 12 | 25 | 75 |
| 1 | 147 | 160 | 166 |
| 4 | 306 | 333 | 453 |
| 10 | 306 | 347 | 507 |

temperature for tillering in temperate grasses is around 18–24°C and tillering is favoured by high light intensities. These are circumstances which favour net photosynthate accumulation by the plant (Langer, 1979).

Nitrogen is the nutrient most likely to restrict tillering. An increase in nitrogen supply to deficient grass increases both leaf growth and tillering. However, the stimulation of the production of new tillers declines at levels of available nitrogen well below those which continue to stimulate topgrowth in general (Adams *et al.*, 1974).

Height of cut also affects tillering. When turf is allowed to grow tall, young tillers cannot survive without the support of the parent shoot because of shading. This consequence is more apparent with tall-growing broadleaved species such as perennial ryegrass than with fine-leaved fescues which produce a high density of tillers even when mown infrequently. Lax maintenance of ryegrass-dominant winter games pitches in the summer results in an open sward with low tiller density predisposing the surface to a rapid loss of grass cover in the playing season.

Within the normal range of turfgrass mowing heights, the potential rate of new tiller production decreases with a decrease in mowing height in agreement with the general principle that tillering is favoured by a good positive energy balance (Table 2.4). Whilst temperature, light intensity, soil moisture and nutritional conditions may have a dominant effect, in essence for every cultivar there is an optimum cutting height above and below which tillering decreases. Maintenance at this optimum, which ranges from less than 10 mm for fine turfgrasses to over 25 mm for most turf-type perennial ryegrasses, ensures the maximum rate of infilling of an open sward caused by wear or sparse germination.

Tiller density (i.e. the number of tillers per unit area) is a characteristic which has been selected for in turf-type perennial ryegrasses. The Aberystwyth cultivar 'S23' was the pasture ryegrass best suited to winter games sportsturf in the UK before major breeding and selection for amenity and sportsturf use began in the early 1970s. This cultivar has a typical maximum tiller density of around 350 shoots $dm^{-2}$, which compares with around 450–500 shoots $dm^{-2}$ of similar weight per shoot for the most compact turf ryegrasses (Shildrick and Peel, 1984).

In a tight turf of perennial species there is a continual turnover of tillers with old tillers dying and being replaced by young ones. This turnover of tillers in a mixed sward provides the opportunity for species or cultivars better adapted to the cultural or environmental conditions to gain dominance. Changes in botanical composition through variations in response to fertilizer input or mowing height can be quite dramatic, even within a single growing season (Adams, 1977).

Some grasses are capable of limited growth at temperatures barely above freezing. Indeed, an accepted criterion for climatic suitability for herbage grasses is the sum of day degrees above $0\,^{\circ}C$ for the first six months of the year (Peacock, 1976). The use in breeding programmes of naturally selected cultivars of species from different climatic areas has enabled the creation of cultivars with a diversity of seasonal growth patterns not exhibited by the native ecotypes.

# Root development

On germination, grasses produce a small number (less than eight) of seminal roots. These are much more highly branched and exploitive of soil than the adventitious roots which succeed them (Langer, 1979). Seminal roots are effective for a few months only and may in part account for the particular vigour of perennial grasses during their first year. Adventitious roots, which are the roots of perennial grasses, arise at the nodes of basal stems, stolons and rhizomes. They are frequently produced above ground in mown turf at the nodes of prostrate tillering stems. The growth form of a grass plant is illustrated in Fig. 2.1.

Much of the root system of perennial grasses is annual or shorter in persistence. There is a turnover of roots during the growing season depending upon maintenance regime and mowing in particular (Evans, 1973). However, there is a major loss of roots over winter which varies between species. The annual turnover of the root system of perennial turfgrasses is likely to be in excess of 70% (Weaver and Zink, 1946; Troughton, 1981).

Active root growth occurs at lower temperatures than leaf growth and substantial growth of new roots occurs in early spring before significant leaf growth is apparent. As the season progresses changes in light input and temperature in addition to maintenance regime influence the amount of live root material. For example, close mowing has a less adverse effect on root mass under high than under low light intensities (Troughton, 1957).

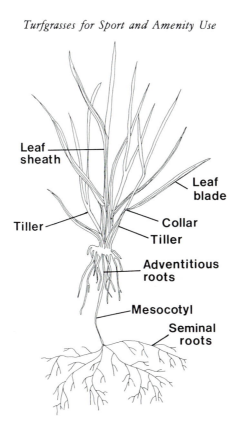

**Fig. 2.1.** The growth form of a grass plant.

Root development requires an input of energy. This is provided in spring through the mobilization of energy stored over winter in the form of non-structural polysaccharides in stem bases and through the translocation of sugars fixed by leaves. In the growing season, roots obtain all their energy from sugars produced in the leaves. Thus whilst stems and leaves depend on roots for their supply of essential mineral nutrients, roots require energy derived from photosynthesis for maintenance and growth. Although internal control is affected by plant growth regulators the interdependence of tops and roots affects their relative performance. One predictable consequence is that severe mowing, which decreases photosynthesis, reduces the energy available for root growth and maintenance. In consequence there is a rapid loss of roots following severe defoliation (Fig. 2.2). Regular close mowing causes a less extreme decrease in photosynthesis but in general closer mowing brings about a decrease in living root mass (Table 2.5). The range of cutting heights over which a decrease in rooting occurs differs between species and cultivars of the same species.

The level of plant nutrient supply affects root development and the

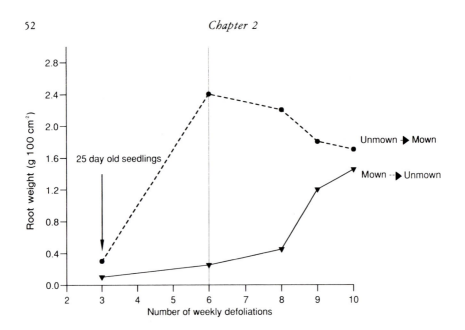

**Fig. 2.2.** Effect of mowing regime on the root weight of perennial ryegrass cv. 'S23'. Grass subplots were either mown weekly for six weeks or left uncut. At that point the treatments were exchanged and the experiment continued up until ten weeks. Root weight was determined by destructive harvesting on five occasions. (Adapted from Bryan, 1969.)

**Table 2.5.** Root mass (mg plant$^{-1}$) of 100-day-old *Lolium perenne* cv. 'S23' in relation to nitrogen supply level and height of cut.

| Nitrogen level (mM) | Height of cut (mm) | | | |
|---|---|---|---|---|
| | 6 | 13 | 25 | 50 |
| 1 | 5 | 10 | 21 | 40 |
| 4 | 3 | 9 | 26 | 60 |
| 10 | 5 | 13 | 17 | 36 |

relative growth of roots and shoots. The plant nutrients most important in this are phosphorus and nitrogen. It has been shown that roots proliferate in soil zones which have higher levels of available phosphorus (Russell, 1977). This implies a potential problem because phosphate through its low mobility tends to accumulate at or near the surface of turfgrass soils where significant soil mixing through cultivation is not possible. Such accumulation may induce shallow rooting although definitive evidence for this is not available.

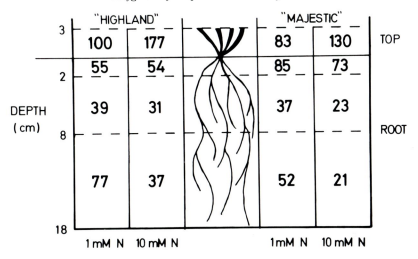

**Fig. 2.3.** Effect of nitrogen supply level on root distribution with depth (g m$^{-2}$) in 100-day-old *Agrostis castellana* 'Highland' and *Lolium perenne* 'Majestic' maintained at 30 mm height of cut (Adams, 1982).

## Nitrogen supply

The nitrogen supply level to turf has major effects on both shoot/root ratio and on rooting pattern. There is also an interaction with height of cut. When a turf cultivar is grown at low nitrogen supply levels, and when nitrogen supply alone controls growth, topgrowth is small relative to the mass of roots maintained. An explanation for this is that nitrogen supply to the tops is insufficient to permit the potential rate of photosynthate utilization for leaf growth. Some surplus carbohydrate is stored and some is used for the maintenance and growth of roots. As nitrogen supply is increased, more photosynthate is used for topgrowth and, though the amount of roots may increase, the ratio of topgrowth to maintained root mass falls (Adams, 1971; Canaway, 1984). As nitrogen supply is increased further, topgrowth continues to rise but a point is reached where root quantity either remains the same or decreases. This critical nitrogen level decreases with decrease in cutting height which affects the potential growth response of the turf. At very high nitrogen inputs root quantity decreases through a decrease in rooting depth (Fig. 2.3). Roots produced at very high nitrogen levels are thicker than those produced at lower levels (Table 2.6).

**Table 2.6.** The effect of nitrogen supply level on the mean root diameter of six turfgrass cultivars. (After Adams *et al.*, 1985.)

| Species and cultivar | Low N (1.0 mM) | High N (50 mM) |
|---|---|---|
| *Agrostis castellana* 'Highland' | 130 | 190 |
| *Festuca rubra* subsp. *commutata* 'Koket' | 120 | 215 |
| *Lolium perenne* 'Barclay' | 135 | 160 |
| *Lolium perenne* 'Loretta' | 80 | 215 |
| *Lolium perenne* 'Majestic' | 120 | 175 |
| *Lolium perenne* 'S23' | 165 | 175 |

## TURFGRASS ECOLOGY

There are five main factors that affect the performance and persistence of grasses in turf. They are soil nutrient status, soil pH, soil moisture regime, height of cut and intensity of use. The relative sensitivity/tolerance of the key sportsturf grasses to the first two of these factors is illustrated in Fig. 2.4. Soil moisture may exert an effect by being in excess and thus causing waterlogging, or by being in short supply and causing drought stress. In natural and semi-natural grassland both conditions exert selective pressure in different sites at different times, but in maintained turf neither should be critical although they often are.

## Effect of nutrient status and soil pH

With regard to soil nutrient status any one of the 13 essential mineral nutrients could be in critically short supply, but in practice the most likely to be limiting are either phosphorus or nitrogen. In natural ecosystems both are often close to being limiting to growth, although in maintained turf it is usually nitrogen which controls growth. This is because phosphate fertilizers have a high residual value and fertilizer applied over a number of years builds up levels in the soil which prevent critical limitation.

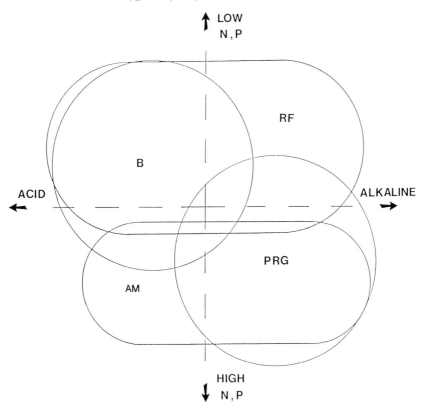

**Fig. 2.4.** General adaptations of four turfgrass species to soil pH and fertility status, B-*Agrostis capillaris*, PRG-*Lolium perenne*, RF-*Festuca rubra* subsp. *rubra* and *commutata*, AM-*Poa annua*.

Most semi-natural grassland in the UK is on acidic soils in upland areas where it is grazed by sheep. The most acidic soils have a pH in the range 3.7–4.0. Earthworms cannot tolerate such acidity and the absence of biological mixing in these uncultivated soils results in the accumulation of surface organic matter. Dominant or co-dominant grasses in the most nutrient-poor sites are moor matgrass (*Nardus stricta*) and purple moor grass (*Molinia caerulea*), with the former predominant on freely drained sites and the latter on wet peaty soils. When nutrient availability is greater (especially of phosphate) but the soil pH is still very low, bents and fine-leaved fescues increase in importance. Cultivars of species within these genera are the ones used in the production of fine turf for amenity and sport. Although acid tolerant, bent and fine-fescue cultivars perform well on neutral soils, but often they are in competition with the much more vigorous broadleaved turfgrasses such as perennial ryegrass, which does not tolerate a soil pH below about 5.0.

An increase in the level of available aluminium is the main selective pressure on grasses and other plants at soil pHs below 5 (Foy *et al.*, 1978). Indeed, a fall in soil pH from 5 to 4 is much more important ecologically than a fall from pH 7 to pH 6. In addition to grasses, several common turfgrass weeds are sensitive to aluminium. White clover is a good example for, although it is the most acid tolerant of common *Trifolium* species, it survives with difficulty in soils more acidic than pH 5.

An important turfgrass found on acidic soils is annual meadowgrass. It has long been recognized that whilst this grass is not found generally in tracts of upland pasture, it accounts for much of the vegetation on and near sheep paths. These sites differ from the extensive pasture areas in two ways; firstly, they are subjected to compaction and, secondly, they receive an abnormally high proportion of animal excreta. These conditions of wear and high fertility are somewhat similar to those on intensively used sportsturf so one should not be surprised at the success of the species across the range of sportsturf uses.

## Competition between turfgrass species

The level of fertilizer nitrogen input and height of cut are the main factors affecting competition between grasses in most maintained turf. Relative responsiveness (in terms of leaf growth) to fertilizer nitrogen and tolerance of close mowing are most important. Of the widely used temperate turfgrass species, perennial ryegrass is the most responsive to nitrogen. It is for this reason that it is aggressive towards the slower-growing bents and fescues when the height of cut does not cause it undue stress. The vigour of the older cultivars of perennial ryegrasses such as Aberystwyth 'S23' used for turf was associated with broad leaves and rapid shoot extension. In consequence only a small amount of leaf area was produced close to the ground and mowing at heights less than 30 mm caused severe weakening.

Efforts worldwide in the breeding and selection within perennial ryegrass have resulted in the production of cultivars which tiller profusely and produce a large leaf area per unit of soil surface area (leaf area index) when mown closely. Many cultivars retain the natural vigour necessary to recover from intensive use for sport and shade out unwanted species. Others are better suited to the less rigorous demands of the ornamental lawn. Most modern turf-type perennial ryegrasses produce a tight turf when mown at a height of 15–20 mm or less. Provided vigour is retained this allows them not only to survive at closer mowing heights, but also to compete with other species, notably annual meadowgrass, which roots less deeply and is more susceptible to fungal diseases.

Fine-leaved fescues and bents tolerate mowing at heights of less than 8 mm. Perennial ryegrasses cannot tolerate regular mowing at this height but annual meadowgrass can. It is unwise to generalize about the attributes of grasses within the bents and fine fescues because they cover such a wide spectrum. Nevertheless, several points justify a mention. Bentgrasses tend to be more responsive to nitrogen whereas fine-leaved fescues are more tolerant of drought stress and low nutrient availability. This difference in tolerance accounts for much of the variation in the relative success of bents and fescues in fine sportsturf, ornamental turf and on hill pastures. Bentgrass alone or more commonly bentgrass/chewings fescue seed mixtures are normally used to establish ornamental lawns and fine turf sports areas such as golf greens and bowling greens. In established fine turf used for sport, fine-leaved fescues are rarely dominant or co-dominant. This contrasts with semi-natural upland pastures and some ornamental turf where fine-leaved fescues are often co-dominant. In golf and bowling greens, naturally selected strains of annual meadowgrass take the place of fine-leaved fescues and a very high proportion of the sward is usually accounted for by bents and annual meadowgrass (Adams, 1975).

The success of annual meadowgrass in both fine sportsturf and winter games pitches results from a combination of its tolerance of close mowing, growth response to high soil fertility and ability to spread both vegetatively and through seed produced over much of the year. Naturally selected strains of annual meadowgrass are moderately tolerant of wear on winter games pitches and the treading wear on fine turf, provided soil moisture and nutrient levels are good. Bryan and Adams (1971) showed the persistence of annual meadowgrass on soccer pitches at the end of the season. A pattern of decrease in the relative proportions of perennial ryegrass compared with annual meadowgrass has been shown in several studies. Changes with time in the proportions of the two species are illustrated in Fig. 2.5, which also shows how end of season overseeding of high wear areas with perennial ryegrass prevents the domination by annual meadowgrass which is evident on low wear areas.

Because of its intolerance of both nutrient and drought stress it is, at least in theory, possible to prevent annual meadowgrass becoming dominant in fine turf areas by cultural means. The problem is that there is a conflict between the maintenance regime necessary to improve appearance and recovery from wear and that required to discourage annual meadowgrass. Low nutrient input (especially nitrogen but also phosphorus) and allowing drought stress to occur, each reduce the vigour of annual meadowgrass more than bents and fescues. This type of cultural control is practicable in ornamental lawns and fine sportsturf with low intensities of use. However, as the intensity of use increases more wear is inflicted and there is an increasing need to stimulate grass growth by using inputs of fertilizer and irrigation water. These inputs increase the competitiveness of annual meadowgrass and make its cultural control impossible. Cultural control of annual meadowgrass is possible on top-class cricket outfields provided they are not used for winter games.

The vigour of regrowth is an important factor in determining persistence

*Chapter 2*

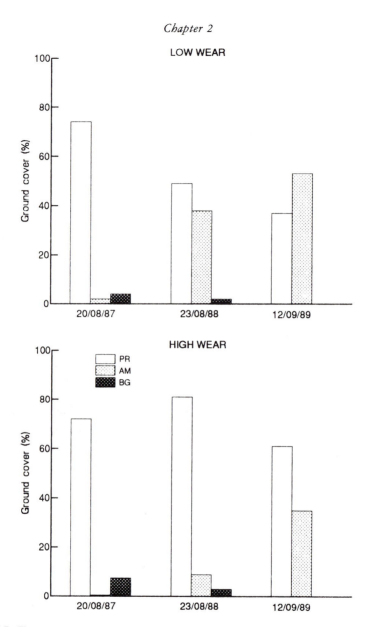

**Fig. 2.5.** The proportions of *Lolium perenne* (PR), *Poa annua* (AM) and bare ground (BG) at the beginning of three consecutive playing seasons on the high and low wear areas of a suspended watertable soccer pitch. Overseeding with *Lolium perenne* was carried out at the end of each playing season. The first playing season was 1987/88.

and an ability to recover rapidly accounts largely for the success of perennial ryegrass. Less vigorous species, particularly the fine-leaved fescues, cannot be maintained under intensive use because recovery is slow and uncompetitive. In rough grazing pastures of the uplands the problem of slow recovery by fescues and other species is shown in the persistent scars which show both the route of vehicles driven by 'off-road' fanatics and the paths of hill-walkers if numbers are sufficiently large.

## TURFGRASS NUTRITION

Grasses, like other green plants, are photosynthetic autotrophs. This means they use the energy in light to convert carbon dioxide into organic carbon compounds which are used as a source of chemical energy and provide the building blocks for the diverse compounds plants contain. The healthy growth of grasses requires an adequate supply of 13 essential mineral nutrients which are obtained from soil (Table 1.9). If the supply of an individual nutrient is inadequate the deficiency may give rise to diagnostic visual symptoms but, in general, grasses do not display these as clearly as some other plants. A deficiency has to be quite severe before symptoms are shown and often a growth-limiting nutrient reduces growth with no visual detrimental effect. There is, for example, a wide range over which the supply of nitrogen affects growth with little or no effect on the apparent health of the turf. Thus when nitrogen fertilizer is applied the increased amount of nitrogen removed in clippings is mainly due to an increase in the yield of clippings rather than an increase in the concentration of nitrogen within them.

Low temperature or drought stress may reduce growth but when nutrient supply restricts growth it is usually attributable to one of a few of the essential nutrients. The nutrients which most frequently control growth are nitrogen, phosphorus and potassium and this is why they are included in fertilizers. To a certain extent this is because these elements are required in the greatest amounts. For example, grasses require over ten thousand times as much nitrogen as they do molybdenum. However, other factors which may be important include large differences between soils in the natural reserves of nutrients and chemical transformations which decrease nutrient availability (see Chapter 1).

## Ratio of nutrients in turfgrasses

The concentration of individual nutrients in grasses varies to a rather limited extent. In the case of the major fertilizer nutrients, the range rarely exceeds two to three-fold, but it may be as much as ten-fold in the case of some minor nutrients. This has two important implications. The first is that the ratio between the concentrations of nitrogen, phosphorus and potassium (and indeed other nutrients) in turfgrasses is fairly constant especially when grass is mown regularly and much of the leaf tissue is of similar physiological age. The second is that when a particular nutrient which is restricting growth is supplied as a fertilizer it will not only increase the uptake of that nutrient but will also increase the demand for the uptake of other nutrients.

The ratio between the concentrations of nitrogen, phosphorus and potassium in the clippings of turfgrasses is around 8.5:1:6 (Skirde, 1974; Wray, 1974; Adams, 1977), which transposed to the normal fertilizer nutrient expression of $N:P_2O_5:K_2O$ approximates to a weight ratio of 4:1:3. This ratio defines the relative amounts of fertilizer nutrients required by turf and thus, simplistically, this is the ratio which should be applied in fertilizers if clippings are removed. The situation is more complicated because of differences between nutrients in leaching losses and in chemical immobilization within soil. Nevertheless, it is a guideline for turfgrass nutrition which becomes more relevant the closer to 'sand culture' turfgrass nutrition becomes.

The quantitative need for the different plant nutrients by turfgrasses is interrelated, and healthy growth and development require nutrient balance. Clichés such as 'phosphate stimulates rooting' should be avoided because they imply that individual nutrients act independently. Under certain circumstances an increase in phosphate supply will affect root development, but changes in nitrogen supply affect root development more often.

Nitrogen is the plant nutrient which most frequently controls growth. From the foregoing section it is evident that an increase in the input of fertilizer nitrogen will increase potential growth and thus the uptake of all other nutrients. Since the plant demand for a particular nutrient depends upon the level of supply of the nutrient which is controlling growth, then there is no precise level of availability which can be considered 'adequate'. One may contrast an intensively used sports area where fertilizer input to stimulate growth and recovery is vital with parkland or amenity grassland where slow growth is acceptable and where clippings may be allowed to fly and the nutrients within them returned to the soil.

Fertilizers have become so cheap and generally available over the last half century that they are assumed to be essential. Fundamentally fertilizers are required to rectify inherent soil deficiencies and to compensate for nutrients removed in clippings or lost to the air or in drainage water. Agriculture is exploitive in the sense that the crop which is removed contains plant nutrients. In turf maintenance, grass clippings are the crop and their removal is equivalent to removing a crop. This particular component of maintenance is the one which more than any other determines fertilizer requirement. Where usage of turf and damage to turf are low and clippings are not removed, there may be no need for fertilizer; indeed, there are many areas of amenity turf and infrequently used sportsturf which have a perfectly adequate sward with no fertilizer input. Intensively used sportsturf is a complete contrast where, in the absence of high fertilizer inputs a satisfactory turf cannot be sustained.

## Functions of essential nutrients

The essential plant nutrients serve many different types of function. Virtually all of the nitrogen in plants occurs in organic compounds and indeed more than half of the compounds within living cells contain nitrogen. In contrast to nitrogen, potassium is not present in any organic compounds. It is very mobile in plants and is translocated to young tissues when in short supply. This is why visual symptoms of a deficiency are the chlorosis and dying back of older leaf tips. Potassium has many roles in plants. It is essential for the action of many enzymes although the bulk of the potassium is involved in the regulation of osmotic pressure and pH control within plant cells. Phosphorus functions in compounds which trap and transfer chemical energy the best known of which is adenosine triphosphate (ATP). In addition, phospholipids, which occur in membranes and nucleic acids which contain the genetic code, contain phosphorus. Three amino acids contain sulphur, and therefore many proteins are sulphur containing. Some proteins contain sulphur in association with iron where the former is not within amino acids. These iron–sulphur proteins include ferredoxin, which is a key compound in the trapping of light energy.

Despite the fact that the minor or trace elements are needed in small quantities, some serve many essential functions. Small amounts are needed because they are either activators or constituents of enzymes where one or a few atoms are required for the functioning of a large and active molecule. The roles of the individual essential plant nutrients are described in detail by Mengel and Kirkby (1987).

## Nutrient uptake

Turfgrass roots take up the nutrients they need from the soil solution and the immediate 'availability' of a nutrient in a soil is related to its concentration in the soil solution. Whilst the availability of nutrients to turfgrasses increases as their concentration in the soil solution increases, the amount of any given nutrient in the soil solution at any one time is small compared with the quantity turfgrasses need over a growing season. In essence, therefore, adequate supply of plant nutrients depends upon the soil system being able to replenish the soil solution when nutrients are taken up by the turf. The mechanisms for this were examined in Chapter 1 and range from the decomposition of organic matter releasing available nitrogen and other nutrients to the equilibria between ions in the soil solution and inorganic solids. The nature of the equilibria between the soil solution and the solid phase is complex and differs between nutrients.

Without even considering the mechanisms by which plant nutrients move to roots, it is obvious that assessing the 'availability' of a nutrient to turfgrasses is difficult. Since the amount of a nutrient in the soil solution is not an adequate guide to availability it is necessary to 'extract' from the soil the reserve of a given nutrient which will become available over a growing season. This involves an empirical extraction which, to have any predictive value, must be seen in the light of correlations between extractable values and plant performance observed in field experiments.

Soil analysis for available nutrients will never enable the precise assessment of the amount of fertilizer to apply but should be used primarily to assess whether, in the long term, fertilizer input is excessive (resulting in unusually high levels of available nutrients) or the opposite. Aspects of soil analysis and interpretation of data are examined in Chapter 5.

Leaf analysis has been proposed and is used as a means of assessing the adequacy of nutrient supply to plants. Such analyses are valuable in identifying deficiencies of trace elements where soil analyses may be less reliable. A limitation with leaf tissue analysis is that concentrations in plant tissues are not related directly to fertilizer response, as is the case with soil analyses. There are other factors which have to be taken into account. Nutrients differ in the extent to which they are translocated within plants. Iron tends to move little whereas potassium is translocated to young tissues, resulting in K sparing when it is in short supply. Thus the physiological age of the tissues sampled may require close control. In practice, with regularly mown turf this is not a serious problem. An additional factor is that tissue analysis may fail to identify a deficiency when it is caused by the interaction between elements. Under such circumstances a physiological deficiency may occur when concentrations in leaf tissue are normal. An example here is that Fe deficiency can be induced by high levels of each of several transition elements including Co, Ni, Cu and Zn, but in such deficiencies Fe concentrations in leaf tissue may be quite normal.

## IDENTIFICATION OF GRASS SEEDS AND TURFGRASSES

The seeds of different turfgrasses vary with species. It is usually possible to identify the seeds of an individual species on morphological appearance alone, but sometimes it is only possible to identify seeds as belonging to a group of species (e.g. fine-leaved fescues). It is not possible to identify seeds of cultivars within a species from morphology alone and more sophisticated techniques are required such as electrophoretic separation of plant proteins. Nevertheless, the ability to recognize the seeds of species is useful because it enables adulteration of seed mixtures by unwanted species to be recognized (e.g. perennial ryegrass in a ryegrass-free mixture). Figure 2.6 shows not only the differences in morphology between the seeds of the main turfgrass species but also differences

| Species | | Number of seeds per gram |
|---|---|---|

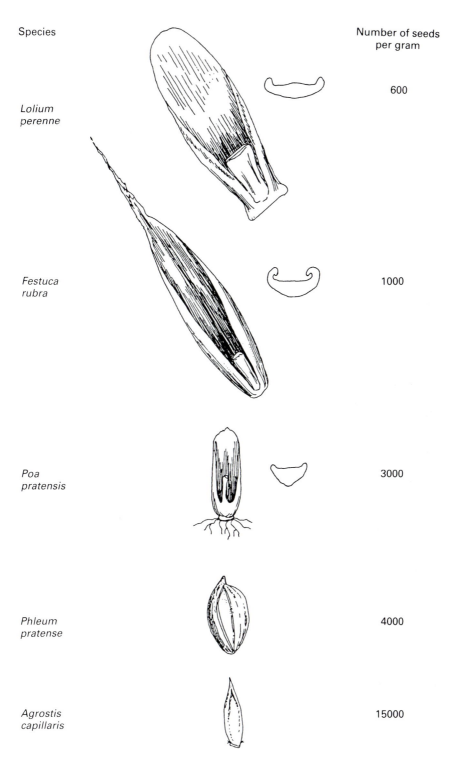

Lolium
perenne

600

Festuca
rubra

1000

Poa
pratensis

3000

Phleum
pratense

4000

Agrostis
capillaris

15000

**Fig. 2.6.** Morphology of the seeds of common turfgrass species and seed number per gram.

in seed size. The latter is important because seed mixtures are described on the basis of proportion of species by weight not seed number. Earlier in this chapter it was pointed out that because of differences between turfgrass species in their speed of establishment, very high seeding rates favour a rapidly establishing species even when it constitutes a small percentage of the seed mixture. We also need to recognize that the size of seeds differs markedly between species so that the same percentage by weight in a mixture of species may provide very different numbers of seeds per unit weight. For example a typical bowling green seed mixture containing 80% Chewings fescue and 20% bentgrass would contain four times the number of bentgrass seeds as fescue seeds (Fig. 2.6). Even if the germination rates of the two species were identical the more rapid establishment of Chewings fescue (Table 2.3) would give it a competitive advantage, countering the difference in seed numbers. Evidently in the case of this particular seed mixture, because of the difference in speed of establishment between the two species, very low seeding rates would favour the predominant establishment of bentgrass and conversely very high seeding rates would ensure competitive dominance by fescue. In short, proportion in the seed mixture, seed size, seeding rate and establishment rate are all factors affecting the species composition of turf established from seed mixtures.

The identification of grasses is much simpler when seedheads are present. In turf which is mown frequently and closely the only species able to produce seed is annual meadowgrass. Seedheads are produced in some amenity grassland situations such as roadside verges when mowing is carried out only once or twice per year. Not only are seedheads absent from most turf but the amount of vegetative growth may be very short. Nevertheless the number of grass species which are likely to be present is quite small and it is usually possible to make an accurate identification from vegetative characteristics. A glossary of morphological features used in the identification of grasses is given in Fig. 2.7 and a key for identification follows. Examples of the use of the key are provided. The morphological features used to identify grasses are often physically small and a small magnifying lens is essential. The ability to identify the common turfgrasses is important, for without this it is not possible to appreciate the effectiveness of overseeding, the relative persistence of particular species and their susceptibility to disease or indeed to monitor changes in turfgrass composition in response to wear or maintenance.

## KEY FOR NAMING SOME COMMON GRASSES

This key is based on Hubbard's classification on naming the grasses by their vegetative characteristic.

  1 Leaf expanded (easily flattened out)` ................................. go to 2
  – Leaf bristle-like (difficult to flatten out) .......................... go to 19

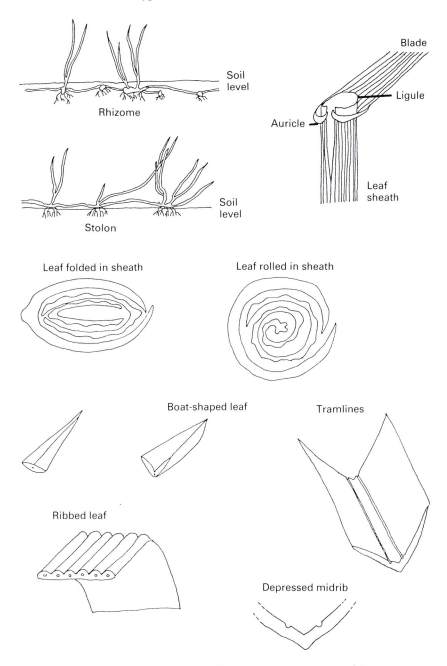

**Fig. 2.7.** Turfgrass characters used in identification in the vegetative condition.

**2** Ligule a dense fringe of short hairs ................................... go to 3
**–** Ligule present ......................................................... go to 4

**3** Auricles absent, leaf blade rolled in shoot, leaf blade slightly hairy. Shoots erect and dark green. Roots tough and cord-like. Leaves deciduous .......................... *Molinia caerulea* (purple moorgrass or flying bent)
**–** Auricles absent, leaf blade folded in shoot, conspicuous spreading hairs at junction of sheath and blade. Leaves have 'tramlines', greyish-green in colour, sparsely hairy. Leaves spreading ........................................ ............................................. *Danthonia decumbens* (heath grass)

**4** Young leaf blade rolled in shoot (convolute) ...................... go to 5
**–** Young leaf folded in shoot (conduplicate). Plant quite glabrous (without hairs) ................................................................. go to 16

**5** Leaves with auricles ............................................... go to 6
**–** leaves without auricles ............................................. go to 8

**6** Plant with conspicuous fleshy creeping rhizomes. Leaves often slightly hairy, leaves dull green in colour ................................................. ........................................ *Agropyron repens* (couch grass or twitch)
**–** Plant without rhizomes .............................................. go to 7

**7** Auricles and marginal junction of sheath and blade glabrous, leaves bright green and glossy beneath. Shoot bases often pink, basal buds rounded. Decaying remains in base dark brown ............................................ .............................................. *Festuca pratensis* (meadow fescue)
**–** Auricles and marginal junction of sheath and blade fringed with a few small hairs. Plant dark green, large and rather coarse to touch. Decaying remains in base tough, whitish, sometimes scale-like sheaths in base ............................................... *Festuca arundinacea* (tall fescue)
**–** Plant quite glabrous (without hairs). Leaf blade glossy beneath. Shoots at base often pink or red, basal buds pointed. Decaying remains in base not conspicuously fibrous .............. *Lolium multiflorum* (Italian ryegrass)

**8** Leaves aromatic. Characteristic smell and taste to leaves. Plant with spreading hairs at junction of sheath and blade; blade slightly hairy ................................ *Anthoxanthum odoratum* (sweet vernal grass)
**–** Leaves not aromatic .................................................. go to 9

**9** Shoots thickened at base, more or less bulbous .................. go to 10
**–** Shoots not thickened at base, not bulbous ........................ go to 11

**10** Lower sheaths yellow or orange in colour. Leaves sparsely hairy or hairless. Basal internodes short, swollen ................................................. ........................ *Arrhenatherum elatius* var. *bulbosum* (tall oat grass)
**–** Basal internodes short. Lower sheaths not yellow or orange, but whitish. Leaves light green with a characteristic twist. Plant quite glabrous (without hairs) ..................... *Phleum pratense/bertolonii* (timothy)

11 Plants with rhizomes or stolons ................................... go to 12
– Plants without rhizomes or stolons ................................ go to 13

12 Rhizomes:
– Ligule very short, blunt. Plant glabrous, sheath smooth. Leaves conspicuously ribbed above and tapering from base to tip. Plant often greyish-
green. Rhizomes small and cork-like ...........................................
................ *Agrostis capillaris/castellana* (common bent or browntop)
– Ligule blunt. Basal leaf sheaths possess faint purple veins. Leaves softly
hairy. Nodes covered with a fringe of downwardly protecting hairs. Plant
has large rhizomes ...................... *Holcus mollis* (creeping softgrass)
– Ligule prominent, blunt. Plant erect, sparsely hairy or hairless. Leaves green
or yellowish-green ...............................................................
.................. *Brachypodium pinnatum* (chalk false brome or tor grass)
– Stolons:
– Ligule rounded, often torn, larger than *A. capillaris*. Plant glabrous,
sheaths mostly smooth, leaves finely pointed. A very variable grass .....
............................................ *Agrostis stolonifera* (creeping bent)

13 Leaf sheath not hairy ................................................. go to 14
– Leaf sheath hairy ..................................................... go to 15

14 Ligule very long and pointed. Leaves dark green. Leaf blade very promi-
nently ribbed, plant stiff and harsh to touch .................................
...................................... *Deschampsia caespitosa* (tufted hairgrass)
– Ligule very blunt. Living basal leaf sheaths often streaked with bright
yellow. Leaf blade trough-shaped at junction with sheath. Upper leaf
surface ribbed, lower leaf surface smooth and glossy .......................
.......................................... *Cynosurus cristatus* (crested dogstail)

15 Basal leaf sheaths possess pink or purple veins. Leaves softly hairy. Leaves
greyish-green ............................... *Holcus lanatus* (Yorkshire fog)
– Leaf sheaths densely and softly hairy. Leaves greyish-green in colour, very
hairy. Sheath split near the tip ...............................................
...................................... *Bromus mollis* (soft brome or lopgrass)
– Leaves hairy but not softly so, rough to the touch, foliage drooping, of pale
sap-green colour. Leaf blade narrowing towards base and tapering to tip
.................. *Brachypodium sylvaticum* (slender or woodfalse brome)

16 Leaves with small auricles (when present). Leaf blade dark green, glossy
and smooth beneath, dull and distinctly ribbed above. Base and sheaths
reddened .............................. *Lolium perenne* (perennial ryegrass)
– Leaves without auricles .............................................. go to 17
– Leaves without auricles, and leaves possess characteristically depressed
mid-rib and a boat-shaped tip. 'Tramlines' often visible ....... go to 18

**17** Ligule very blunt. Living basal leaf sheaths often streaked with bright yellow. Leaf-blade trough-shaped at junction with sheath. Upper leaf surface ribbed, lower leaf surface smooth and glossy ........................
.......................................... *Cynosurus cristatus* (crested dogstail)
 – Ligule pronounced. Living basal leaf sheaths whitish, sticky. Shoots broad, strongly laterally compressed. Leaves green, greyish-green .................
................................................ *Dactylis glomerata* (cocksfoot)

**18** Ligule conspicuous and membranous. Young leaves often transversely crinkled. Leaves yellow-green in colour. Fibrous often shallow root system. Creeping base ........................... *Poa annua* (annual meadowgrass)
 – Ligule long and pointed. Leaf blade green or purplish, very glossy beneath. Base and sheath may be reddened. Plant has creeping stolons ...........
.................................... *Poa trivialis* (rough stalked meadowgrass)
 – Ligule short and blunt. Leaves dull grey-green with blunt apex. Tramlines normally present. Plant has rhizomes .........................................
.................................. *Poa pratensis* (smooth stalked meadowgrass)

**19** Ligule obscure or very short .......................................... go to 20
 – Ligule short but quite distinct. Ligule broader than leaf. Plant pliant, waxy to touch. Leaves dark green ..... *Deschampsia flexuosa* (wavy hairgrass)

**20** Roots cord-like. Shoots very closely packed on short rhizomes. Outer leaf blades horizontally spreading. Basal sheaths tough and shining ..........
................................................ *Nardus stricta* (moor matgrass)
 – Roots fine and slender ............................................... go to 21

**21** Sheath split near the top, plants quite glabrous. Shoot bases sometimes slightly pink .................................. *Festuca ovina* (sheep's fescue)
 – Sheath not split near top. Base of shoots usually pink or red in colour. Leaf sheaths usually minutely pubescent. In luxuriant forms leaf can be flattened out. Rhizomes present .............. *Festuca rubra* (red fescue)

## EXAMPLES USING THE KEY

**1** To find *Lolium perenne*:

| Question | Answer | Result |
|---|---|---|
| 1 | Leaf expanded ................................................ | go to 2 |
| 2 | Ligule present ................................................ | go to 4 |
| 4 | Young blade folded ....................................... | go to 16 |
| 16 | Leaves with small auricles ............................................... | |
| | .................................. *Lolium perenne* (perennial ryegrass) | |

**2** To find *Agrostis stolonifera*:

| | | |
|---|---|---|
| 1 | Leaf expanded | go to 2 |
| 2 | Ligule present | go to 4 |
| 4 | Young blade rolled | go to 5 |
| 5 | Leaves without auricles | go to 8 |
| 8 | Leaves not aromatic | go to 9 |
| 9 | Base not bulbous | go to 11 |
| 11 | With stolons | go to 12 |

12   Stolons rounded, long ligule, spreading by leafy stolons, prostrate base roots from nodes ....................................................
................................ *Agrostis stolonifera* (creeping bent)

## REFERENCES

Adams, W.A. (1971) Management of sportsturf. In: Stewart, V.I. and Adams, A.D. (eds) *Lectures on Sportsfield Construction and Management.* Sisis Equipment (Macclesfield) Ltd, Cheshire, pp. 90–94.

Adams, W.A. (1975) Some developments in the selection and maintenance of turfgrasses. *Scientific Horticulture* 26, 22–27.

Adams, W.A. (1977) Effects of nitrogen fertilization and cutting height on the shoot growth, nutrient removal and turfgrass composition of an initially perennial ryegrass dominant sportsturf. In: Beard, J.B. (ed.) *Proceedings of the 3rd International Turfgrass Research Conference.* American Society of Agronomy, Madison, Wisconsin, pp. 343–350.

Adams, W.A. (1982) Planned use of turf – design and maintenance. *Mitterilungen Deutsche Bodenkundliche Gesellschaft* 33, 215–223.

Adams, W.A. and Bryan, P.J. (1974) *Poa pratensis* L. as a turfgrass in Britain. In: Roberts, E.C. (ed.) *Proceedings of the 2nd International Turfgrass Research Conference.* American Society of Agronomy, Madison, Wisconsin, pp. 41–47.

Adams, W.A., Bryan, P.J. and Walker, G.E. (1974) Effects of cutting height and nitrogen nutrition on the growth pattern of turfgrasses. In: Roberts, E.C. (ed.) *Proceedings of the 2nd International Turfgrass Research Conference.* American Society of Agronomy, Madison, Wisconsin, pp. 131–144.

Adams, W.A., Tanavud, C. and Springsguth, C.T. (1985) Factors influencing the stability of sportsturf rootzones. In: Lemaire, F. (ed.) *Proceedings of the 5th International Turfgrass Research Conference.* Institut National de la Recherche Agronomique, Paris, pp. 391–399.

Bryan, P.J. (1969) Effect of cutting and nutrition on the distribution of growth in young *Lolium perenne* cv. 'S23'. Unpublished Honours Dissertation, University of Wales.

Bryan, P.J. and Adams, W.A. (1971) Observations on grass species persisting on English League soccer pitches in spring 1970. *Rasen Turf Gazon* 2, 46–51.

Canaway, P.M. (1984) The response of *Lolium perenne* (perennial ryegrass) turf grown on sand and soil to fertilizer nitrogen. II. Above-ground biomass, tiller numbers and root biomass. *Journal of the Sports Turf Research Institute* 60, 19–26.

Canaway, P.M., Carr, L., Bennett, R.A. and Isaac, S.P. (1986) The effect of renovation on the ground cover of swards of eight turfgrass species grown on sand and soil and subjected to football-type wear. *Journal of the Sports Turf Research Institute* 62, 118–132.

Evans, P.S. (1973) The effects of repeated defoliation to three different levels on root growth of five pasture species. *New Zealand Journal of Agricultural Research* 16, 31–34.

Foy, C.D., Chaney, R.L. and White, M.C. (1978) The physiology of metal toxicity in plants. *Annual Review of Plant Physiology* 29, 511–566.

Hubbard, C.E. (1984) *Grasses*, 3rd edn revised by J.C.E. Hubbard. Penguin Books Ltd, Middlesex, 476 pp.

Langer, R.H.M. (1979) *How Grasses Grow*, 2nd edn. Studies in Biology No. 34, Edward Arnold, London, 66 pp.

Lewis, G.C. (1988) Fungicide seed treatments – their effect on grass seedling emergence and growth. In: Gibbs, R.J. and Adams, W.A. (eds) *Proceedings of the 6th Discussion Meeting of Amenity Grass Research*. University College of Wales, Aberystwyth, pp. 32–45.

Mengel, K. and Kirkby, E.A. (1987) *Principles of Plant Nutrition*, 4th edn. International Potash Research Institute, Berne, Switzerland.

Peacock, J.M. (1976) Temperature and leaf growth in four grass species. *Journal of Applied Ecology* 13, 225–232.

Peel, C.H. (1982) Review of the biology of *Poa annua* L. with special reference to sports-turf. *Journal of the Sports Turf Research Institute* 58, 28–40.

Russell, R. Scott (1977) *Plant Root Systems*. McGraw-Hill (UK) Ltd, Maidenhead, Berks, 298 pp.

Shildrick, J.P. (1986) Mowing regimes and turfgrass regrowth. *Journal of the Sports Turf Research Institute* 62, 36–49.

Shildrick, J.P. and Peel, C.H. (1984) Preliminary trials of perennial ryegrass cultivars (Trials A1, B2 and B3) 1980–3. *Journal of the Sports Turf Research Institute* 60, 73–95.

Skirde, W. (1974) Nahrstoffgehalt und Nahrstoffentzug von Rasen bei verschieden hoher Dungung und verschiedenem Bodenaufbau. *Rasen Turf Gazon* 5, 68–73.

Troughton, A. (1957) *The Underground Organs of Herbage Grasses*. Bulletin, Commonwealth Bureau of Pastures and Field Crops No. 44, Hurley.

Troughton, A. (1981) Length of life of grass roots. *Grass and Forage Science* 36, 117–120.

Weaver, J.E. and Zink, E. (1946) Length of life of roots of ten perennial range and pasture grasses. *Plant Physiology* 19, 201–217.

Wray, F.J. (1974) Seasonal growth and major nutrient uptake of turfgrasses under cool wet conditions. In: Roberts, E.C. (ed.) *Proceedings of the 2nd International Turfgrass Research Conference*. American Society of Agronomy, Madison, Wisconsin, pp. 79–88.

**PLATE 1.** Profile of an acidic upland pasture soil with natural development of a surface organic horizon (see Chapter 1, p.10).

**PLATE 2.** A selection of 0–20 mm surface sections of county cricket squares in 1967 showing the widespread occurrence of layering (see Chapter 5, p.186).

**3**

**4**

PLATE 3.  Section of a 'Cell system' sand profile golf green at Myerscough College, Preston, UK, showing a black layer of sulphide accumulation (35–55 mm) underlain by a thin zone of iron oxide accumulation (see Chapter 5, p.188).

PLATE 4.  Grass roots proliferating in hollow tine holes in an upturned section of a compacted golf green (see Chapter 5, p.189).

**PLATE 5.** Typical 'diamond' shaped wear pattern on a soccer pitch (see Chapter 6, p.212).

**PLATE 6.** The wetting up after three hours of dry soil cores, one from a 'normal' fine turf area and one from an adjacent 'dry patch' area. Grass was cut to ground level and cores inverted onto a tension table at zero tension (see Chapter 7, p.289).

**7**

**8**

**9**

PLATE 7.   Bradgate Park, Leicestershire, UK, a designated Country Park, is a
deer park which includes grass, heath, woodland and aquatic habitats (see
Chapter 10, p.354).

PLATE 8.   Iffley Meadows, Oxford, UK, is managed to encourage populations of
the fritillary butterfly *(Fritillaria meleagris)*, which attract many visitors in late
spring each year (see Chapter 10, p.357).

PLATE 9.   A 3 m-wide grass and wildflower strip sown at the edge of a cereal
field creates an attractive feature (see Chapter 10, p.371).

# 3 | SPORTSTURF DRAINAGE SYSTEMS

## INTRODUCTION

From Chapter 1 it should be evident that the growing potential of a soil is related to it being able to provide two important divergent physical characteristics: that of drainage to allow aeration and that of water retention for use by plant roots. From the plant point of view, the physical framework of a soil must therefore contain a system of pores stable enough to preserve its porosity, but not so strong that the individual plant roots cannot enlarge and explore pores within the soil system. The continuity of pores is as important as the quantity.

A sufficient quantity of large diameter, stable and continuous macropores will ensure satisfactory infiltration of water, drainage and aeration within the soil profile, and unimpeded root extension; a sufficient quantity of small diameter, stable micropores will ensure that gravitational drainage is restricted although these micropores must still be large enough to release water to plant roots. The size and quantity of pores that confer these characteristics have been discussed in Chapter 1. However, most soils only possess the above characteristics when the primary particles of sand, silt and clay are bound together into a fragile system of water-stable aggregates.

The extent to which this fragile physical system can function for amenity purposes in its natural state without being adversely affected depends not only on the climate and intended use, but also on the soil type. Only in sand and loamy sand soils (with a particle size composition of at least 70% sand) can sufficient macroporosity be maintained in the absence of particle aggregation to guarantee rapid transmission of water and a satisfactory state of aeration. However, since the great majority of native soils in the UK and elsewhere in

**Table 3.1.** Pore size distribution in a sandy soil at bulk densities of 1.24 g m$^{-3}$ and 1.52 g m$^{-3}$. (Adapted from Schuurman, 1965.)

| Pore diameter ($\mu$m) | % porosity (v/v) for soil bulk density (g m$^{-3}$) | |
|---|---|---|
| | 1.24 | 1.52 |
| > 1200 | 2.4 | 1.3 |
| 100-1200 | 15.0 | 0.3 |
| 6-100 | 23.6 | 23.9 |
| < 6 | 9.5 | 14.2 |
| | 50.5 | 39.7 |

the world have silt plus clay contents in excess of 30%, adequate drainage and aeration are possible in these soils only through particle aggregation.

## THE POTENTIAL FOR POOR DRAINAGE IN AMENITY SOILS

The physical properties of most natural soils are vunerable to major changes when mechanical stress disrupts their aggregated structure. Such disruption is most likely to occur when soils are at or wetter than their naturally drained condition (i.e. 'field capacity'). Thus the indiscriminate use of loamy or clayey soils for sport either during or immediately after rain invariably results in the degradation of surface soil structure. All soils will become compressed to some extent but the end consequence (namely a plastic poached surface and a very compact and dense soil) depends upon the particle size distribution of the soil. In any compaction event whether by animals, people, machines or natural forces, it is always the largest pores which are most readily destroyed (Table 3.1). Loamy or heavier soils, when compacted, are therefore likely to be made up almost exclusively of micropores because the soil contains too little sand to avoid the macropore system between the sand particles being blocked by silt and clay.

Surface degradation of soil structure causes a marked decrease in water infiltration rate and hydraulic conductivity, which in turn leads to surface waterlogging or ponding. These conditions not only provide an unfavourable environment for root growth, but may also cause a reduction in soil strength. If play continues, the surface quickly becomes slippery and plastic as ground cover is destroyed. The likelihood of these conditions occurring is also greatly increased if the soil has been moved (e.g. in cut-and-fill operations) or has been stored, because both processes destroy stable aggregation. On most natural soils used for sportsturf, it is saturation of the compacted surface layer and the

**Table 3.2.** Percentages of pitches with drainage problems. (After Ward, 1983a.)

| | Major problem [a] | Moderate problem [b] | Total |
|---|---|---|---|
| School | 13.0 | 18.4 | 31.4 |
| Public | 17.1 | 34.2 | 51.3 |
| Overall | 20.3 | 23.7 | 44.0 |

[a] Defined as a pitch where matches were frequently cancelled during November-February due to waterlogging.
[b] Defined as a pitch where cancellations occurred occasionally during November-February due to waterlogging.

inability of its micropore systems to transmit rainfall rapidly to the subsoil which cause poor playing conditions or cancellation of games.

The extent and severity of poor drainage in amenity soils was estimated in a comprehensive review carried out by Ward (1983a). In this review, questionnaires were sent to Football League clubs, local authorities and golf clubs. The worst problems of poor drainage were associated with the latter two groups. Only 12.5% of local authorities had no problems with drainage of their pitches. In the public sector as a whole, 44% of pitches had a drainage problem and 20% of pitches regularly had matches cancelled between November and February (Table 3.2). Even the drainage of newly constructed pitches was often inadequate.

For golf, Ward (1983a) found that the number of days a course was closed during the monitoring period (1981) varied from zero to 45 days. On average, 18 hole golf courses were closed for 5 days per year due to waterlogging. Greens were more prone to poor drainage than fairways or tees and over 80% of the courses monitored had poor drainage on at least one of their greens.

For horse racing, Wyatt (1990) estimated that somewhat less than 5% of fixtures had been lost in the last 20 years of which 73% had been due to waterlogging and frost, the rest being lost because of snow, high winds and fog.

## THE TRADITIONAL AGRICULTURAL APPROACH TO DRAINAGE

The art and practice of field drainage for combating the problem of wet soils has been an important technique of soil husbandry in agriculture since at least Roman times and is needed on about 60% of lowland Britain. The history of agricultural field drainage contrasts markedly with that of sportsturf drainage since it is only from about 1965 in Britain that the drainage of sportsturf has been recognized to present problems not normally associated with the agricultural use of soils and has received specific attention. Prior to this time and on many more recent occasions the principles of agricultural drainage have been applied to sportsturf drainage resulting in a catalogue of failures.

## Objective and benefits of agricultural drainage

The objective of draining agricultural land has always been to increase the proportion of the year when cultivations or grazing can be carried out without risk of damaging the soil and to create a more favourable physical environment for crop growth. Most systems achieve this objective through the removal of water from the subsoil more quickly than would happen naturally, thus speeding the return of a soil to its field capacity state after rain. Drainage is needed in the UK because much of the agricultural land is on loamy or clayey soil and because the climate provides long periods in the year when temperature and light conditions are favourable for growth, but excessive soil wetness reduces the potential for successful seedbed preparation, germination, establishment and growth of crops.

Although the adverse effects of waterlogging on plant performance, growth and development are the primary reason for installing agricultural drainage, the ability of soil to carry machinery or indeed withstand any physical stress without damage to soil structure decreases as soil moisture content increases. Although a soil's wettest drained condition or field capacity varies somewhat in practical terms, it can be defined objectively (Chapter 1). A soil cannot be drained to a condition drier than field capacity and the period of time in a year when soils are in this condition or wetter is used as a criterion for assessing the number of days when a soil is too wet to cultivate (field capacity days). When field capacity is exceeded, the risk of causing damage to soil structure from both farm machinery and grazing animals is severe.

Field capacity is often used to imply a soil condition when aeration is adequate for healthy root growth. This is not necessarily the case because the expression merely describes the situation when loss of water by gravitational drainage has effectively ceased. The expression does not give information on the balance of air-filled and water-filled soil pores. Thus whereas a well-aggregated soil contains a large amount of air-filled pore space at field capacity, a compact soil containing only micropores would be saturated at field capacity and incapable of admitting air into the profile once surface water had been cleared. This latter situation is typical of sportsturf soils where degradation of surface soil structure has arisen through indiscriminate use in wet conditions.

## Basic design criteria of agricultural drainage systems

Climatic data are valuable when designing agricultural drainage systems and for determining the extent to which soil moisture surpluses are likely to limit the safe use of land as well as the extent to which soil moisture surpluses are likely to damage plant growth through the adverse effects of waterlogging. For example, in Agroclimatic Area 50 (Dyfed, England and Wales) there is an average of approximately 700 mm of excess winter rainfall annually (rainfall minus potential evapotranspiration) which could create considerable water-logging problems in the absence of satisfactory drainage (Smith, 1984). In the south Derbyshire/north Leicestershire region, this excess reduces to 250 mm and in the Vale of Kent to 210 mm. Similarly, in Dyfed, field capacity is exceeded on average for 257 days per year, compared with 168 days for the south Derbyshire/north Leicestershire region and 146 days per year for the Vale of Kent. The benefit of agricultural drainage increases the greater the number of field capacity days.

Meteorological data also provide information on the risk of the soil being too dry for grass growth, a consideration just as important as that of the need for drainage. Whilst in Dyfed there may only be an average of five drought-stressed days for grass per year (with a maximum summer soil moisture deficit of 57 mm), in the Vale of Kent these figures extend to 44 days per year and 109 mm respectively (Smith, 1984). As Stewart (1980) pointed out, for those growing amenity grass, the indications are therefore that the major problem in the east and south of the UK is likely to be summer drought, with waterlogging being the main problem in the north and west of the country. Despite this general comment, the meteorological figures are based on averages of widely varying seasons, and drought conditions do occur in the west and north of the UK and waterlogging can be a problem in the south and east. Overall this suggests that climatic data based on averages is unlikely to be particularly useful for those designing sportsturf drainage schemes and a different approach is required (see Chapter 4).

Davies *et al.* (1982) cited four different conditions which gave rise to waterlogging problems in agricultural soils which could be reduced by the installation of drains. These were: (i) lowland soils where it was necessary to lower the watertable using a network of ditches supplemented by pipe drainage; (ii) clayland areas where the permeability of clay soils was too slow to give efficient natural drainage and a system of deep subsoiling or mole drainage was needed together with pipe drainage in conjunction with gravel backfill to act as a connector between the surface treatments and the pipes; (iii) spring sites where permeable soils were usually overlying impermeable clays and the precise location and depth of pipe drains was vitally important to keep water moving laterally at the junction of the permeable and impermeable strata; (iv) impermeable pans either of natural origin or man-made where the usual remedy was deep subsoiling perhaps in conjunction with pipe drainage.

The economic benefit of agricultural drainage is often small and for this reason priority must be given to the cheapest effective system. The spacing of an underdrain system can be greatly increased for example if a second tier of a non-permanent carrier network of mole drains is superimposed (Fig. 3.1).

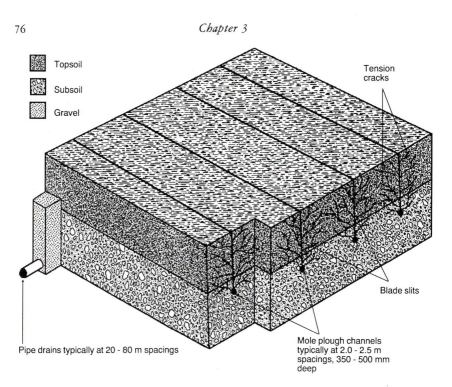

Topsoil

Subsoil

Gravel

Tension cracks

Blade slits

Pipe drains typically at 20 - 80 m spacings

Mole plough channels typically at 2.0 - 2.5 m spacings, 350 - 500 mm deep

**Fig. 3.1.** Closely spaced non-permanent mole drains connecting with a widely spaced underdrain system.

Whether water is accumulating in a soil because of subsoil impermeability or because the site is a receiving one, the purpose of the pipe drains is: (i) to cut off or contain the flow of extraneous water; (ii) to control the ground watertable level; (iii) to carry away surface water directed to them. The pipe drains act as a conduit to transport water off the land. They do not draw water out of the soil. Water has to move through the soil profile or down a slope under the influence of gravity until it reaches a pipe drain.

It is a common misconception to think that drainage can be approached as if all the characteristics of poor drainage relate to ground watertable problems. In fact, the range of circumstances where agricultural drainage improves the productivity and versatility of cropping is wide. It is, however, possible to make the generalization, as stated earlier, that the improvement is usually achieved by the more rapid removal of subsoil water thus increasing the time when the soil is at field capacity or drier.

The agricultural benefit from a pipe drainage system is progressive, not instantaneous. The gradual lowering of the watertable and removal of excess rainfall allows a greater degree of soil drying and cracking than would be possible without drainage. Whilst some consideration is given to a soil's hydraulic characteristics in designing an agricultural drainage system especially in relation to drain spacings, the emphasis is on the capacity of the pipe drains to transport the water received rather than on controlling the rate at which water moves to the pipes.

The design drainage rate in agricultural systems is computed from pipe diameters, their spacings and the slope. Typical values range from 25 to 50 mm per day (Castle *et al.*, 1984). These values are reasonable in relation to anticipated daily rainfall and the rate at which rain is transmitted to pipe drains. Agricultural drains often continue to flow for up to a week of dry weather following rain, and soil efficiently drained by an agricultural system is unlikely to reach field capacity in less than 48 hours. Temporary ponding of surface water is likely to occur frequently and, in the wetter areas of the UK, the soil is also likely to be wetter than field capacity for much of the winter period.

In sportsturf the emphasis has to be on the rapid transport of water through the soil to the pipe drains because sporting events are timetabled and played irrespective of weather (rainfall). Simplistically the design drainage rate for sportsfields could be increased by decreasing the drain spacing so as to match the maximum anticipated rate of rainfall rather than that averaged over a daily period. Indeed this approach used to be common practice. For example, the original pipe drainage system at Derby County's ground had 75 mm diameter pipe drains at less than 2.5 m spacings and a design drainage rate of approximately 1100 mm per day. The fact that the pitch had to be reconstructed in 1975 illustrates that provision for rapid removal of excess water in the subsoil is not the solution to problems of sportsfield drainage. During the period 1979–1981, most drainage schemes installed on local authority playing fields were still restricted to pipe drainage systems (Ward, 1983a). Invariably, drainage problems manifested themselves by showing poached areas in between drain lines with the benefit of improved drainage extending only a few metres either side of the drain (Fig. 3.2).

The illustration in Fig. 3.2, where permeable fill over the drains was brought to 150 mm below the surface does not imply that pipe drainage schemes are inadequate, merely that the movement of water to the pipes is inadequate. A comprehensive pipe drainage scheme is usually a prerequisite for schemes designed to improve the drainage of sportsfields but that itself is not the solution.

The rate-limiting step in the pathway of water flow in sportsturf is the component between soil surface and subsoil. In most situations, the critical component is the top 70 mm or so of soil. The reason for this is that macropore systems created by aggregates are destroyed and the pore size distribution comes to be determined by the close packing of primary particles. Therefore solutions to sportsturf drainage must be sought independent of the benefits conferred by the aggregation of soil particles.

Recognition of the basic characteristic of the problem dates back some time, but in the 1960s, it was the observant and experienced groundsmen who

**Fig. 3.2.** The ineffectiveness of a pipe drainage system for intensive sportsturf use. The benefit of the drains extends only a metre or so either side of the drain. (Photograph courtesy of K. McAuliffe.)

began to initiate solutions. A typical example was Bert Bond of Aston Villa who used to regularly cultivate fine clinker into the topsoil to improve its openness. Although not an ideal solution, it did ensure that the pitch at Aston Villa was one of the best Football League grounds at that time.

## APPROACHES TO INTENSIVE SPORTSTURF DRAINAGE

Since the beginning of the 1970s there has been a large amount of research on sportsturf drainage, ranging from detailed monitoring of field situations to comprehensive laboratory studies on the physical properties of soil materials for construction and amelioration. Although solutions to sportsturf drainage problems in amenity soils differ in functional efficiency largely in response to demands for playing quality and intensity of use, the aims and concepts are similar. The key aim is to transmit rainfall rapidly to the subsoil which is achieved by one or a combination of two methods. The first method is a system of close-spaced vertical draining channels filled with permeable material installed to transmit water to a pipe drainage system thus bypassing the bulk soil. The second method is a rootzone whose particle size distribution guarantees an adequate hydraulic conductivity overlying a permeable subsoil or permeable fill linked to a pipe drainage system. Conceptually these are either bypass systems or through-drainage systems. They can be considered as alternative approaches but the commonest high-grade schemes in the UK for winter games pitches combine the benefits of rootzone design with the extra benefit of bypass drainage. A key point is that bypass systems alone create a systematically heterogeneous rootzone which may result in untrueness and/or non-uniform behaviour. In the rest of this chapter, the principles of each of these drainage solutions are illustrated in an approximate order of ascending drainage capability. Fundamental aspects concerning design criteria, selection of materials and installation procedures are dealt with in Chapter 4.

# Slit drainage systems

Slit drainage systems (or bypass systems) consist of vertical channels of highly permeable material connecting with the surface and passing through the top-soil. The slits usually connect with gravel backfill overlying a system of lateral pipe drains situated deeper in the soil profile (Fig. 3.3), but they may connect with a permeable subsoil. Slit drainage systems can be considered the sportsturf equivalent of secondary agricultural drainage treatments such as subsoiling or mole ploughing.

## *Terminology*

There is a wide variety of terminology used to describe slit drainage systems (e.g. sand banding, surface banding, sand slitting, sand grooving, sand place-ment and sand injection) and interpretations of this terminology may differ between individuals, thus leading to some confusion. Nevertheless, there are only two basic types of slit drain which can be installed (Adams, 1986) and these are:

**1.** Excavated slit drains which are installed using a trencher or rotating disc where the spoil is removed and the resulting trench is filled in a separate opera-tion either with sand alone (typically referred to as 'sand bands') or a combina-tion of sand over gravel (typically referred to as 'sand/gravel slits' or 'surface banding'). Some of the earliest excavated slit drainage systems in the early 1970s were filled with sand only and this was impossible to achieve successfully unless dried sand was used (subsequent developments in machinery have largely overcome this problem). Moreover, the lateral conductivity properties of medium or fine sand which was used in the slit drains were often insufficient by themselves to satisfy the drainage rate requirements of early designs, and for efficient function they depended on good subsoil permeability (see Mode of action below).

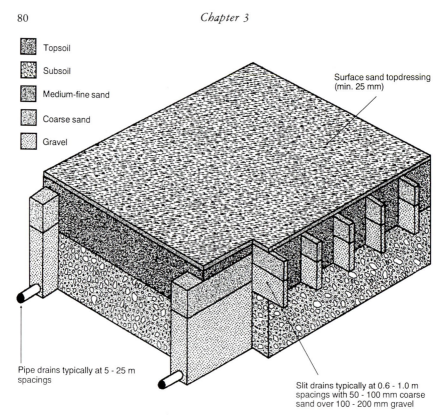

Topsoil

Subsoil

Medium-fine sand

Coarse sand

Gravel

Surface sand topdressing
(min. 25 mm)

Pipe drains typically at 5 - 25 m
spacings

Slit drains typically at 0.6 - 1.0 m
spacings with 50 - 100 mm coarse
sand over 100 - 200 mm gravel

**Fig. 3.3.** Slit drainage system of highly permeable excavated vertical channels filled with gravel and coarse sand with gravel backfill over the pipe drains.

2.   Injected slit drains which are installed using a modified mole plough utilizing a cutting blade which forces the soil apart and fills the resulting fissure in a single pass either with sand alone (typically referred to as 'sand injection' or 'sand placement'), or sand and gravel.

A sub-category of injected slit drains is also recognized, and is commonly termed 'surface grooving' (Baker, 1988), an expression used to describe very narrow and shallow slits of sand which are injected into the top 100–150 mm of the surface, and typically used to reconnect an existing slit drainage system which has become capped with soil (Fig. 3.4).

Excavated slit drains are generally 50–75 mm wide and up to 350 mm deep with a large capacity for water transmission and as such are normally used where they provide the major route for water flow from the surface to a pipe drainage system. Injected slit drains containing only sand are normally narrower (20–35 mm) and shallower (typically between 100 mm and 250 mm deep) and are often installed on a more intensive basis than excavated slit drains because of their lower cost per unit length and because they are not as efficient as

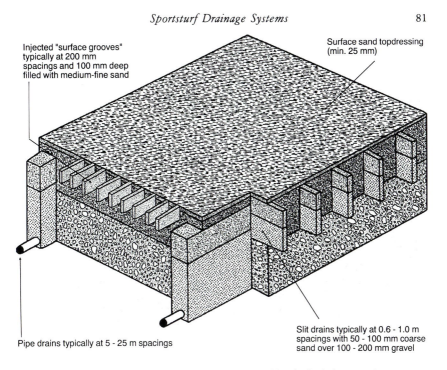

Injected "surface grooves" typically at 200 mm spacings and 100 mm deep filled with medium-fine sand

Surface sand topdressing (min. 25 mm)

Pipe drains typically at 5 - 25 m spacings

Slit drains typically at 0.6 - 1.0 m spacings with 50 - 100 mm coarse sand over 100 - 200 mm gravel

**Fig. 3.4.** As Fig. 3.3, but with close-spaced injected sand-filled slit drains to act as a secondary slit drainage tier, or to reconnect existing slit drains with the surface.

excavated slit drains for transmission of water. In some slit drainage systems, a combination of excavated and injected slit drains is used to provide a very intensive matrix for surface water removal.

A number of mini-pipe and fibre-wrapped honeycomb or waffle structures ('fin' or 'strip' drains) have been developed which are capable of being placed in the bottom of excavated slit drains. Although there has been little investigation on the effectiveness of these developments for sportsturf, one potential advantage in adopting this type of system is that the need for traditional lateral pipe drains is removed since the mini-pipe or honeycomb structures fulfil the function of underdrains. However, these slit drains usually need to be at much closer spacings than traditional lateral pipe drains and therefore inevitably there are many more pipe junctions to install (Fig. 3.5). A secondary tier of slit drains (typically filled with sand only) is often installed at right angles to the mini-pipe slit drains to provide a more intensive method for water interception (Fig. 3.6). It is important that mini-pipe and fin/strip-type slit drains are installed deep enough to avoid any subsequent maintenance or drainage operations puncturing the pipes. For this reason, the depth of installation is usually greater than traditional sand/gravel slit drains.

The final component of any slit drainage system, regardless of design, is

**Fig. 3.5.** A fibre-wrapped honeycomb structure type of drain pipe being installed in excavated slit drains on a winter games pitch. Such installations remove the need for a traditional pipe drainage system. The corrugated pipe in both pictures is the main drain.

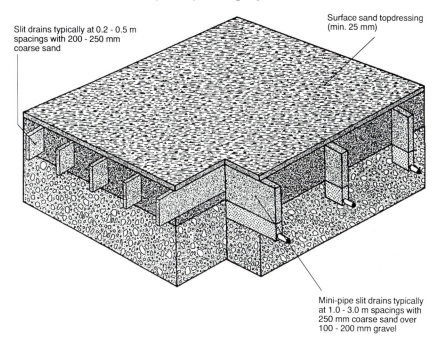

Slit drains typically at 0.2 - 0.5 m spacings with 200 - 250 mm coarse sand

Surface sand topdressing (min. 25 mm)

Mini-pipe slit drains typically at 1.0 - 3.0 m spacings with 250 mm coarse sand over 100 - 200 mm gravel

**Fig. 3.6.** A two-tier slit drainage system consisting of excavated or injected sand-filled slit drains installed at right angles to mini-pipe slit drains.

an adequate layer of sand topdressing (*c.* 25 mm) to protect the integrity of the slit drains. This aspect is dealt with in more detail in Chapters 4 and 6.

## *Mode of action*

Although there are many variants of slit drainage systems, the basic principle of operation is the same. In brief, the system permits an increase in water infiltration rate into the surface without the need for major reconstruction or amelioration of the native soil. This increase is achieved by allowing rainfall to flow across the surface until it enters a highly permeable slit drain which then allows lateral transmission of water until an intersection with a pipe drain is reached.

In a few rare cases, it is unnecessary to include a pipe drainage system with gravel backfill in the design of a slit drainage system because a permeable sub-soil already exists and the slit drains are needed only to move water vertically through a poorly structured topsoil. In this situation, the slit drains need only contain a material sufficiently permeable (usually sand) to allow vertical water entry at a rate consistent with the rainfall intensity if ponding is to be prevented

(Ward, 1983b). However, in most cases, lateral flow is required at the base of a slit drain in order to conduct water laterally to an intersection with a pipe drain. Since sand does not have a saturated hydraulic conductivity to allow this process to take place sufficiently fast enough over large distances (e.g. greater than approximately 3 m), it is normal for the primary drainage slits to be excavated and filled with gravel at the base of the slit to ensure rapid lateral flow, the gravel being topped up with a layer of blinding sand. This combination of gravel and sand has become the most widely used slit drain in common practice (Adams, 1986).

Although slit drainage systems overcome the initial problem of the existing soil being physically unsuitable for rapid surface water removal, these systems do not improve the rootzone properties of the topsoil as a whole because the slit drains usually occupy no more than approximately 5% of the surface area. Conditions for turf growth are improved only with respect to a general reduction in excess surface moisture. The areas between the slit drains are nevertheless still expected to support a vigorous sward despite little or no soil physical improvement. Moreover, because of the nature of the design of slit drainage systems, some form of ponding *has* to occur when overland flow is transmitted to the slit drains (Gibbs, 1988), which is an undesirable characteristic whilst play is actually taking place. Thus slit drainage systems have often been considered as a technique for removing the symptoms of waterlogging rather than as a technique for curing the problem.

### *Benefits, disadvantages and applications of slit drainage systems*

The main advantages of slit drainage systems lie in the economical use of drainage materials (i.e. sand and gravel) and more rapid return to use compared with when the total playing area is disturbed and turf has to be re-established. In consequence many local authorities, private clubs and schools have adopted this technique as a solution to drainage problems.

However, there are several disadvantages of slit drainage systems. In soils that swell and shrink on wetting and drying (generally soils with more than approximately 20% clay), the lines of the slit drains act as planes of weakness as the soil dries out with the result that the slits open up and the level of permeable material in the slit drain sinks. This can result in a surface which is dangerous to play on because of the risk of ankle or leg injury. On rewetting, a corrugated surface develops which leads to a deterioration in surface evenness. In addition, soil is likely to spread over and cap the slit drains thus decreasing their effectiveness. Slit capping by soil occurs in time with all bypass systems and this disadvantage is the single most important factor governing the effective life time of slit drainage systems. Other disadvantages are:

1.   It is difficult to establish turfgrass from seed over the lines of the slit drains as the permeable material in the slit drains has very poor moisture and nutrient retention properties.

**2.** Once ground cover has been eroded by play over the lines of the slit drains, they are prone to erosion especially towards the end of a winter season when the surface begins to dry out, resulting in an uneven playing surface.

**3.** The areas between the slit drains become compacted with time which in turn leads to a shallow-rooted sward unless some remedial action is undertaken.

The above aspects are considered in more detail in later chapters.

Slit drainage systems have been used mainly on soccer and rugby pitches, but because of their undesirable effect on surface trueness (especially on heavier soils), they are not so successful for hockey pitches and cricket outfields. Similarly, slit drainage systems are not recommended for golf and bowling greens unless narrow slit drains can be installed without adversely affecting surface evenness. Slit drains may be used to improve the drainage of golf fairways and tees and horse racing tracks.

# Sand/soil drainage systems

Like slit drainage systems, sand/soil drainage systems are commonly used where the physical characteristics of the existing soil prevent satisfactory drainage and aeration especially after the effects of compaction. However, unlike slit drainage systems, the purpose of sand/soil drainage systems is to provide and maintain a free-draining and well-aerated rootzone material despite the effects of compaction, instead of relying on a small amount of the surface area for rainfall interception and removal.

## *Terminology*

Sand/soil drainage systems can be constructed in a variety of different ways, some methods being specific to certain designs. In general, the following types of sand/soil drainage systems are recognized:

**1.** A system where a predetermined quantity of sand is mixed with a specified or existing soil either on-site or off-site to produce the required depth of sand/soil mix (typically approximately 150 mm in depth for winter games pitches and 200 mm for golf greens). A closely spaced pipe drainage system should underlie the modified sand/soil rootzone with permeable fill up to the undersoil/rootzone interface to intercept water moving through the rootzone (Fig. 3.7).

**2.** A system where soil amelioration with sand is carried out as above to ensure acceptable hydraulic conductivity of the rootzone, but where slit drains are also installed after amelioration to increase the rate of water flow to the drains (Fig. 3.8).

**3.** A system where a prescribed sand/soil mix is created off-site and is brought in for the rootzone component of a total profile reconstruction (Fig. 3.9).

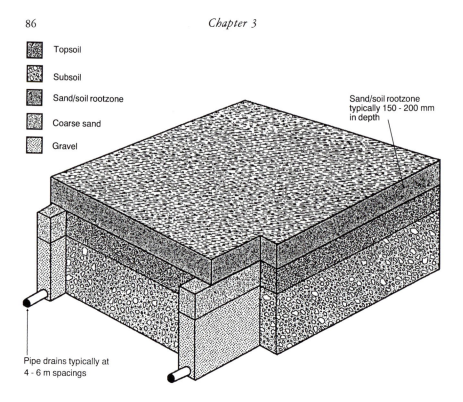

Topsoil

Subsoil

Sand/soil rootzone

Coarse sand

Gravel

Sand/soil rootzone
typically 150 - 200 mm
in depth

Pipe drains typically at
4 - 6 m spacings

**Fig. 3.7.** A sand/soil drainage system where the modified rootzone overlies a closely
spaced pipe drainage system with permeable fill up to the base of the rootzone.

One of the best known sand/soil drainage systems under this category is that
of the United States Golf Association (USGA) who have provided profile
design criteria for the construction of golf greens since 1960 (Radko, 1974).
Many successful bowling greens have also been built in the UK utilizing the
components of the above design (Evans, 1988).

## Mode of action

The success of sand/soil drainage systems depends on the use of a uniform
sand, for which the particles will not interpack, and achieving the correct
mixing ratio of soil to sand (Baker, 1988). With sensible management, the
whole amended rootzone profile can then be expected to withstand intensive
foot traffic, maintain the ability to drain excess water quickly, yet still retain an
adequate supply of water for plant growth. In addition it may be expected to
possess a small cation exchange capacity sufficient to help buffer against times
of nutrient depletion (Brown and Duble, 1975). These properties in turn will
be determined by the final particle size distribution of the sand/soil mix, the

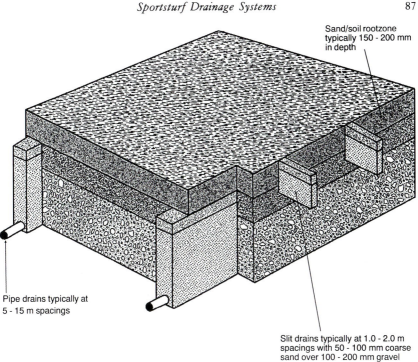

Sand/soil rootzone typically 150 - 200 mm in depth

Pipe drains typically at 5 - 15 m spacings

Slit drains typically at 1.0 - 2.0 m spacings with 50 - 100 mm coarse sand over 100 - 200 mm gravel

**Fig. 3.8.** As Fig. 3.7, but where slit drains have also been added to the sand/soil drainage system. The drain spacing for the pipe drains can be greater in the above type of system than where the slit drains are absent.

type and amount of clay in the mix and the quality and amount of organic matter (Taylor and Blake, 1979).

The priority for any sand/soil mix is therefore to dilute the silt and clay fraction to below prescribed limits in order to produce a material whereby the primary single grain structure still confers a satisfactory hydraulic conductivity and balance of air and water-filled pores. It takes far less of the fine component of a soil to dominate its physical characteristics than the coarse component. For example, all soils behave as clay soils when the clay content exceeds 35%, but sand-dominant soils are required to contain at least 70% sand (Hodgson, 1974).

This means that large proportions of sand are needed for amelioration purposes with the overall objective of ending up with between 75% and 90% by weight of the final mix in the sand fraction depending on the precise requirements of the turf (Adams *et al.*, 1971; Taylor and Blake, 1979). Adams (1976) demonstrated that a sand/soil mix should theoretically contain no more than 12% (w/w) of particles less than 50 $\mu$m in diameter in order to maintain a rootzone saturated hydraulic conductivity greater than 10 mm h$^{-1}$. A

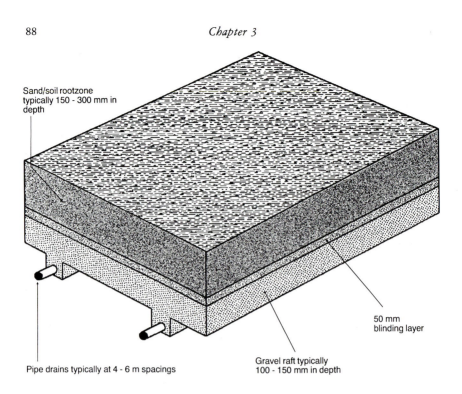

Sand/soil rootzone typically 150 - 300 mm in depth

50 mm blinding layer

Pipe drains typically at 4 - 6 m spacings

Gravel raft typically 100 - 150 mm in depth

**Fig. 3.9.** A total profile design utilizing a prescribed sand/soil mix over a gravel raft. The above system is also known as a suspended or perched watertable construction.

minimum practical specification set in the UK by Baker (1985) is that the final mix must contain less than 20% fines (particles <0.125 mm in diameter), less than 10% silt and clay (particles <0.05 mm in diameter) and less than 5% clay (particles <0.002 mm in diameter), with proportionally more sand being required for faster drainage rates.

Adding insufficient quantities of sand to a soil may actually worsen the soil's physical properties because the silt and clay in the existing soil will simply fill the pore spaces between the sand particles. In this situation, total porosity may be of the order of only 30% (v/v) unless the sand particles are frequent enough to be in contact with each other in all directions and still leave some unfilled space between them. Many of the English Football League club pitches in the early 1970s contained insufficient quantities of sand in their rootzones, and they were of a poor quality by today's standards. Out of a survey of 17 pitches sampled in 1970, only Aston Villa, Blackpool, Ipswich and Manchester United approached the criteria which are currently acceptable (Thornton, 1978). In contrast, the three worst pitches in the survey of 1970 (those of West Bromwich Albion, Nottingham Forest and West Ham United, all teams then in the First Division), had an average rootzone analysis of only 38% in the

125–500 μm particle size range and an average of 48% of particles less than 125 μm in diameter.

The pathway for water flow for a sand/soil mix over a pipe drainage system differs from the sand/soil mix as part of a total profile construction. In the former system, lateral flow is required through the ameliorated layer in order for water to intercept permeable fill over a lateral pipe drain (assuming that the native soil is essentially impermeable). In the latter system, only vertical water flow is required through the rootzone since the sand/soil mix is underlain by a complete gravel drainage raft. This raft also allows a better control of water retention in the sand/soil mix than where the layer is absent (see Mode of action of all-sand drainage systems).

The ratios of soil and sand required to produce a mix of defined particle size distribution can be calculated but the final mix should be checked not only for its particle size distribution but also for other properties including: saturated hydraulic conductivity, total and air-filled porosity and organic matter content. This testing is not practicable when on-site mixing is carried out and this type of mixing requires a high level of technical expertise. Mixing off-site is generally preferable and should be considered essential in golf green construction (USGA Green Section Staff, 1993).

## Benefits, disadvantages and applications of sand/soil drainage systems

The main advantage of a sand/soil drainage system is that, unlike a slit drainage system, the entire rootzone can theoretically be expected to remove rainfall even after the effects of compaction. Moreover, the physical properties within the rootzone are more consistent and surface firmness is maintained even when wet. Unlike the 'all-sand' constructions, sand/soil rootzones also contain a moderate amount of silt and clay and indigenous soil organic matter, all of which can exert benefits in terms of biological activity and nutrient retention/ availability. There is thus some buffering against environmental extremes and, to a certain extent, poor management. Surface trueness is also not disrupted, which has particular advantages for certain types of sport (e.g. golf and bowls).

Despite the undoubted merits of sand/soil drainage systems, there have been failures of this type of system, some spectacularly so (e.g. Radford *et al.*, 1985). The main reasons for problems appear to be either inadequate designs or failure to implement the precise design requirements. In a drainage survey carried out by Ward (1983a), over 40% of sand-ameliorated local authority pitches still had inadequate drainage; this led her to conclude that sand/soil drainage systems were not a good choice of provision for local authorities in the UK. A more recent survey of well-constructed and well-maintained winter games pitches in the UK presented a different picture (Adams *et al.*, 1992). The sand/soil designs included in the survey provided high quality playing surfaces when used for two to three matches per week throughout the season. Three pitches were still performing well 15 years after construction. Nevertheless, the survey identified a situation where sand top-dressing was inadequate to maintain a desirable rootzone composition in the long term.

Sand/soil drainage systems are more expensive to install than slit drainage systems and a greater proportion of their cost is taken up by the construction materials (see Chapter 6). Sand/soil amelioration requires complete re-establishment of turf, unlike slit drainage systems which can be installed without destroying the turf. The improved drainage of a sand-dominant rootzone is gained at the expense of a decrease in both water and nutrient retention. Thus sand/soil systems require greater fertilizer input, especially of N and K, and irrigation in summer is normally necessary. Benefits in both nutrient and water economy are achieved if the sand/soil rootzone overlies soil rather than a sand or gravel base.

With winter games, when ground cover is completely destroyed on intensively used areas, the fine and coarse materials in sand/soil mixes tend to separate out at the surface (especially under wet conditions). This segregation reduces the infiltration rate of water dramatically. If this type of deterioration occurs the segregation must be destroyed by disturbing the surface layer by mechanical treatment at the end of the season.

## All-sand drainage systems

A number of construction techniques have been developed which use sand without soil as the rootzone material in the general belief that these systems provide the 'ultimate' technique for improving playing quality and usage potential. By omitting soil from the rootzone, the problems of incorrect mixing methodology and segregation of fine particulate material associated with sand/soil mixes can be avoided, although other problems in turn are created. However, the term 'all-sand' is somewhat misleading because some form of amendment (usually peat) is often added to the sand during the construction stage to increase nutrient/water retention of the sand and aid grass establishment.

### *Terminology*

Like the sand/soil drainage systems, all-sand drainage systems aim to provide and maintain a homogeneous, free-draining and well-aerated rootzone material even after the effects of compaction. The way in which this is achieved varies considerably and Baker (1988) recognized three basic types of all-sand drainage systems:

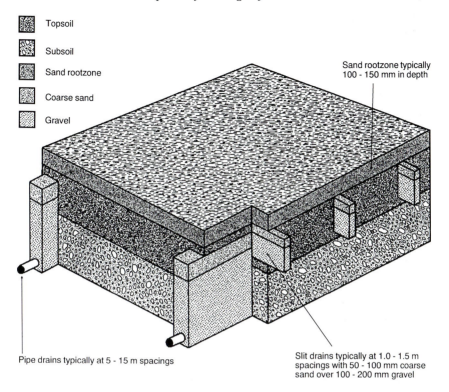

Topsoil

Subsoil

Sand rootzone

Coarse sand

Gravel

Sand rootzone typically
100 - 150 mm in depth

Pipe drains typically at 5 - 15 m spacings

Slit drains typically at 1.0 - 1.5 m
spacings with 50 - 100 mm coarse
sand over 100 - 200 mm gravel

**Fig. 3.10.** A 'Prunty-Mulqueen' sand carpet drainage system consisting of a 100-150 mm layer of specified sand over a matrix of slit drains (UK Patent No. 1217409).

1. A system where a 100–150 mm layer of specified sand overlies the existing soil in which an intensive matrix of slit drains and pipe drains has been installed (Fig. 3.10). In the UK, this type of approach is commonly referred to as a sand carpet system or a 'Prunty–Mulqueen' system as it was originally developed by J.P. Prunty and J. Mulqueen of Northern Ireland in 1968 (Prunty, 1970). Less intensive sand carpet drainage systems have also been built in the UK which consist of a shallower depth of sand (typically 30–50 mm) over a pipe drainage system without secondary slit drains. In this latter case, it is important that the permeable fill over the pipe drains connects directly with the sand carpet, as in the sand/soil drainage system.

2. A system where a rootzone layer of specified sand (typically 250–300 mm deep) overlies a coarse blinding layer and gravel drainage raft to create a suspended watertable construction (see Mode of action below). An impermeable plastic membrane may also enclose the whole construction or just the sides of the construction. No use is made of the existing soil as the technique involves total profile reconstruction (Fig. 3.11). The USGA originally developed the

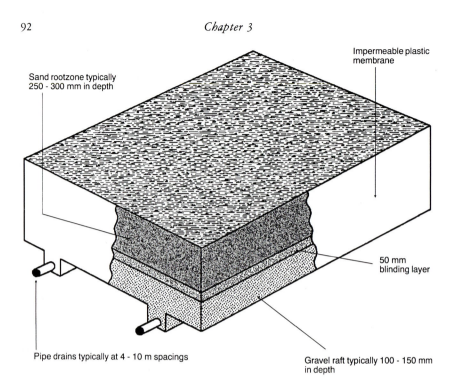

Sand rootzone typically
250 - 300 mm in depth

Impermeable plastic
membrane

50 mm
blinding layer

Pipe drains typically at 4 - 10 m spacings

Gravel raft typically 100 - 150 mm
in depth

**Fig. 3.11.** A suspended watertable construction where a specified layer of rootzone sand overlies a gravel drainage raft. Some designs of this type also utilize an impermeable plastic membrane around the entire construction (as here) or around the sides of the construction only.

principle of this system in 1960 where the rootzone material was a sand/soil mix (Fig. 3.9), but sand is now often used as an alternative.

3.    A system where only one basic construction material is used for the profile reconstruction, which is surrounded by an impermeable plastic membrane to gain complete control of the watertable (Fig. 3.12). Drainage pipes, which are placed at the base of the profile, are fitted with controllable outlets to allow removal, or conservation of water. Two basic variants of this system have been developed. The first variant is where the water is removed by gravity alone and the construction is typically divided up into several drainage bays or cells which in some systems may be individually controlled. Examples of this first variant include the 'Cellsystem' (Moesch, 1975) and 'Purr-Wick' (Daniel, 1969) systems. The second variant is where drainage is assisted by pumps for the very rapid removal of excess water. These pumps also allow for controlled and automatic sub-irrigation. Examples of the second variant include the 'Prescription Athletic Turf' (Daniel *et al.*, 1974) and 'Vacudrain' (Chrystie, pers. comm.) systems. For reasons of cost, only the 'Cellsystem' has been installed in the UK.

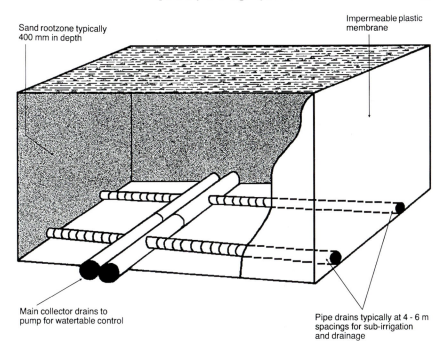

Sand rootzone typically
400 mm in depth

Impermeable plastic
membrane

Main collector drains to
pump for watertable control

Pipe drains typically at 4 - 6 m
spacings for sub-irrigation
and drainage

**Fig. 3.12.** An all-sand drainage system utilizing one construction medium enclosed within an impermeable plastic membrane. In some highly specialized designs of this type, pumps are attached to the main drain for the very rapid removal of excess water, as shown in the above Prescription Athletic Turf System (USA Patent No. 3908385).

## *Mode of action*

Due to the absence of clay and silt in the rootzone, all-sand drainage systems should provide the fastest drainage rates of all intensive drainage solutions for sportsturf. However, it must be emphasized that like the sand/soil drainage systems, the success of all-sand drainage systems depends on the correct selection of sand to provide the desired hydraulic conductivity, air/water balance and physical stability. These requirements are to some extent contradictory and this considerably narrows down the selection of suitable sands (see Chapters 1 and 4). Where unsuitable sands have been used (e.g. where they have been too coarse), there have been some notable failures of all-sand drainage systems (Kamp, 1985).

In contrast to sand/soil drainage systems, all-sand drainage systems should be relatively free of clay particles and organic matter which means that they will not possess the same buffering qualities as sand/soil mixes. Whilst the

**Table 3.3.** Percentage saturation of the pore space in the surface 20 mm of different depths of sand[a] overlying a silty clay loam soil. (Adapted from Adams *et al.*, 1971.)

| Construction details | Equilibration time following saturation of the construction (hours) | | | | |
|---|---|---|---|---|---|
| | 0 | 0.5 | 2.0 | 24 | 48 |
| 80 mm of sand on 500 mm of local soil | 97 | 95 | 94 | 72 | 57 |
| 230 mm of sand on 500 mm of local soil | 84 | 79 | 72 | 44 | 31 |
| 380 mm of sand on 500 mm of local soil | 96 | 35 | 22 | 14 | 10 |

[a] Sand used: $D_{10} = 140 \, \mu$m; $D_{90} = 280 \, \mu$m.

absence of soil may be beneficial from the drainage point of view, all-sand root-zones present a far harsher environment for establishment and development of grass. As a compromise to this harsh environment, sand carpet drainage systems aim to provide the best aspects of both soil and sand rootzones by allowing uniform free drainage at the surface with consequent high quality playing conditions, together with the possibility of storage and retention of nutrients in the topsoil layer below (Canaway and Hacker, 1988).

Sand carpet drainage systems do not behave like slit drainage systems where water infiltration is confined to the lines of the slits. Instead water entry into the surface is uniform and lateral flow through the sand layer is necessary for water to reach the slit drains (Gibbs, 1988). Ideally roots should also penetrate into the topsoil layer below the sand carpet which itself should not only be in a non-compacted state at the time of construction, but should also remain uncompacted during play.

As yet there is little scientific evidence to suggest the optimum depth of sand that is necessary to prevent compaction of the underlying soil. In some sand carpet constructions in the UK, roots have failed to penetrate through the sand/soil interface because the underlying soil has become compact and anaerobic. Soil compaction leading to waterlogging may have been caused during construction or may have developed later. However, depths of root-zone sand greater than 150 mm cannot be recommended, not only for finan-cial reasons, but also because any capillary or small pore continuity between the sand rootzone and the underlying soil may dry the sand out too much (Table 3.3). A similar argument applies to sand/soil mixes overlying soil with a pipe drainage system, although, in this case, a depth of 200 mm is probably acceptable before the underlying soil dries out the rootzone too much due to the greater critical tension of the sand/soil mix.

**Table 3.4.** Percentage saturation of the pore space in the surface 20 mm of different depths of sand[a] overlying a 150 mm gravel drainage raft. (Adapted from Adams *et al.*, 1971.)

| Construction details | Equilibration time following saturation of the construction (hours) | | | | |
|---|---|---|---|---|---|
| | 0 | 0.5 | 2.0 | 24 | 48 |
| 120 mm of sand on a 150 mm gravel layer | 96 | 95 | 93 | 79 | 75 |
| 280 mm of sand on a 150 mm gravel layer | 94 | 90 | 85 | 68 | 62 |
| 430 mm of sand on a 150 mm gravel layer | 97 | 55 | 42 | 32 | 24 |

[a] Sand used: $D_{10} = 140\ \mu m$; $D_{90} = 280\ \mu m$.

The phenomenon of fine pore systems removing water from systems with coarser pores can also present problems in situations where water movement is predominantly lateral. Where sand or sand-dominant rootzones adjoin 'natural' soil, for example at the edge of a golf green, a dry soil will extract water from the rootzone near the perimeter of the green unless lateral movement of water is prevented by surrounding the construction with an impermeable plastic membrane.

The gravel raft in a suspended watertable construction helps prevent the drying out of an overlying rootzone. The large pores in the gravel raft drain under the force of gravity so creating a break in the continuity of water-filled pores at the interface of the rootzone with the gravel. The interface is called a capillary break and it behaves like a watertable in the way it controls the retention of water in the rootzone (see Chapter 1). Provided the depth from the soil surface to the capillary break is a little greater than the critical tension of the rootzone, the gravel raft can be used to improve water retention in the rootzone whilst ensuring air entry at the surface. However, if the rootzone layer is too deep, too much water will be removed from the surface (Table 3.4). If the layer is too shallow, critical tension will not be overcome and the rootzone will remain saturated after drainage has ceased.

Although the capillary break affects water retention in a rootzone, it cannot supply water to the overlying rootzone in the same way as a normal watertable. It is of course possible to envelop the complete construction within a plastic membrane (Fig. 3.11) so that by controlling the watertable within the construction, the benefits of sub-irrigation and a capillary break layer can be enjoyed (Adams *et al.*, 1971). Whilst a few schemes of this type have been constructed in the UK, there are severe practical difficulties in preventing the puncturing of the plastic membrane.

The third category of all-sand drainage systems has no capillary break, but utilizes an impermeable plastic membrane to create a watertable which also allows sub-irrigation. This category allows greater control of watering require-ments than the traditional suspended watertable construction without a plastic membrane, because the height of the watertable may be lowered or raised either manually or automatically depending on the weather and growing conditions. At times of excessive rainfall, the outlets of the drainage pipes are opened and gravitational forces (perhaps supplemented by vacuum pumps) remove surface water. When soil moisture is in deficit, water may be pumped through the drainage pipes and it then moves upwards through the rootzone sand by capillary rise. The capillary rise properties of the rootzone sand should match the greatest expected rates of potential evapotranspiration so that the rate of water rise is not limiting. Both the 'Cellsystem' and 'Purr-Wick' (Plastic Under Reservoir Rootzone with Wick action) systems work on this principle of drainage and watering.

There is however a danger in relying on capillary rise only for the method of watering for two reasons. Firstly, the height of the saturated zone above a watertable is about 30% greater if the profile is initially saturated from above compared with wetting only by capillary rise from below (Fig. 1.8). The inter-pretation of this is that for a fixed watertable depth at the base of a rootzone over a plastic membrane, if the rootzone depth was specified only in relation to its capillary rise properties, it would be likely to remain waterlogged if saturated by rainfall. Using a greater depth of rootzone to guarantee air entry at the surface after rainfall would mean that the capillary rise properties of the sand would not ensure adequate water supply to the surface. In these situa-tions, the rootzone must be deeper than its critical tension, but also it must be possible to raise and lower the watertable within the rootzone.

Secondly, all-sand enclosed drainage systems are frequently recommended for use in arid regions of the world where water conservation is at a premium, and where sub-irrigation can considerably reduce watering costs. However, in these climates, potential evapotranspiration always greatly exceeds rainfall with the inevitable result that salts accumulate at the surface. The only way to remove an accumulation of salts is by regular flushing with high quality irriga-tion water. This is only possible if an overhead irrigation system is available. There have been some severe problems with salinity in these types of construc-tion in arid areas such as the Middle East in situations where an overhead irriga-tion facility has not been installed.

The 'Prescription Athletic Turf (PAT)' and 'Vacudrain' systems employ extra features in addition to those used by the 'Cellsystem' and 'Purr-Wick' systems. Here suction pumps are attached to the drainage pipes so that if heavy rain falls immediately before a match or during play, the pumps can be used to increase the hydraulic gradient within the sand rootzone and prevent ponding (Daniel *et al.*, 1974). In addition, the rootzone moisture providing optimum playing conditions can theoretically be achieved by utilizing the pumps to either remove or add water. Moisture sensors within the rootzone can be used to automatically activate the pumps as necessary.

## *Benefits, disadvantages and applications of all-sand drainage systems*

The major attraction of all-sand drainage systems is that they provide a firm and free-draining playing surface. Because of these features, playing quality and wear tolerance are potentially better than can be achieved with other systems. Nevertheless, they are expensive to construct and maintain and they demand a greater technical knowledge in management than other systems.

Although all-sand drainage systems overcome the problem of waterlogging and cancellation of sporting fixtures because of rain, the use of a sand rootzone creates several major disadvantages over even sand/soil rootzones. These disadvantages are summarized below:

1. Control of moisture in the rootzone is difficult especially when establishing turf from seed, but also in general maintenance. Irrigation provision is essential.
2. Sand rootzones are 'hungry' and the use of 'slow release' fertilizer sources of N is virtually essential. Leaching of nutrients, especially $NO_3$, may be a problem.
3. Disease and pest susceptibility is often increased in turf grown on sand rootzones because of the lack of beneficial microorganisms and the ease with which nutrient and moisture levels can be depleted.
4. Because of the small specific surface area and poor adsorption characteristics of sands, the dosages of biocides used must be precisely controlled to avoid residual toxicity.
5. The surface becomes unstable once ground cover has been destroyed (especially applicable to winter games constructions) leading to totally unacceptable playing conditions (Fig. 3.13).
6. When all-sand constructions are on sloping sites (e.g. mainly golf greens), water is transferred rapidly from the high areas to the low areas by siphoning, with the result that the high areas become drought stressed, or if this is corrected, low areas remain waterlogged. Control of this problem is very difficult, but the 'Purr-Wick' system involves installing divided compartments in the construction to equalize water retention within the green (Fig. 3.14).

The problems associated with all-sand drainage systems can be overcome and the reward for skilful management is a superior playing surface. However, it is important to be aware at the outset of the consequences of choosing this type of drainage system in terms of the management input and precise construction requirements.

All-sand drainage systems are appropriate for sports where play takes place during the worst weather of the year (e.g. golf, soccer and rugby), and

**Fig. 3.13.** Unacceptable deterioration of surface stability of a suspended watertable sand rootzone soccer pitch, arising from excessive use in the goal area.

**Fig. 3.14.** Internal dividers in a 'Purr-Wick' golf green prevent siphoning of water from high areas to low areas (photograph courtesy of Martyn Jones).

especially where the quality of the playing surface takes precedence over cost. All-sand drainage systems are not appropriate where soil binding characteristics are required (tennis and cricket).

## Other drainage systems

The range of drainage options illustrated in this chapter mainly reflect British and USA approaches to sportsturf drainage (with the exception of the 'Cellsystem' and 'Vacudrain' designs). Most other sportsturf drainage designs in the world have originated from Sweden, Denmark, The Netherlands and Germany, and although the type of material used (particularly the grade of sand) and depths of various profiles may differ to some extent from those covered in this book they employ no unique concepts. For example, one method of draining sportsfields which is popular in The Netherlands is to use an artificial sand carpet over the existing soil (van Wijk, 1980). This is similar in principle to the 'Prunty–Mulqueen' system although slit drains are usually not included. German, Swedish and Danish designs in the past have all used a gravel drainage raft topped by a constructed rootzone (e.g. the German 'DIN Standard' systems, the Swedish 'Weigrass' system and the Danish 'Petersen' system).

Priority has been given in this chapter to drainage solutions for areas of turf which are used intensively because it is primarily the intensity of use that dictates the approach to drainage. For extensive areas which receive a low intensity of use (e.g. open parks, airport fields, golf fairways), traditional solutions to soil drainage may be adequate. The key factor in deciding the approach to drainage in these areas is whether the controlling factor in drainage through the soil profile is the permeability of the topsoil or the permeability of the subsoil.

## REFERENCES

Adams, W.A. (1976) The effect of fine soil fractions on the hydraulic conductivity of compacted sand/soil mixes used for sportsturf rootzones. *Rasen Turf Gazon* 7, 92–94.

Adams, W.A. (1986) Practical aspects of sportsfield drainage. *Soil Use and Management* 2, 51–54.

Adams, W.A., Stewart, V.I. and Thornton, D.J. (1971) The assessment of sands suitable for use in sportsfields. *Journal of the Sports Turf Research Institute* 47, 77–85.

Adams, W.A., Gibbs, R.J., Baker, S.W. and Lance, C.D. (1992) Making the most of natural turf pitches. A national survey of winter games pitches with high quality drainage designs. *Natural Turf Pitch Prototypes Advisory Panel Report No 10*, The Sports Council, London, 20 pp.

Baker, S.W. (1985) Topsoil quality: relation to the performance of sand–soil mixes. In: Lemaire, F. (ed.) *Proceedings of the 5th International Turfgrass Research Conference*. Avignon, France, pp. 401–409.

Baker, S.W. (1988) Construction techniques for winter games pitches. In: Reilly, T. *et al.* (eds) *Proceedings of the 1st World Congress of Science and Football*. E. & F.N. Spon, London, pp. 399–405.

Brown, K.W. and Duble, R.L. (1975) Physical characteristics of soil mixtures used for golf green construction. *Agronomy Journal* 67, 647–652.

Canaway, P.M. and Hacker, J.W. (1988) The response of *Lolium perenne* L. grown on a Prunty–Mulqueen sand carpet rootzone to fertilizer nitrogen. I. Ground cover response as affected by football-type wear. *Journal of the Sports Turf Research Institute* 64, 63–75.

Castle, D.A., McCunnall, J.M. and Tring, I.M. (1984) *Field Drainage, Principles and Practice*. Batsford Academic, London, 250 pp.

Daniel, W.H. (1969) The "Purr-Wick" rootzone system for compacted turf areas. In: The Sports Turf Research Institute (ed.), *Proceedings of the 1st International Turfgrass Research Conference*. Harrogate, England, pp. 323–325.

Daniel, W.H., Freeborg, R.P. and Robey, M.J. (1974) Prescription Athletic Turf system. In: Roberts, E.C. (ed.) *Proceedings of the 2nd International Turfgrass Research Conference*. American Society of Agronomy, Madison, Wisconsin, pp. 277–280.

Davies, D.B., Eagle, D.J. and Finney, J.B. (1982) *Soil Management*. Farming Press Ltd, Ipswich, 287 pp.

Evans, R.D.C. (1988) *Bowling Greens: Their History, Construction and Maintenance*. The Sports Turf Research Institute, Bingley, 196 pp.

Gibbs, R.J. (1988) The influence of winter sports pitch drainage systems on the measurement of water infiltration rate. *Journal of the Sports Turf Research Institute* 64, 99–106.

Hodgson, J.M. (1974) *Soil Survey Field Handbook Technical Monograph No. 5*. Soil Survey, Harpenden, 99 pp.

Kamp, H.A. (1985) Dutch experience with a Cellsystem pitch. *Zeitschrift für Vegetationstechnik* 8, 6–10.

Moesch, R. (1975) Be- und Entwässerung von Rasenflächen nach dem Cellsystem. *Rasen Turf Gazon* 6, 83–85.

Prunty, J.P. (1970) Improvements in or relating to lawns, greens or playing fields and a method of forming same. *British Patent 1217409*, The Patent Office, London.

Radford, C., Dury, P.L.K. and Skinner, N. (1985) *The Kirkby Project – A Study of Outdoor Joint Use Sports Facility Provision at Kirkby Kingsway Park*. Education Department, Nottinghamshire County Council, 103 pp.

Radko, A.M. (1974) Refining Green Section specifications for putting green construction. In: Roberts, E.C. (ed.) *Proceedings of the 2nd International Turfgrass Research Conference*. American Society of Agronomy, Madison, Wisconsin, pp. 287–297.

Schuurman, J.J. (1965) Influence of soil density on root development and growth of oats. *Plant and Soil* 22(3), 352–373.

Smith, L.P. (1984) *The Agricultural Climate of England and Wales*. MAFF Reference Book, 435, HMSO, London, 149 pp.

Stewart, V.I. (1980) Soil drainage and soil moisture. In: Rorison, I.H. and Hunt, R.

(eds) *Amenity Grassland: An Ecological Perspective.* John Wiley and Sons Ltd, Chichester, pp. 119–124.

Taylor, D.H. and Blake, G.R. (1979) Sand content of sand–soil–peat mixtures for turfgrass. *Soil Science Society of America Journal* 43, 394–398.

Thornton, D.J. (1978) The construction and drainage of some specified sports field playing surfaces. Unpublished PhD Thesis, University College of Wales, Aberystwyth.

USGA Green Section Staff (1993) USGA recommendations for a method of putting green construction. *USGA Green Section Record* Mar/Apr, 1–3.

van Wijk, A.L.M. (1980) *A Soil Technological Study on Effectuating and Maintaining Adequate Playing Conditions of Grass Sports Fields.* Agricultural Research Report 903, Centre for Agricultural Publishing and Documentation, Wageningen, 124 pp.

Ward, C.J. (1983a) Drainage of sports turf areas: results of a questionnaire survey. *Journal of the Sports Turf Research Institute* 59, 29–45.

Ward, C.J. (1983b) Sports turf drainage: a review. *Journal of the Sports Turf Research Institute* 59, 9–28.

Wyatt, N.J.S. (1990) Special requirements for turf reinforcement for horses. In: Shildrick, J.P. (ed.) *National Turfgrass Council Workshop Report No. 19,* pp. 53–56.

# 4 | Design Criteria and Sportsturf Drainage Installation

## Introduction

The different approaches to sportsturf drainage outlined in Chapter 3 involve modifications of the soil profile either overall or locally, for example with slit drains, to create systems with defined hydraulic properties. This makes it possible to calculate, using standard drainage theory, the capacity of different drainage designs to conduct rainfall falling on the surface to a drain outfall.

Although the design rate of a sportsturf drainage system will be governed by the anticipated demands made of that system and its eventual cost, ideally the best quality sportsturf drainage system should be able to cope with the greatest local rainfall intensity of at least 30 minutes duration that is likely to occur over a one year period. Typically in Britain, such a rainfall intensity will be as large as 20 mm h$^{-1}$ (Adams, 1981) and, at this design rate, it should be possible for rainwater to be removed as rapidly as it falls so that a sporting fixture need not be put off because of the weather.

In practice, drainage rates of 20 mm h$^{-1}$ or greater can only be guaranteed using costly sand-based constructions. Other systems such as those involving slit drains alone have a lower capacity and it is usual to find published design drainage rates for sportsturf integrated over a 24 hour period rather than a one hour period. Typical values are in the range of 24–100 mm d$^{-1}$ for the UK (Thornton, 1978).

With slit drainage systems, it is the rate of lateral flow through the base of the slits which is the rate-limiting step to water flow. Baker (1982) considered it essential for the design drainage rate to be assessed for a time period that is critical for this type of flow. He suggested that the maximum design drainage rate should reflect the greatest winter rainfall that is likely to occur in a 24 hour period once every two years, but that a ten year return period should be used for top quality sportsturf areas. As might be expected, there is a large geographical variation in the maximum 24 hour winter rainfall (Fig. 4.1), which for a return period of two years, can be as low as $13 \text{ mm d}^{-1}$ in parts of Cambridgeshire to over $50 \text{ mm d}^{-1}$ in the upland areas of Britain (Baker, 1982).

Three other design features are important in considering design drainage rates, particularly for slit drainage systems. Firstly, it is advisable to assume that the drainage system is required to cope with *all* the incident rainfall. This is especially so when the drained area is being used during the winter months when evapotranspiration is small and where there may be very poor permeability into the topsoil between slit drains.

Secondly, it is better to assume that sportsturf drainage systems overlie an impermeable subsoil especially when this is the 'natural' subsoil (Adams, 1981). Calculations which assume an impermeable subsoil provide minimum values for drainage rate and may greatly underestimate the benefits of sportsturf drainage systems (particularly slit drainage systems) because many sportsturf subsoils are much more permeable than compacted topsoils. When a permeable subsoil does exist, which is common with old designs where a topsoil was replaced over a clinker or stone base, then even narrow sand slits can have a dramatic effect on the theoretical drainage capacity of the system (Table 4.1).

Thirdly, it should be appreciated that the macropore systems of gravel or stone used as permeable fill or layers in constructions have the ability to store rainfall before this water has to be transmitted to a pipe drain. For example, a typical sand/gravel slit drainage system may be able to store approximately 4 or 5 mm of rainfall in the permeable material itself.

**Table 4.1.** Theoretical capability of two slit drainage systems to transfer surface water to pipe drains (in $\text{mm h}^{-1}$), compared with similar slit drainage systems which make direct contact with a permeable subsoil (from Adams, 1981).

|  | 50 mm wide slit drains 300 mm deep at 1 m spacings filled with fine sand | 50 mm wide slit drains 300 mm deep at 1 m spacings filled with 200 mm gravel and 100 mm coarse blinding sand |
| --- | --- | --- |
| Transfer capability to pipe drains with impermeable subsoil | 0.4[a] | 4[b] |
| Transfer capability direct to permeable subsoil | 25 | 25 |

[a] Pipe drains at 5 m centres.
[b] Pipe drains at 20 m centres.

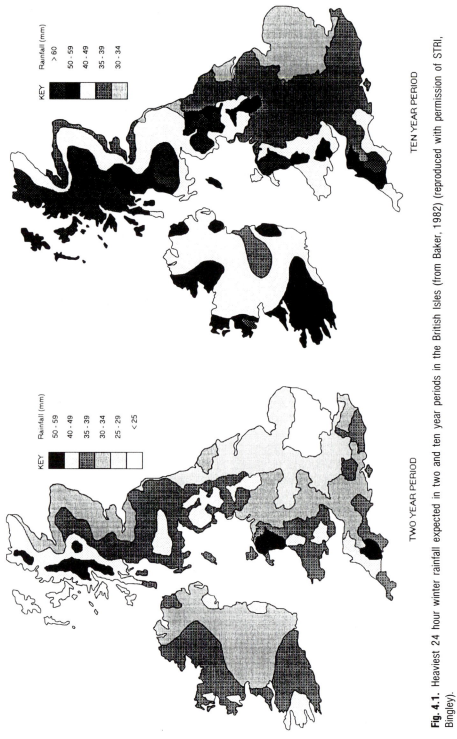

**Fig. 4.1.** Heaviest 24 hour winter rainfall expected in two and ten year periods in the British Isles (from Baker, 1982) (reproduced with permission of STRI, Bingley).

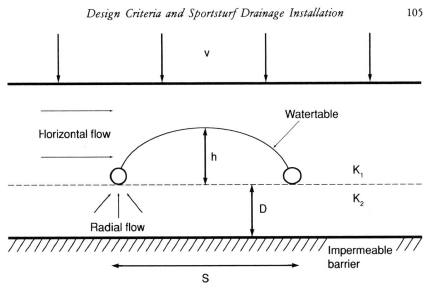

**Fig. 4.2.** Illustration of Hooghoudt's Drainage Equation. See text for explanation of symbols.

## THEORETICAL DESIGN OF SPORTSTURF DRAINAGE

To apply a design drainage rate, it is first essential to relate the rainfall characteristics of the site to the layout and dimensions of the proposed drainage scheme.

## Hooghoudt's Drainage Equation

Probably the most widely used drain-spacing equation in Europe is the Hooghoudt Equation, where drain discharge is assumed to equal the incoming rainfall, and the watertable at mid-drain spacing is maintained at a steady height above the drain level (Castle *et al.*, 1984):

$$S^2 = (8K_2 dh + 4K_1 h^2)/v \qquad (4.1)$$

where $S$ = drain spacing (m); $K_1$ = hydraulic conductivity of the layer above the drains (m d$^{-1}$); $K_2$ = hydraulic conductivity of the layer below the drains (m d$^{-1}$); $h$ = height of the watertable above the drain level midway between the drains (m); $d$ = equivalent depth – this is related to the distance ($D$) below the drains to an impermeable barrier and the drain spacing ($S$) in metres; $v$ = drain discharge or daily rainfall (m d$^{-1}$).

The Hooghoudt Equation takes into account not only horizontal flow, but also the radial flow caused by the convergence of flow lines over the drains (Fig. 4.2). This is accomplished by reducing the depth of the flow layer $D$

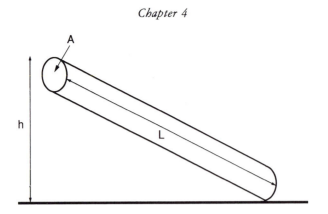

**Fig. 4.3.** Illustration of Darcy's Law. See text for explanation of symbols.

below the drains to the hypothetical depth $d$ of an 'equivalent layer' where $d$ depends on $D$ and $S$ (drain spacing) and the radius of the drain, the values of which can be found in standard drainage textbooks (e.g. Castle *et al.*, 1984).

The first part of the right-hand side of Equation 4.1 relates to flow below the drains and can be ignored when the drains are on top of an impermeable layer where $D$ effectively becomes zero (Raadsma, 1974). This situation is assumed for most types of sportsturf drainage and in fact with certain designs (e.g. the 'Cellsystem') the assumption is completely valid because of the use of an impermeable plastic membrane. However, in practice, the layer below the drains can be considered impermeable if $K_2 < 0.1K_1$ (Thomasson, 1975). Thus, for sportsturf applications, the Hooghoudt Equation reduces to the second part of the equation which describes flow above the drains:

$$S^2 = \frac{4Kh^2}{v} \qquad \text{or} \qquad S = \sqrt{\frac{4Kh^2}{v}} \qquad (4.2)$$

To calculate $v$ (now termed the design drainage rate), Equation 4.2 is rearranged to:

$$v = \frac{4Kh^2}{S^2} \qquad (4.3)$$

It can be shown that Equation 4.3 is a direct derivation of Darcy's Law of flow of water in a saturated medium. Darcy's Law states that the flow of water is proportional to rate of change in hydraulic head along the average flow line. It is illustrated conceptually in the sloping tube in Fig. 4.3 and can be expressed as follows:

$$Q = \frac{KAh}{L} \qquad (4.4)$$

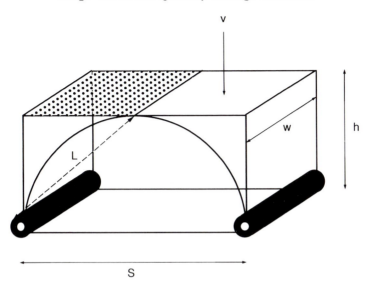

**Fig. 4.4.** Section of sportsturf containing drainage pipes. See text for explanation of symbols.

where  $Q$ = the rate of discharge of water through a tube of cross-sectional area $A$ (m³ d⁻¹); $L$ = the average length of the flow line (m); $h$ = the pressure head (m); $K$ = the hydraulic conductivity of the medium through which the water flows (m d⁻¹).

As with the Hooghoudt Equation, various units for the individual parameters can be used, but when real values are substituted into the equation, the same order of magnitude for each unit of measurement should be used (e.g. metres or millimetres, but not a combination of the two).

The 'tube' illustrated in Fig. 4.3 can be drawn as a section of sportsturf containing drainage pipes (Fig. 4.4), where the cross-sectional area $A$ of Darcy's Law becomes $hw$. Since the drain depth will be small compared with the drain spacing, $L$ (the average length of flow) equals $S/2$ (e.g. for a 10 m drain spacing, water falling on the surface will have to travel approximately 5 m to reach the pipes). When $hw$ and $S/2$ are substituted into Equation 4.4, it modifies to:

$$Q = \frac{Khwh}{S/2} \qquad (4.5)$$

thus:

$$Q = \frac{2Kh^2w}{S} \qquad (4.6)$$

Equation 4.6 now describes a volume of water flowing in a given time to one of the pipe drains. If the rainfall intensity or design drainage rate ($v$) is known, $Q$ can be described by:

$$Q = \frac{vwS}{2} \tag{4.7}$$

where $(wS)/2$ is the shaded area in Fig. 4.4.
Substituting into Equation 4.6 gives:

$$\frac{vwS}{2} = \frac{2Kh^2w}{S} \tag{4.8}$$

thus:

$$v = \frac{4Kh^2}{S^2}$$

(Hooghoudt's Equation for impermeable layers beneath the drain pipes.)

## Modification of Hooghoudt's Equation for slit drainage

With a slit drainage system, it is assumed that all rainfall is intercepted and removed via the slit drains. However, the slit drains only occupy a small fraction of the surface area to be drained, so the design drainage rate ($v$) should be modified by a factor ($D/W$) where $D$ is the slit spacing and $W$ is the slit width:

$$v = \frac{4Kh^2W}{S^2D} \tag{4.9}$$

The actual value used for the hydraulic conductivity ($K$) of the permeable material in the slit drain depends on whether there is only one conducting material in the slit or two (i.e. sand over gravel). Where two materials are used, the combined saturated hydraulic conductivity should be used:

$$K = \frac{K_1h_1 + K_2h_2}{h_1 + h_2} \tag{4.10}$$

where $K_1$ and $h_1$ are the hydraulic conductivity and depth of the first material; and $K_2$ and $h_2$ are the hydraulic conductivity and depth of the second material.

Usually only the hydraulic conductivity and depth of the laterally conducting layer (i.e. the gravel) need to be considered because the hydraulic conductivity of the gravel is very large compared with the sand above it. Nevertheless the drainage capacity of the system could be seriously overestimated if vertical water flow through the sand blinding layer above the gravel restricted the lateral flow through the gravel. Thus it is better to include the combined saturated hydraulic conductivities of the blinding sand and gravel components especially if the blinding sand is of poor quality. Baker (1982) pointed out that since slit drains do not necessarily intercept the pipe drains at right angles, $S$ should be redefined as being the length of slit drains between pipe drains.

## Using Hooghoudt's Equation

### Slit drainage systems

Consider a slit drainage system as in Fig. 3.3, with 50 mm wide slit drains at 1 m spacings containing 150 mm gravel ($K = 10^5$ mm h$^{-1}$) topped by 100 mm coarse blinding sand ($K = 500$ mm h$^{-1}$) intercepting lateral pipe drains at 15 m centres. The combined saturated hydraulic conductivity of the permeable material in the slit drains must first be calculated:

$$K = \frac{(150 \times 10^5) + (100 \times 500)}{250} = 60,200 \text{ mm h}^{-1}$$

given:

$$v = \frac{4Kh^2W}{S^2D}$$

then:

$$v = \frac{4 \times 60,200 \times 250 \times 250 \times 50}{15,000 \times 15,000 \times 1,000} \times 24 = 80 \text{ mm d}^{-1}$$

There are innumerable permutations for slit drainage systems, but the most common methods of achieving the desired design drainage rate are usually by modifying the distance between the slit drains or the lateral drains or the depth of gravel in the slit drains or some combination of all three. The most commonly used spacing of 50 mm wide sand/gravel slit drains is around 1 m, and 2 m spacing seems to be the minimum intensity worthy of installation (Adams, 1986). The design drainage rates of a typical range of permutations for traditional sand/gravel slit drainage systems (see Fig. 3.3) are given in Table 4.2. It may be noted that halving the slit spacing doubles the design rate whereas halving the drain spacing quadruples the design rate. This is because $v$ is inversely proportional to $D$ but is inversely proportional to $S^2$. In the examples given, slit depth and width are constant. As a general rule, however, since $v$ is proportional to $W$ but also to $h^2$, increasing the depth of slits causes a much greater increase in their ability to transmit water than increasing their width.

### Sand/soil drainage systems

A typical sand/soil drainage system overlying a pipe drainage system will have an ameliorated rootzone depth of 150 mm for a winter games pitch (Fig. 3.7).

**Table 4.2.** Design drainage rates (mm d$^{-1}$) for different combinations of lateral drain spacings, slit drain spacings and depth of gravel in the slit.

| Slit drain spacing (m) | Type of slit drain[a] | Lateral drain spacing (m) | | | | |
|---|---|---|---|---|---|---|
| | | 5 | 10 | 15 | 20 | 25 |
| 1.0 | 1 | 484 | 121 | 58 | 30 | 19 |
| | 2 | 722 | 181 | 80 | 45 | 29 |
| | 3 | 961 | 240 | 107 | 60 | 38 |
| 1.5 | 1 | 322 | 81 | 36 | 20 | 13 |
| | 2 | 482 | 120 | 54 | 30 | 19 |
| | 3 | 641 | 160 | 71 | 40 | 26 |
| 2.0 | 1 | 242 | 60 | 27 | 15 | 10 |
| | 2 | 361 | 90 | 40 | 23 | 14 |
| | 3 | 481 | 120 | 53 | 30 | 19 |

[a] Slit width 50 mm with the following depth combinations: 1, 100 mm gravel, 150 mm blinding sand; 2, 150 mm gravel, 100 mm blinding sand; 3, 200 mm gravel, 50 mm blinding sand. $K$ for gravel $= 10^5$ mm h$^{-1}$; $K$ for blinding sand $= 500$ mm h$^{-1}$.

If the design drainage rate is set at 240 mm d$^{-1}$ (10 mm h$^{-1}$), and the sand/soil mix has a saturated hydraulic conductivity of 200 mm h$^{-1}$, the drain spacing ($S$) is calculated as follows:

$$S = \sqrt{\frac{4 \times 200 \times 150 \times 150}{10}} = 1341 \text{ mm or } 1.34 \text{ m}$$

Such a close spacing for lateral pipe drains would be uneconomic to install as it is rarely viable to install lateral pipe drains at spacings closer than 4 m. If the laterals were put in at 4 m instead of 1.34 m, this would give a design drainage rate of only 27 mm d$^{-1}$, considerably smaller than the original 240 mm d$^{-1}$ proposed above but nevertheless adequate in some situations.

Two options exist to increase the design drainage rate. The first is to increase the depth of amended rootzone, which may be uneconomical for a winter games pitch, but practicable for a golf or bowling green. If the depth was increased to 300 mm instead of 150 mm, then this would give a design rate of 108 mm d$^{-1}$ (5 mm h$^{-1}$) with laterals at 4 m spacings, which would guarantee play under most rainfall conditions.

The second option, which is used very often, applies to winter games pitches only and involves superimposing a slit drainage system onto the ameliorated sand/soil rootzone (Fig. 3.8). If the ameliorated depth was kept at 150 mm, and slit drains were installed at 1 m spacings, these would serve two purposes by acting as interceptors of surface water and as lateral drains.

Clearly, from the calculation above slits acting as laterals alone would more than satisfy the design requirement provided the slits themselves could transmit water at an adequate rate to the underdrain system. Slits which were 50 mm wide and 250 mm deep with 150 mm of gravel and 100 mm of blinding sand could deliver water at the rate of 1130 mm d$^{-1}$ to a drain system at 4 m centres. Were this system adopted, therefore, the spacing of the laterals would be more than doubled and still meet the design rate of 240 mm d$^{-1}$.

### Sand carpet drainage systems

In the sand carpet design, sand replaces the sand-dominant soil in the previous system, otherwise they have common features. The patented 'PM pitch' (Prunty-Mulqueen) comprises 100 mm medium-fine sand (150 mm for a 'PM green'), sand/gravel slit drains at 1.5 m spacings and lateral pipe drains at 5 m spacings (Fig. 3.10). Assuming a saturated hydraulic conductivity for the medium-fine sand of 500 mm h$^{-1}$, the design drainage rate through the sand layer is:

$$v = \frac{4 \times 500 \times 100 \times 100}{1500 \times 1500} \times 24 = 213 \text{ mm d}^{-1}$$

where the primary interceptor drains for the water travelling through the rootzone are the slit drains, not the lateral pipe drains. Secondary calculations would need to confirm that the lateral pipe spacing between the slit drains would be able to cope with a design drainage rate of 213 mm d$^{-1}$ (8.9 mm h$^{-1}$). Assuming the slit drains were of type 2 in Table 4.2 the lateral pipe drains would need to be at 9.2 m centres or less to satisfy the requirement.

### Designed profile drainage systems

In the case of a complete profile construction such as the sand/soil rootzone over a gravel drainage raft, the Hooghoudt Equation cannot be used because the critical component in drainage is the downward movement of water through the rootzone into the gravel layer through which movement of water is extremely rapid to the pipe drains (Fig. 3.9). Thus the design drainage rate is regulated by the saturated hydraulic conductivity of the rootzone and the blinding layer. The apparent permeability downwards is calculated using the equation of Luthin (1966):

$$\text{Apparent permeability} = \sum l \Big/ \left( \frac{l_1}{K_1} + \frac{l_2}{K_2} \dots \text{etc.} \right) \qquad (4.11)$$

where $\Sigma l$ = the total depth of all the layers (mm); $l_1$ = the depth of layer 1 (mm); $K_1$ = the saturated hydraulic conductivity of layer 1 (mm h$^{-1}$); $l_2$ = the depth of layer 2 (mm); $K_2$ = the saturated hydraulic conductivity of layer 2 (mm h$^{-1}$), etc. for more layers.

Assuming a sand/soil rootzone depth of 300 mm ($K = 50$ mm h$^{-1}$), and a blinding layer depth of 50 mm ($K = 100$ mm h$^{-1}$), the apparent permeability is 54 mm h$^{-1}$. Thus water flowing at a rate of 54 mm h$^{-1}$ (maximum) could enter a gravel layer typically 100 mm deep ($K = 10^5$ mm h$^{-1}$). The Hooghoudt Equation can now be applied to show that laterals at over 8.5 m centres could cope, provided they were of sufficient diameter. A key point here is that when dealing with a gravel bed or other subsoil material of high permeability, the rootzone can be of much lower hydraulic conductivity than when water has to move laterally through it to a drainage system.

### Enclosed drainage systems

Although most enclosed systems are patented, and thus have predetermined pipe spacing in accordance with the manufacturer's instructions, the Hooghoudt Equation can still be applied to test the soundness of the design. Since there is only one construction medium, there is no need first to calculate the apparent permeability through the rootzone. Therefore, assuming a profile depth of 400 mm with a saturated hydraulic conductivity of 500 mm h$^{-1}$ and a design drainage rate of 20 mm h$^{-1}$ to cope with the maximum rainfall intensities,

$$S = \sqrt{\frac{4 \times 400 \times 400 \times 500}{20}} = 4000 \text{ mm or } 4 \text{ m}$$

which is the typical lateral pipe spacing that many of these types of system use.

## Soil amelioration

Soils which possess physical characteristics unsuitable for adequate drainage and aeration after the effects of play, can be physically improved by changing the particle size class through the addition of sand. It is crucial that the correct type of sand is used and this was detailed in Chapter 1. In Chapter 3 it was stated that the overall objective is to end up with between 75% and 90% by weight of the final mix in the desired sand fraction range (125–500 $\mu$m).

The following equation provides a practical guideline for the amount of a *suitable* sand required to be mixed with a soil of known mechanical composition to create a free-draining rootzone material:

$$A = \frac{(R - B)}{(C - R)} \times 100 \tag{4.12}$$

where $A$ = the weight of sand necessary to add to 100 weight units of the original soil; $B$ = % of original soil in the 125–500 $\mu$m size range; $C$ = % of 125–500 $\mu$m particles in the sand used for amelioration; $R$ = the desired % of 125–500 $\mu$m sand in the final mixture.

The weight of soil in 1 hectare to 10 mm depth is approximately 100 tonnes (i.e. a field bulk density of $1.0 \, t \, m^{-3}$. Therefore $A$ in the above equation becomes the amount of sand in tonnes needed to ameliorate 10 mm of the original soil per hectare.

## An example using the sand/soil mixing equation

The particle size distributions of soil and sand are given in Table 4.3.

**Table 4.3.** Particle size distribution (%) of soil and sand.

| | Particle size range (mm) | | | | | | | |
|---|---|---|---|---|---|---|---|---|
| | 10-5 | 5-2 | 2-1 | 1-0.5 | 0.5-0.25 | 0.25-0.125 | 0.125-0.063 | <0.063 |
| Soil | 3 | 10 | 17 | 13 | 14 | 11 | 9 | 23 |
| Sand | 0 | 0 | 2 | 5 | 53 | 37 | 2 | 1 |

For this example, assuming $R = 80\%$:

$$A = \frac{(80 - 25)}{(90 - 80)} \times 100$$

i.e. 550 t of sand per 100 t of soil (or 10 mm of soil depth per hectare) would be required to raise the 125–500 $\mu$m percentage to 80% in the final mix.

If the sand and soil were mixed off-site it would be more appropriate to mix them on a volume basis. The bulk density of loose air-dry soil is approximately $1.0 \, t \, m^{-3}$ and moist sand is approximately $1.5 \, t \, m^{-3}$. Therefore to achieve a volume ratio for mixing, divide the weight of sand required per 100 t of soil by the sand/soil bulk density, i.e. $1.5 / 1.0 = 1.5$. Thus, if $A = 550$ t the volume (v/v) ratio of soil to sand in the mixture would be: $100 : 550 / 1.5$ or $1 : 3.7$.

## Determining the depth of a sand/soil rootzone mixed on-site

The depth which a sand/soil mix occupies is usually somewhat less than the summed individual depths of soil and sand in the mix because of interpacking of fine soil into the sand. Depending on the degree of compaction, a mix with about 80% in the 125–500 $\mu$m range will have a bulk density in the range of 1.2 to greater than $1.6 \, t \, m^{-3}$. In order to get an estimate of the depth of a rootzone mix it is reasonable to assume a final bulk density of $1.4 \, t \, m^{-3}$ or, looked at in another way, 14 t per hectare millimetre. In the calculation earlier, 550 t of sand were required per 100 t of soil. The total quantity of mix (650 t) would therefore occupy a depth of $650/14 = 46.4$ mm.

In order to create a 150 mm depth of mix one would need a total weight of mix of $14 \times 150 = 2100\,t\,ha^{-1}$. In the example given, $550/650 \times 2100 = 1777\,t$ of sand would be needed and $100/650 \times 2100 = 323\,t$ of soil would be needed. On-site mixing would require therefore the mixing of $1777\,t\,ha^{-1}$ of sand with 32.3 mm depth of native soil. The weight of sand would be scaled up or down from 1 ha depending upon the actual area involved but the depth of soil is independent of the area.

## Depth of ameliorated rootzones over a gravel raft

The primary consideration for both sand and sand/soil constructed rootzones over a gravel drainage raft (or impermeable plastic membrane) is that air entry is guaranteed at the surface, to prevent saturated conditions occurring at the surface after drainage has ceased. Depths of these rootzones (whether sand or sand/soil) will fall between 200 and 400 mm (Chapter 3) assuming that the predominant particles are in the medium to fine sand category (see Table 1.5 for relationship between particle size and critical tension/air-entry pressure).

Laboratory measurements of critical tension are therefore useful to help determine depths of constructed rootzones over a gravel drainage raft or impermeable plastic membrane. However, critical tension is not the only measurement of importance once a sand or sand/soil mix has been selected. Water storage is equally important and therefore the selection of a coarser material to reduce the construction depth could have serious consequences for the water management.

## Depth of ameliorated rootzones over indigenous soil

For ameliorated rootzones which overlie existing topsoil, it is usual to specify a depth in the range 100–250 mm for the reasons outlined in Chapter 3.

It should be noted that minimum rootzone depths may be dictated by factors other than the nature of the rootzone material. For example, if irrigation heads are installed within the construction, then they must be adequately covered. On golf greens, the rootzone should ideally be sufficiently deep to accommodate the flag post and cup (a minimum of 200 mm depth). When sportsfields are constructed on old refuse tips it is essential that the rootzone or capping is sufficiently deep to avoid the risk of glass or metal becoming exposed at the surface.

## GRADIENTS AND ORIENTATION OF SPORTSTURF FACILITIES

Gradients and cambers on sportsturf facilities, in particular winter games pitches, have traditionally been used to assist surface water removal. Slopes over long distances, for example goal to goal on winter games pitches, give no practical benefit to the flow of surface water off the pitch but a slope across the pitch or better still a slope to each touchline assists the shedding of surface water. For crown bowling greens and golf greens the general topography and slope are important aspects of the nature of the game. In the case of golf greens a poorly thought out topographical design can adversely affect drainage.

A modest, preferably diagonal slope across a playing field area enables pipe drains to be installed at a constant depth and lateral surface flow of water to slit drains is facilitated. A diagonal slope of between 1% (1:100) and 1.5% (1:67) is a good rule of thumb. The British Department for Education's directive for school pitches is that they should have a maximum slope of 1% (1:100) along the pitch and 2% (1:50) across the pitch (DES, 1982). When pitches are on sloping land they should be orientated so that the greatest slope is across the pitch when it will have less of an adverse effect on play. Even so, slopes in excess of 2.5% are undesirable.

With improved soil and soil profile design for sportsturf a sloping surface is less important for winter games pitches because of the better infiltration of water over the entire surface. Indeed suspended watertable and enclosed systems need to be flat to allow satisfactory control of the watertable or sub-irrigation. The terraced impermeable membrane in a Purr-Wick system (Chapter 3) helps cope with a sloping surface.

Provided the surface is even, a general slope on a cricket outfield does not affect the game adversely. Lord's outfield for example has a general slope of around 2%. Ideally an outfield should be flat and no strong case can be made for the need for a slope to aid drainage in summer. Slopes on a cricket square are more critical. It is desirable for pitches to be level between stumps with a slight crossfall of about 1.25% (1:80) to assist drainage. It is undesirable for the gradient to exceed 1.25% (1:80) in the direction of play or 2.5% (1:40) across the direction of play (County Cricket standards are more restrictive). Similarly, grass tennis courts should be level in the direction of play, although a crossfall of as much as 1.67% (1:60) may be necessary on heavy soils. A minimum fall of 0.4% (1:240) is desirable.

Attention should be paid to the orientation of pitches to minimize the problem of glare due to the sun. Hockey, Association and Rugby Football pitches, grass tennis courts and cricket pitches should run in a roughly north–south direction if possible. Orientation is most important in tennis and cricket where a player receives the ball consistently from the same direction.

## Fundamental Elements of a Pipe Drainage System

All sportsturf drainage systems use some type of pipe drains and it is important to select appropriate pipe type and diameter for each part of the drainage system. Selecting a pipe that is too small for the purpose is more serious than selecting one that is too large in diameter. An over-sized pipe is uneconomic and it is a misconception to think that a larger pipe will absorb and remove more water. An under-sized pipe may become surcharged at peak rainfall intensities which could result in an increase in the height of the watertable and create the risk of a blow-out at the surface especially in unstable soils. Surcharging occurs when a pipe drain receives inflow of water along its length, and where the pipe is filled to capacity at any one point.

## Terminology

Traditionally, pipe drains have had a variety of functions for agricultural use (MAFF, 1982) and, in principle, these same functions apply to sportsturf use since the main function of a pipe drain is to provide a means for water to be removed from the soil profile. However, there are different types of pipe drain available for different functions and the categories generally recognized for sportsturf use are detailed below.

### Lateral pipe drains

These are subsurface pipe drains which collect water continuously along their length usually through slots in the pipe wall. Most of the pipe drains in the drainage systems illustrated in Chapter 3 come in this category. Lateral pipe drains are generally the ones whose primary function is the eventual removal of excess moisture from the playing surface and lowering of the watertable if the latter is too high.

### Main carrier pipe drains

This is the name usually given to a drain which collects water from lateral pipe drains and carries the water from the drained area to the outfall. Although called restricted inlet drains in agricultural terms, these drains may have slots in the pipe wall so that they can act as primary water interceptors. Main carrier pipe drains are larger in diameter than lateral pipe drains, typically being between 115 and 180 mm internal diameter for the former, and between 65 and 85 mm internal diameter for the latter.

*Catchwater or cut-off pipe drains*

This is the name commonly given to drains that are of first priority in draining a sportsturf area. These drains prevent extraneous water from entering the area. Extraneous water can enter a site as surface or subsurface runoff from upslope. It is especially common where batters have been formed around pitches or banks around golf greens. Cut-off pipe drains are usually placed in the base of trenches filled to the surface with permeable fill. They have a particularly important role in preventing erosion on newly constructed playing fields and golf courses.

*Interceptor pipe drains*

Extraneous water may arise locally in the form of underground springs. Interceptor drains are slotted pipes placed to drain these sites and prevent water from getting on to the playing area.

*Discharge pipe drains*

These are generally the pipe drains taking water from a silt trap or inspection chamber to the final discharge point, which may be in the form of a soakaway, a storm water drain, ditch or stream/river. Because the function of these drains is simply to transport water from one source to another, they are usually solid wall pipe drains without permeable fill.

The above categories need not necessarily be considered as discrete independent functions. For example, a catchwater pipe drain or restricted inlet slotted main drain may equally have a role in watertable control and spring-water interception. Moreover, in addition to the various types of pipe drain listed above, open drains and ditches are often of value to sportsturf drainage, particularly on golf courses where they can act as outlets for numerous localized pipe drainage systems or even in their own right as cut-off drains. If open ditches affect play seriously, they may need to be piped and filled in, but if they act as cut-off drains they must be filled to the surface with permeable fill.

# Pipe materials

Traditional clayware pipes (tile drains) have been almost completely replaced in the UK by plastic pipes which are now generally recognized as being more appropriate for sportsturf drainage work. Clayware pipes are cumbersome to lay, are subject to breakages and have large transport and handling costs. A key disadvantage relative to plastic pipes is that a minimum gap when clay tiles

**Table 4.4.** Flow rate ($l\,s^{-1}$) of lateral corrugated plastic pipe drains over a range of gradients commonly used in sportsturf construction (from MAFF, 1982).

| Internal pipe | Gradient | | | |
|---|---|---|---|---|
| diameter (mm) | 1:300 | 1:200 | 1:100 | 1:50 |
| 65 | 1.3 | 1.5 | 2.0 | 2.6 |
| 75 | 1.9 | 2.2 | 2.8 | 3.7 |
| 85 | 2.6 | 3.0 | 3.9 | 5.1 |
| 100 | 4.0 | 4.5 | 5.8 | 7.7 |
| 115 | 5.6 | 6.4 | 8.0 | 11.0 |
| 135 | 8.4 | 9.5 | 12.5 | 16.5 |
| 155 | 12.0 | 13.5 | 17.5 | 23.0 |
| 180 | 17.0 | 19.5 | 26.0 | 34.0 |

are butted cannot be guaranteed and this prevents the use of fine grades of permeable fill.

Perforated plastic pipes can be used in almost any pipe drainage role except at the final outlet where it is advisable to have a stronger solid walled plastic or concrete pipe. A range of supplementary fittings including end stops, reducers and purpose-made junctions is available. It is also recommended to use unperforated solid plastic pipes if these pipes are likely to be in the close proximity of trees (whose roots may block up pipes), and if pipes follow a steep bank (these pipes are less subject to the effect of land displacement). On very stony ground, it may be necessary to place a layer of gravel in the excavated trench base on which to bed and secure the pipes.

## Selection of pipe size

The quantity of water which a pipe can carry depends on its diameter, the gradient to which it is laid, and detailed pipe geometry including the 'roughness' of the pipe material. A range of design charts supplied either by the pipe manufacturers or the Ministry of Agriculture, Fisheries and Food provides a recognized method of calculating pipe size (e.g. MAFF, 1982). These charts cover all types of drainage pipe, both clayware and plastic, smooth and corrugated. For sportsturf drainage, the most commonly used drainage pipe is the corrugated plastic type. Table 4.4 summarizes pipe flow rates for corrugated plastic pipes for a typical range of internal diameters and gradients used in sportsturf construction.

# Gradients for pipe drains

The gradients or falls that pipe drains are laid to typically range from 2% (1:50) down to 0.33% (1:300), the latter of which should be considered an absolute minimum for sportsturf use. If possible a minimum of 0.5% (1:200) should be aimed for. The gradient should preferably be uniform although a drain may be laid in increasingly steeper falls. On no account should the gradient become less as this will slow down the rate of flow, causing possible surcharge and even silting up.

Modern drain-laying methology utilizes a laser beam to ensure a smooth gradient. This technique is extremely valuable when permitted gradients are very shallow and/or the surface topography is uneven. When pitch surfaces are graded to an even slope drain laying is simplified by using a constant drain depth.

# Maximum length of pipes to discharge point

Once the basic drainage scheme has been designed, the information needed to determine the maximum length of pipe to a discharge point will be the drain spacing (m) and the design drainage rate (mm h$^{-1}$). The appropriate chart for the particular pipe drain in question is consulted and the manufacturer's flow rate in litres per second for the appropriate gradient is read off. The following equation is then applied:

$$M = \frac{F}{vS} \qquad (4.13)$$

where $M$ = maximum length of pipe (m); $F$ = pipe manufacturer's flow rate (l h$^{-1}$); $v$ = design drainage rate (mm h$^{-1}$); $S$ = drain spacing (m).

For example, a 65 mm internal diameter plastic corrugated lateral pipe drain installed at a gradient of 1% (1:100) would have a manufacturer's flow rate of 2 l s$^{-1}$ (7200 l h$^{-1}$). Assuming a design drainage rate of 20 mm h$^{-1}$ for the whole scheme and a lateral pipe drain spacing of 15 m, Equation 4.13 gives:

$$M = \frac{7200}{20 \times 15} = 24 \text{ m}$$

Thus for a suspended watertable soccer pitch of size 100 m by 70 m and a design drainage rate of 20 mm h$^{-1}$, 65 mm diameter lateral pipe drains at 15 m centres would be grossly inadequate for carrying water at the design drainage rate across the width of the pitch. The alternatives are to: (i) use a bigger diameter pipe; (ii) reduce the lateral pipe drain spacing; or (iii) increase the pipe gradient. Main carrier pipe drains must have sufficient diameter at the installed gradient to remove water from the entire drained area at the design drainage rate in addition to any water that may enter them from extraneous sources.

## Depth of installation of pipe drains

With Hooghoudt's Equation modified for systems which overlie an impermeable subsoil, the pipe drains are in effect laid in trenches cut into or lying on the impermeable barrier. Here the main consideration for depth of placement is that the pipe drain is sufficiently protected against surface traffic, whether this be construction traffic or maintenance traffic. Maintenance machinery that has deep spiking implements or mini-mole ploughs/subsoilers attached needs to be able to be operated well above a pipe drain. Similarly, pipe drains need to be deep enough to avoid being damaged by subsequent installation of irrigation pipe work or slit drains. Typically lateral pipe drains, which are the nearest to ground level in sportsturf drainage systems (excluding mini-pipe slit drains), should be no shallower than 450 mm from the finished surface height. Gooch and Escritt (1975) estimated that 600 mm was probably a good compromise in terms of cost of excavation, efficiency and cost of backfill.

Main carrier pipe drains that laterals connect to, are likely to be deeper (typically 700 mm from the finished surface height). There is little benefit in installing lateral or main carrier pipe drains deeper than 750 mm in impermeable subsoils as long as sufficient fall can be achieved above this depth without the pipe drain coming too near to the surface. Catchwater, cut-off and interceptor pipe drains vary in depth depending on the specific problem. It is not uncommon for cut-off drains to be 1000–1500 mm deep.

## Watertable control

Sportsturf drainage focusses mainly on getting water away from the surface to the pipe drains because this is where the key problems arise. However, problems of a shallow watertable can be a complicating issue. It is generally recommended that the watertable is kept below 500 mm from the surface and the depth of placement of the pipe drains will control the height to which the watertable can rise so long as the capacity of the pipe drains to transmit water is not exceeded (Ward, 1983). The full Hooghoudt Equation (Equation 4.1) may be needed to calculate the appropriate drain spacing in exceptional situations, but since sportsturf pipe drain spacings are generally much less than agricultural drain spacings, watertable problems are unlikely to arise.

It is not uncommon for sportsfields to be in very low lying sites where water

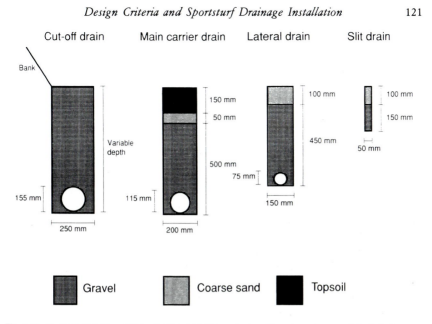

**Fig. 4.5.** Typical trench widths, depths and internal pipe diameters for sportsturf drainage systems.

levels in adjacent rivers or ditches may be shallower than pipe drains in wet weather in winter. This problem can normally be overcome by isolating the ground from floodwater and creating a sump for infield drainage. The sump can be emptied by automatic pump and the water lifted into a ditch, stream or river. (The National Stadium in Cardiff is an example of this situation.) If the sump is large enough it can serve as a source of irrigation water in summer.

## Width of pipe drain trenches

Drain trenches should be as narrow as possible with the proviso that they allow placement of the pipe drain without risk of forcing the pipe into the trench and smearing the pipe with soil. A range of appropriate trench widths and pipe sizes is shown in Fig. 4.5. As a general rule the bottom of a trench should be at least 50 mm wider than the outside diameter of the pipe.

In unstable soils and soils containing moderate or large stones, drain trenches will inevitably be wider than specified. In these situations it is impossible to estimate with accuracy the quantities of permeable backfill needed to fill the trench (Fig. 4.6). Trenches which are left as open drains must be wider than trenches for pipe drains and they should have sloping sides. The narrowest

**Fig. 4.6.** An irregularly cut lateral drain trench. Quantities of backfill are difficult to estimate in these situations.

trenches for pipe drains are achieved with a 'trenchless' system where the pipe is placed, in a single operation, in a mole plough channel.

## Backfilling over pipe drains

All pipe drains used for draining sportsturf areas require permeable fill over the pipe drains to provide drainage continuity with superimposed drainage layers or rootzones. The key function of permeable fill is to facilitate the rapid flow of water from the soil surface to the pipe drains. In agricultural drainage systems permeable backfill is used less frequently but is needed for schemes which utilize mole ploughing or subsoiling as secondary operations or where the control of spring water or seepage lines is important.

## Outfalls, inspection chambers and pipe blockages

The essential requirement of any drainage scheme is the provision of a suitable outfall. This may take the form of an existing land or pipe drain or may be a river, ditch, stream or municipal storm water drain. Some soils have sufficiently permeable subsoils to allow the use of constructed soakaways (e.g. sites over permeable chalk).

In general, drainage outlets should be protected, provide ample fall and not be subject to backing up. Outlets into ditches, streams and rivers should be at least 150 mm above peak levels unless tide flaps are used. Steel vermin grids are also recommended and it is important that the final 1.5 m of buried outfall drain is rigid with any protected portion being frost resistant.

All sportsturf drainage schemes should specify the provision for the inspection of the outfall from main drains, and for silt collection and removal. On sites where drainage maintenance is likely to be a regular feature, purpose-built inspection chambers will be needed at regular points in the drainage scheme.

Silting up of drains is a problem on poorly structured silty soils in particular, but poor drainage design can be a prime factor. Any feature which is likely to result in the slowing down of drain water flow or even backing up of water in drains must be avoided.

Drains may become blocked by iron ochre but incidents are related to specific circumstances. The situation can arise when drains are introduced into a severely waterlogged and reducing soil where iron is mainly in its reduced, ferrous state. In its ferrous form, iron is mobile but if air is introduced it is oxidized to the ferric form and becomes immobile as a gelatinous oxide. Specific bacteria are usually involved in the oxidation. When a pipe drain is placed in the soil, air can enter and the inside and immediate surrounds of the pipe act as a sink for the ferrous iron which moves towards it as the orange-red ochre builds up. Ochre will only continue to build up whilst parts of the subsoil are anaerobic and reducing. Once the subsoil becomes fully aerated no further build-up will occur.

## Pipe drainage maintenance

There are two basic techniques for clearing blocked drains. The most sophisticated sportsturf drainage systems have inspection holes over the pipe drains to allow regular checking of the drainage system by endoscopy. Rodding is the best method for clearing blocked drains as long as there is sufficient drainflow to assist washing out, but normally only straight, unbranched pipe drains with individual accessible outlets can be cleared in this way. A variety of probes, scrapers or brushes can be used to break up and clean out most deposits although any ochre blocking drain slots is not removed by rodding. An alternative technique is to use drain jetting equipment which may assist in removing ochre from drain slots, a process which involves the use of high pressure water jets.

## Organization of pipe drainage

The organization of pipe drainage depends primarily on the nature of the drainage problem and intended use of the drained area. For all drainage schemes, the main objective is to keep the layout as simple as possible for ease of laying out the scheme, machinery operation and future work on the system (e.g. superimposing a slit drainage system at a later date). For most sportsturf drainage systems, there are only three basic types of organization of pipe drainage that are generally appropriate for intensive sportsturf use and one system for larger, less intensively used areas.

### Herringbone system

A herringbone organization of pipe drains has been used for many years. As its name suggests, it consists of a central main carrier pipe drain with lateral pipe drains branching off alternately at an angle of approximately 30° (for which purpose-made junctions are available) (Fig. 4.7). Junctions must be used to connect the lateral pipe drains with the main carrier pipe drain and it is important that there is no levelling off of the gradient at this junction as serious sedimentation can potentially occur at these points. Ideally steeper gradients should be introduced in the last 1–2 m of lateral run before the connection is made. Where practical, the steeper gradient line of the lateral pipe drain

Large playing field                                    Golf green

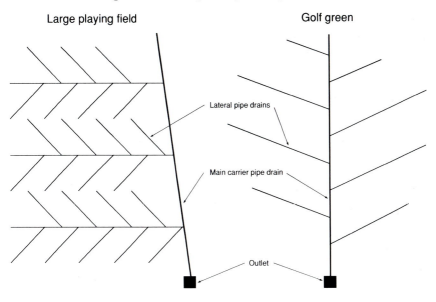

Lateral pipe drains

Main carrier pipe drain

Outlet

**Fig. 4.7.** Examples of the herringbone system of organization of pipe drainage.

should be at least 60 mm to 90 mm higher than the gradient elevation necessary for making the connection.

Herringbone systems are regularly recommended for large playing field areas and golf greens where the area being drained is not of a regular shape. They are particularly advantageous for large areas because thay can consist of many relatively short lateral drainage lengths which means that, on flat land, large depths of trench excavation are not needed to obtain the necessary gradient (Gooch and Escritt, 1975). However, herringbone systems suffer from the disadvantage that they are relatively complicated to install and difficult to locate once installed. Slit drainage superimposed on herringbone systems is difficult to orientate so that the slits cut the laterals at close to 90°. On graded land, herringbone systems have the advantage that both lateral and main carrier pipe drains can utilize the same general slope without needing trenches excavated at increasing depths.

*Parallel systems*

Parallel or grid systems are better suited for regularly shaped areas like winter games pitches and bowling greens (Fig. 4.8). Moreover, because of the layout, lateral drains can be located easily. Parallel systems also involve fewer junctions and are therefore cheaper to install (Gooch and Escritt, 1975).

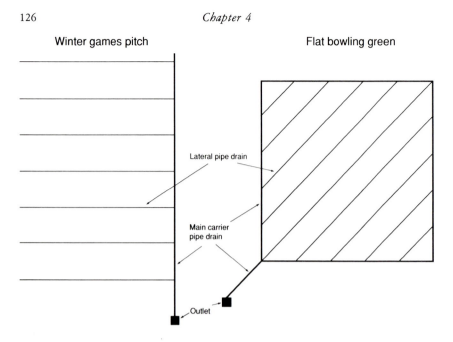

**Fig. 4.8.** Examples of parallel or grid systems of organization of pipe drainage.


Parallel systems can be installed with the lateral pipe drains connecting to the main carrier pipe drain at right angles or at a more acute angle using purpose-built connectors. For soils particularly prone to silting problems, right-angled junctions are inadvisable as the junction can act as a trap for silt if installed in this way.

### Fan-shaped system

In a fan system, laterals radiate from a single outfall (Fig. 4.9). The connections at the outlet may be complex but all laterals can be examined at one point. The system is suitable for some golf greens but generally not for other areas.

### Natural or random systems

These systems are appropriate for undulating land. They involve an informal layout where drains are laid to drain water from perceived wet areas and depressions. They are potentially efficient, low intensity layouts which, for their success, depend upon high quality site assessment. This type of system may be used on, for example, golf fairways and country parks (Fig. 4.9).

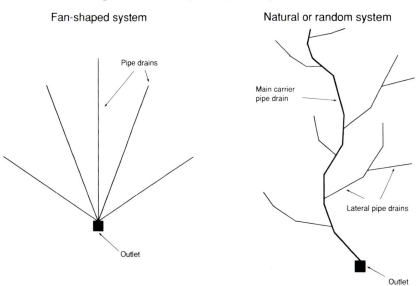

**Fig. 4.9.** Examples of the fan-shaped and natural/random systems of organization of pipe drainage.

### *Interceptor/cut-off systems*

Cut-off or interceptor pipe drains are sited strategically to serve their function. They may feed into or serve as the main carrier pipe drain accepting water from the lateral pipe drains. Caution must be exercised because the quantity of water collected by a cut-off drain is impossible to predict and may be much greater than imagined.

## SELECTION OF MATERIALS FOR SPORTSTURF CONSTRUCTION

## Permeable fill

Permeable fill is used to increase the rate of vertical and/or lateral water movement over pipe drains, in slit drains and as a layer in construction profiles. Gravel or stone may be used as permeable fill.

The term 'gravel' implies a material of particle size intermediate between sand and stone. The upper limit of sand sized particles is 2 mm but there is no universally accepted upper limit for gravel although 10 mm is used widely (Table 1.2). The hydraulic conductivity of gravel sized material is very large so there is no reason to use larger particles to achieve an adequate flow rate of water (Fig. 1.7).

There are two principle reasons why gravel sized particles are used in sportsturf drainage. The first is that the drainage layer of gravel or stone is usually overlain with other materials, often sand. In order for the overlying sand to produce a natural filter and not weep progressively into the drainage layer it must contain a moderate proportion of particles no smaller than one sixth of the diameter of the particles in the drainage layer. Widely available concreting sands can be used to overlie 5–10 mm gravel and these are acceptable to players when they continue right to the surface. However, drainage layers containing particles greater than 20 mm must be overlain by materials which contain some quite large and abrasive particles. The second reason, which applies to slit drains and permeable fill over laterals, is that with gravel there is much less tendency for soil to move into and occlude the pore space than when coarser stone is used.

Gravel is the normal material to use but there are two exceptions. One is that larger stone is more appropriate to create a stable base on soft ground. Also for cut-off drains, stone brought to the surface is more likely to remain 'clean' and free from vegetation.

Gravel sized materials may be rounded or angular. Naturally rounded gravels are derived mainly from shore deposits although some river gravels are rounded. 'Lytag' is an artificial rounded aggregate produced from coal-fired power station waste. Angular gravels are usually derived from crushed limestone or granite rock and their main use is for road surfacing. Hard limestone is quite satisfactory but there have been a few instances in low rainfall areas in Europe where soft limestone aggregate drainage layers have degraded to a cemented mass. The likely explanation for this is that calcium carbonate is dissolved at shallow depths through the influence of natural acidity or acidifying fertilizers but becomes reprecipitated within the $CaCO_3$ matrix at a greater depth. Some industrial 'slags' used as drainage aggregate may be phytotoxic but this is not usually a problem since the material is normally below rooting depth.

From the practical point of view, rounded gravels flow easily and are therefore well suited for use as fill over pipe drains and in slit drains. They are extremely difficult to handle when used to create drainage layers because they are unstable under local loading. Conversely angular gravels lock together and are stable to work on but flow less readily.

With regard to particle size preferences within the 2–10 mm range there are several points. If plastic drainage tube is used there is no danger of gravel getting into the pipes provided particles are larger than 3 mm. Irrespective of particle size the narrower the range in particle sizes the greater the total pore space. The maximum is about 40% for single sized rounded particles but can be over 50% for angular gravel. Particles of 3 mm and greater do not 'clog' and flow easily (accepting an effect of shape). Finally, choice is very often restricted by availability of materials. A 4–8 mm gravel is well suited to all of the prescribed uses but either a 3–6 mm or 6–9 mm gravel could substitute adequately.

**Table 4.5.** Recommended and acceptable limits of sand size (in terms of % passing through sieves) for blinding two types of drainage gravel. (From Baker, 1990a.)

| Sieve size (mm) | 5-8 m gravel | | 5-10 mm gravel | |
|---|---|---|---|---|
| | Recommended | Acceptable | Recommended | Acceptable |
| < 0.063 | 0-1 | 0-2 | 0-1 | 0-2 |
| 0.125 | 0-2 | 0-5 | 0-2 | 0-5 |
| 0.250 | 5-25 | 0-30 | 0-10 | 0-25 |
| 0.500 | 20-50 | 15-60 | 0-35 | 0-50 |
| 1.000 | 45-80 | 40-90 | 10-70 | 0-80 |
| 2.000 | 80-100 | 75-100 | 50-95 | 20-100 |
| 4.000 | 100 | 98-100 | 92-100 | 80-100 |
| 8.000 | 100 | 98-100 | 98-100 | 90-100 |

# Blinding layer

A blinding layer is designed to prevent particle migration from finer textured material overlying a drainage layer into the underlying gavel. It may also avoid taking gravel right to the surface with slit drainage systems.

It is not necessary to prevent particle migration downwards completely, but a natural filter must be achieved without significant migration. In order to achieve this 20% at least of the blinding layer should be of particles greater than one sixth the size of the smallest 20% of particles in the underlying layer. The most common use of a blinding layer is over gravels having particle size ranges within 4 mm to 10 mm. Concreting sands ranging in particle size between 125 $\mu$m and 5 mm are usually used. Table 4.5 summarizes the current Sports Turf Research Institute specifications for recommended and acceptable blinding layers over 5-10 mm and 5-8 mm gravels (Baker, 1990a).

Concreting sands have a high Gradation Index and considerable interpacking of particles can occur. In consequence a blinding sand with particles in the range of 125 $\mu$m to 5 mm and a $D_{90}/D_{10}$ of around 15 has a much smaller hydraulic conductivity than a clean fine sand (125–250 $\mu$m) with a $D_{90}/D_{10}$ of around 2.0. The hydraulic conductivity of concreting sands, typically 300–700 mm h$^{-1}$, is usually adequate. An alternative to using a broad spectrum blinding sand is to use a clean coarse material over gravel and this is specified by the USGA Green Section Staff (1993) who specify 1–4 mm sand over 6–9 mm gravel. Such a blinding sand has a very large hydraulic conductivity (see Chapter 1) but the material is rarely available in the UK. Figure 4.10 illustrates the downward migration of a rootzone sand placed directly on gravel and Fig. 4.11 shows the absence of migration when an appropriate blinding sand is used.

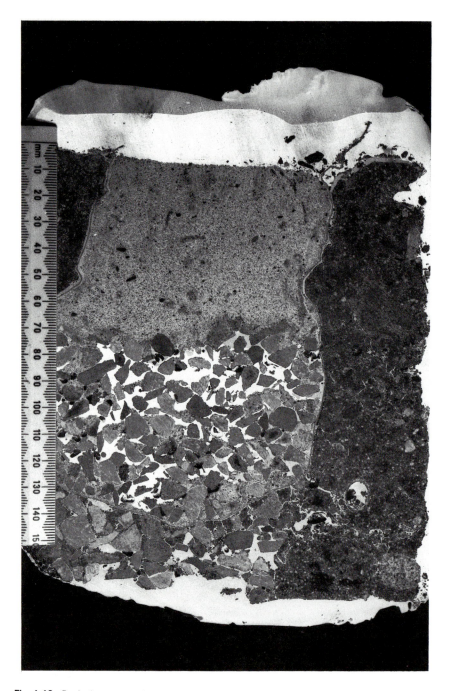

**Fig. 4.10.** Resin impregnated section of a medium-fine sand overlying a 4-10 mm gravel in an experimental slit drain surrounded by a geotextile membrane at STRI, Bingley. The sand has migrated into the gravel layer leading to a reduction in porosity of the gravel. The white areas are large pore spaces within the gravel which remained uncontaminated. The scale is in units of millimetres (from Baker *et al.*, 1991).

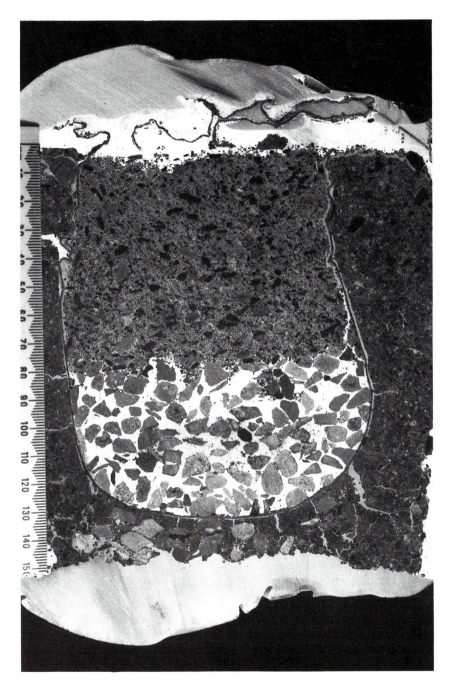

**Fig. 4.11.** Resin impregnated section of a coarse grit sand overlying a 4-10 mm gravel in an experimental slit drain surrounded by a geotextile membrane at STRI, Bingley. The pore space within the gravel has remained uncontaminated. The scale is in units of millimetres (from Baker *et al.*, 1991).

*Geotextile blinding layers*

In addition to sands being used as blinding layers, geotextile membranes are used with varying degrees of success between a rootzone and gravel drainage raft. These membranes should allow water to pass through, but prevent the downward migration of particles. The decision to use these materials often hinges on cost, but the use of a geotextile membrane does not change the requirement for an adequate depth of rootzone to ensure air entry at the surface.

Geotextile membranes have other applications in sportsturf construction. They are particularly useful for lining the bases of constructions on which a gravel raft is to be placed to prevent the underlying soil contaminating the gravel. They have been used frequently to line the base of golf bunkers to prevent contamination of the bunker sand, but unless the bunker maintenance regularly involves replacing lost sand, these membranes inevitably get raked to the surface and cause maintenance difficulties (Fig. 4.12). They are also used to line the fibre-wrapped honeycomb drain pipes (Fig. 3.5) and can be used in general as a filter between soil and gravel in drainage trenches, so preventing soil being washed into drains. However, these materials must be selected in relation to the particle size distribution of the neighbouring soil if blocking up of the membrane is to be avoided. Problems have arisen where these materials have become clogged by fine soil material, so rendering them virtually impermeable. An additional important point is that geotextile membranes at rather shallow depths prevent many operations such as subsoiling, slit drain insertion and pipe drain installation.

# Rootzone sands

The choice of rootzone sands was covered in Chapter 1, which described how the requirements for fast hydraulic conductivity, surface stability, large total porosity and sensible construction depths placed tight limits on the particle size distribution of sand and sand-dominant mixes for sportsturf rootzones. Both inadequate selection of sand and insufficient proportions of sand in sand/soil rootzones have been shown to markedly affect soil physical properties and playing quality characteristics of winter games and fine turf constructions.

For example, Baker and Isaac (1987) noticed that uniform medium-coarse sands ($D_{50} = 550\,\mu m$) gave an unstable surface for a winter games construction using a pure sand rootzone, particularly in dry conditions, with low ball bounce and poor moisture retention. On the same type of construction with a much broader range of particles in the sand ($D_{90}/D_{10} = 24.6$), a hard surface with high ball bounce resulted. Baker (1988) found that on winter games pitches constructed using sand/soil mixes with inadequate proportions of sand

**Fig. 4.12.** Geotextile membranes used to line the base of golf bunkers are easily ripped up by raking unless the bunker is regularly topped up with sand to replace that lost during play.

in the mix, the type of sand became relatively unimportant because the soil had a dominant effect on the physical properties of the rootzone.

With golf, the rootzone composition has an important effect on the hardness of a green and its holding power (the stopping distance of a ball after its initial impact under defined conditions of velocity, backspin and an approach angle). In experimental trials, Baker and Richards (1991) noted that surfaces of rootzones containing too little sand were soft in wet weather but hard in the summer months. In drier conditions the holding power was generally larger for sand/soil mixes than for the pure sand constructions, but this trend was reversed in wet conditions because the sand/soil mixes held more moisture than the pure sand rootzones. Golf balls also tended to stop more quickly on rootzones containing uniform medium-fine sand ($D_{50} = 230 \, \mu$m) than on those containing medium-coarse sand ($D_{50} = 550 \, \mu$m).

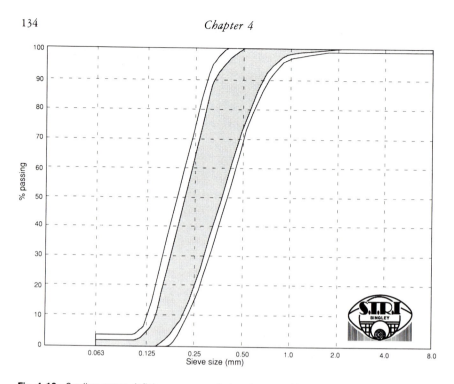

**Fig. 4.13.** Grading curve defining recommended and acceptable limits of sand size for pure sand rootzones for winter games pitches (reproduced with permission of STRI, Bingley).

### Sports Turf Research Institute (STRI) recommendations

In Chapter 1 the 'ideal' particle size distribution of rootzone sands for different uses was derived. In practice the choice of sands is often limited and acceptable deviations from the ideal need to be specified. Baker (1990a) reviewed the relevant literature on sand types used for specific sportsturf applications and produced general specifications on recommended and acceptable limits for sands for different sportsturf uses. It was recognized that sands falling within the acceptable range but not totally within the recommended range cannot be used with complete confidence.

The Sports Turf Research Institute's recommended and acceptable limits of sand size for winter games pitches and for golf greens and bowling greens are reproduced in Figs. 4.13 and 4.14. These specifications apply to pure sand rootzones for winter games pitches but to sand for sand/soil mixes for golf and bowling greens.

### USGA recommendations

In addition to recommending particle size limits for sands for construction of

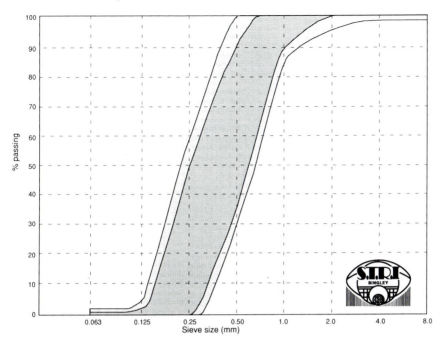

**Fig. 4.14.** Grading curve defining recommended and acceptable limits of sand size for golf and bowling greens (reproduced with permission of STRI, Bingley).

USGA putting greens, the USGA Green Section Staff also recommend a variety of other desirable physical properties of the mix once constructed. One of the shortcomings of earlier USGA guidelines was that they concentrated primarily on the definition of the rootzone properties themselves (measured in the laboratory), rather than on defining the rootzone composition which would actually confer these properties (Adams, 1986). This shortcoming has now been overcome and it is interesting to note that the recommendations for sand/soil mixes which originally favoured relatively coarse materials are now more in line with British recommendations for sands with uniform medium-coarse contents. Specifications are detailed in Chapter 7.

## Calcium carbonate content of blinding and rootzone sands

As with the type of gravel being used for permeable fill over pipe drains, the calcium carbonate ($CaCO_3$) content of sands for sportsturf use requires special appraisal (Adams *et al.*, 1971), especially with fine turf constructions where an increase in soil pH caused by the liming effect encourages earthworm activity and certain turf diseases like take-all patch (*Gaeumannomyces graminis*).

Conversely, acid conditions are unfavourable to the bacteria responsible for sulphide production in anaerobic conditions (Adams and Smith, 1993). Furthermore, turfgrasses differ in their sensitivity to soil pH. For example, most fine turfgrasses are tolerant to pHs below 5.0 but perennial ryegrass is not (see Chapter 2).

Even a very small content of calcium carbonate in a rootzone sand constitutes a substantial liming effect. The lime addition is the more important because of the small cation exchange capacities of sandy rootzones and therefore their poor buffering against pH change (see Chapter 1). A rootzone sand of 150 mm depth weighs over $2000 \, t \, ha^{-1}$ and a $CaCO_3$ content of only 0.25% represents an addition of $5 \, t \, ha^{-1}$ of $CaCO_3$ which is a typical agricultural application of ground mineral limestone. In contrast a sand topdressing of $50 \, t \, ha^{-1}$ with the same $CaCO_3$ content would supply only 125 kg of liming material.

The $CaCO_3$ in sands is usually present as fragments of shell and its liming effect (in terms of speed of effect) depends on particle size which determines the surface area for chemical dissolution. The total free carbonate content of a sand can be determined as weight loss following treatment with M HCl or, more accurately, by measuring the $CO_2$ evolved on treatment with acid. Measurements of pH on sands provide no useful information because of the almost total absence of buffering.

It is not possible to be precise about the liming effect of the $CaCO_3$ in a sand used for a rootzone, as topdressing or as a blinding layer, but guidelines are needed. Baker (1990a) recommended that rootzone and topdressing sands used for fine turf should not contain more than 0.5% (w/w) of $CaCO_3$. For perennial ryegrass and other broadleaved turf larger $CaCO_3$ contents are acceptable and sands with up to 15% $CaCO_3$ have been used successfully. Soils produced from such sands have pHs typically in the range 7.2–7.6. These alkaline conditions favour ammonia loss by volatilization when fertilizers containing urea are used.

For blinding sands over gravel in slit drains, a specification of less than 15% calcium carbonate applies where the slit drains are draining coarse turf areas such as on winter games pitches. However, where slit drainage is used to drain fine turf areas such as cricket outfields or golf fairways, then the fine turf specification of less than 0.5% calcium carbonate applies (Baker, 1990a).

## Selection of soils for sand/soil mixes

The quality of topsoil used in sand/soil mixes can have a marked effect on the overall quality of the mix despite substantial amelioration with sand. It cannot always be assumed that the indigenous soil is appropriate for amelioration purposes and this, coupled with the logistical difficulties in producing a uniform mix, explains why there are now many proprietry sand/soil mixes available on the market.

Baker (1985a) felt that soils to be used for amelioration purposes should be restricted to the textural classes of sand, loamy sand, sandy loam, the lighter sandy clay loams and the lighter loams. Nevertheless, heavier textured soils, for example clay loams, have been used successfully. Other criteria suggested by Baker (1985a) are:

1.   A stone content (i.e. particles > 2 mm) of less than 10% (w/w). Even with 10% stones, provision for stone removal will probably be needed. (Note: machinery capable of burying stones is now available.)
2.   An organic matter content by loss on ignition of between 2% and 12% by weight with the more organic soils (i.e. those with more than 8%) being used only where the sand/soil ratio exceeds 1:2 by volume.
3.   A clay dispersion coefficient (see Chapter 1) of less than 45%. Values greater than 45% suggest that the soil is prone to mechanical breakdown when wet. However this limit can be relaxed when the clay content is less than 15%.
4.   A pH in the range of 4.0–7.0 for coarse turf areas and 4.0–6.0 for fine turf.

## Degree of sand/soil mixing

The degree of mixing will affect any naturally aggregated structure present in the topsoil selected for the sand/soil rootzone. Baker (1985b) found that rootzone mixes containing 20% or more of particles less than 125 $\mu$m will only give satisfactory performance if some soil structure is present, particularly if clay particles form a large percentage of the fines content. In general it is better to adopt a loose form of mixing where aggregates of soil remain as discrete units in the mix, so that continuous pathways of sand exist in the mix, which are supplemented by small 'reservoirs' of water and nutrient storage in the discrete aggregates of soil (Fig. 4.15).

As far as testing a sand/soil rootzone mix in the laboratory is concerned, it is appropriate to homogeneously mix a sample intended for winter games since, in field use, shearing and compaction forces are likely to break down all discrete soil aggregates at the soil surface. This at least then provides a worst-case analysis. For fine turf sports, loosely mixed aggregates of soil in a sand/soil mix are likely to stay as discrete units because the nature of wear on the surface is less severe.

## Soil amendments and conditioners

Whilst the priority for fast drainage rates in sportsturf has led to all-sand and sand/soil systems, this development has in turn led to problems in other areas,

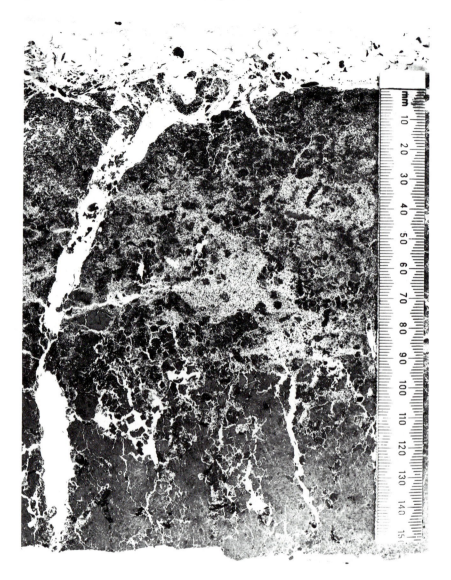

**Fig. 4.15.** Resin impregnated section of a loose sand/soil mix. Aggregates of soil remain as discrete units in the mix leaving tortuous but continuous pathways of water flow through the sand. A large earthworm channel can also be seen. The scale is in units of millimetres.

such as poor surface stability, excessive surface hardness, poor moisture and nutrient retention and diminished soil biological activity. As a result, a range of soil stabilizers and conditioners have been developed to help solve the problems incurred.

## Soil stabilizers or reinforcers

These synthetic products are often marketed for their alleged ability to 'reinforce' turf and to extend the wear tolerance of natural turf surfaces. The number of natural turf surfaces utilizing these materials is growing rapidly. The stabilization of sand by synthetic materials provides a means of maintaining playing quality where turfgrass roots alone are unable to sustain this property. Their principle mode of action is to increase the resistance to shear of turf surfaces. They do not necessarily improve the wear tolerance of the grass itself.

Currently there are three types of reinforcement materials:

1.  Intact fabrics or carpets placed into or a little below the surface.
2.  Fragments of mesh typically around 100 mm by 50 mm mixed randomly into the rootzone.
3.  Individual fibres typically of 30–40 mm length mixed randomly into the rootzone.

Reinforcement of natural turf is a relatively new technology. Several products are being marketed for which there are few data on mode of action and short and long-term benefits and interactions with natural turf.

Intact fabrics have had a chequered history. Most notable was the introduction and rapid discarding of 'Grass fleece' (The Sports Council, 1977). The most widely used material in this category at present is VHAF. This is a needle-punched polypropylene fabric with vertical, horizontal and angular fibres. Baker (1990b) found this to be the most acceptable of the fabric reinforcements available especially in relation to its stabilizing effect and safety in use. Intact fabrics prevent some maintenance or upgrading procedures (e.g. subsoiling) and their long-term persistence in rootzones may be a problem. Adams and Gibbs (1989) concluded that the strategic use of VHAF on small areas (e.g. goal mouths) was to be recommended.

Netlon mesh elements have been used especially for rootzone reinforcement on horse racing tracks and on golf tees where divoting is reduced and the recovery period from damage decreased. The physical properties of the rootzone, in particular total and air-filled porosity, may also be improved (Beard and Sifers, 1989).

Individual fibres, as used for example in 'Fibresand', have probably the greatest potential for use in the reinforcement of sand-dominant rootzones. Provided incorporation rates are low (around 0.2% w/w) maintenance programmes should not be affected adversely. This type of reinforcement material

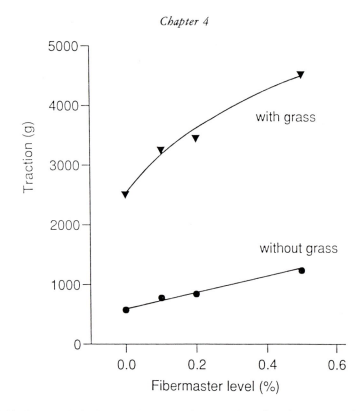

**Fig. 4.16.** Relationship between resistance to shear (traction) of sand rootzone surface and level of incorporation of 'Fibermaster' fibrillated polypropylene fibres both without and with roots of 124-day-old perennial ryegrass (Adams, unpublished data).

is interesting in that the fibres interact positively with turfgrass roots. Thus the increase in stability of a rootzone containing turfgrass roots increases much more rapidly with increase in the rate of fibre incorporation than when roots are absent (Fig. 4.16).

### Improvement in nutrient/moisture retention and playing quality

Amendments added to sand or sand-dominant rootzones to aid nutrient and moisture retention fall into three categories: (i) those which are organic (e.g. peat, seaweed conditioners, sawdust, bark, lignite, slurry, sewage sludge); (ii) those which are mineral based (e.g. calcined clay, vermiculite, perlite); and, more recently (iii) those which contain gel-forming polymers (e.g. starch co-polymers, polyvinyl alcohols, polyacrylamides), with the last group being developed primarily for use in arid climates where suppression of evapotranspiration losses and an increased efficiency in the use of irrigation water are of prime concern.

With the renewed interest in 'organic' products for sportsturf construction and maintenance, products like peat are still widely used especially in all-sand constructions where peat is added primarily to aid germination after completion of the construction. In fact, a sand/soil/peat blend is considered by many to be an ideal mixture of materials for sportsturf rootzones. Peat is not a single uniform product. It varies depending on origin (e.g. sedge or sphagnum moss) and fibre content.

Despite the widespread and successful horticultural use of peat, it is a potentially dangerous substance to add to sportsturf rootzones. This is because of the adverse effects of colloidal organic matter on the physical properties of the rootzone particularly in terms of blocking macropores (Chapter 1).

The method of incorporation of peat is critical in the construction phase because it tends to separate out into a discrete layer within the rootzone. It may become buried by sequential topdressings and finally turn into an anaerobic layer impeding drainage, aeration and vertical root growth. In the suspended watertable, sand rootzone soccer pitch shown in Fig. 4.17 infiltration rates decreased from $100\ mm\ h^{-1}$ to $1\ mm\ h^{-1}$ after only two seasons of use as a result of excessive peat application prior to seedbed establishment (Gibbs *et al.*, 1991). When peat is mixed with sand the level of incorporation should not be greater than to give 5% by weight of organic matter in the top 50 mm of rootzone (Adams, 1986).

Sand rootzones without organic amendment are likely to be hard. For example, Baker and Hacker (1988) found that a sand carpet construction on unamended sand was too hard shortly after being constructed. Nevertheless, a build-up of organic matter with time modifies behaviour. Thus it is probably better to have a surface which is hard to begin with, rather than compromise on the physical properties of the rootzone by the addition of peat. The use of peat can be avoided using controlled release fertilizers in conjunction with a high quality irrigation system.

Of the other organic amendments used at the construction stage, seaweed soil conditioners provide a source of trace elements and may contain rooting stimulants. The low rates of incorporation of these products means they are unlikely to cause the pore blockage associated with peat. Beneficial effects have been reported on ground cover and playing quality (Canaway, 1992). Nevertheless, the effects have not been attributed to specific constituents of commercial products.

Mineral amendments like calcined clay are porous. However, in the review by Ward (1983), there was little to commend these types of product for intensive sportsturf use either because the water they stored was unavailable for plant growth (calcined clay), or because like peat they decreased hydraulic conductivity (vermiculite), or because they disintegrated under compactive forces (perlite).

Polyacrylamide gels increase moisture retention in sand-dominant soils (Johnson, 1984). Moreover, by virtue of their increased moisture retention,

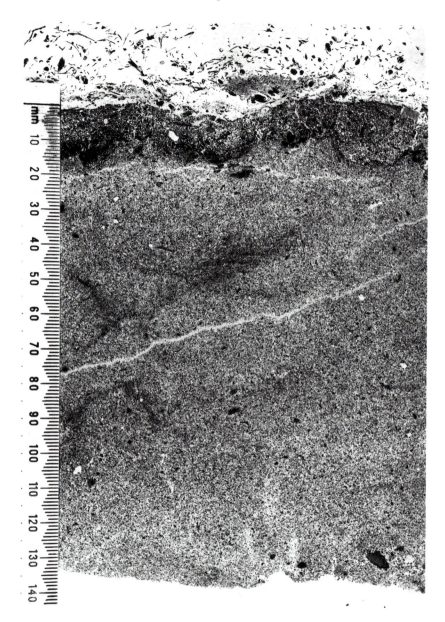

**Fig. 4.17.** Resin impregnated section of an all-sand suspended watertable construction where peat has been added to the surface of the rootzone to aid grass establishment. The peat has formed an impermeable layer restricting infiltration of water to values as low as $1 \, \text{mm} \, \text{h}^{-1}$. The scale is in units of millimetres.

they have other advantages in terms of a reduction in surface hardness in dry weather as well as an apparent increase in ground cover during simulated soccer-type wear (Baker, 1991). Such materials are able to absorb approximately 400–500 times their weight in water, so rates need to be strictly controlled in order to avoid creating a soft, spongy surface. In particular, lack of uniformity of mixing will produce large variations in the playing characteristics of the root-zone surface.

Rates of incorporation vary typically between 0.1% and 0.5% by weight. Moreover, unlike calcined clay, the stored water in polyacrylamide gels is available for plant uptake. For example, Baker (1991) observed that at a water tension of 400 mm, the volumetric water content increased from 26.2% for unamended sand to 41.0% when 0.2% by weight of a polyacrylamide co-polymer was added. However, this initial moisture retention advantage was largely lost after 27 months as the material was biodegraded.

## *Improving soil biological activity and structure*

This is the most controversial area of soil conditioners and it is a widely marketed claim that such products act as catalysts for increasing soil microbial activity with a consequent benefit on soil structure. The seaweed soil conditioners and improvers also fall in this category in addition to providing a valuable source of micro-nutrients. Seaweed soil conditioners contain polyuronides and alginates (in addition to materials with growth-regulating effects), which assist in aggregate formation in soils whilst at the same time providing substrates susceptible to decomposition by soil bacteria and fungi. Other products claim to carry out similar benefits of introducing beneficial microorganisms in all types of soil through the enzyme action they introduce into the soil. Certain products are a cocktail of materials thought to improve soil biological activity and structure including various combinations of peat, lignite, slurry, seaweed soil conditioners and gypsum.

Gypsum (calcium sulphate) is a traditional soil structure improver. It is used to reclaim saline soils and there is no doubt of its benefit in this context. There is interest in using gypsum as a soil ameliorant in sportsturf construction and maintenance work where it may be applied either as a material in its own right or within a sand carrier. The material may have benefits on areas of construction, where indigenous soil is being used, as long as its application can be combined with mechanical means of structure regeneration (e.g. subsoiling).

Soil conditioners are often marketed as cures to any or all soil and turf problems. This they are not. Before using any conditioner one should be clear that it is in order to solve a precisely defined problem. A precise problem is not the desire to increase biological activity – that can be achieved with a spoonful of sugar.

# METHODOLOGY OF SPORTSTURF INSTALLATION

It is beyond the scope of this book to provide a contractor's or golf course architect's step-by-step guide to the practical installation of sportsturf. Instead a framework for practical installation is illustrated, which also summarizes the necessary documentation required for construction work and common pitfalls that must be avoided. Some of the information in this final section of this chapter is reproduced with permission of the Commonwealth of Land Use and Land Reclamation Consultants and the Sports Council.

The provision of good natural turf surfaces has design, construction and maintenance components, each of which is vital in the achievement of the ultimate goal of consistent high quality. Site assessment and the design of a programme of construction or upgrading require specialist knowledge to identify the potential of a site and to produce a design which will satisfy a client's needs and overcome or minimize limitations within defined financial constraints. It is advisable therefore to appoint a consultant to formulate a construction scheme and provide adequate details of materials, quantities and procedures which enable a contractor to tender for the work with a clear understanding of what is required. A consultant should also function as a coordinator so that everyone involved in the development and its subsequent maintenance is kept informed and has an opportunity to contribute views and comments.

It is the consultant's responsibility to identify any unusual or difficult features of the site which may affect the way in which a contractor can operate. However, this does not absolve contractors from the responsibility of making their own assessment of any difficulties in implementing the proposed scheme. A consultant or consultants may be appointed not only to design a scheme, but also to take responsibility for its successful implementation. A common but unwise action is for the provider or client to go directly to a contractor and ask for an area to be constructed or upgraded with no clear design or scheme in mind.

For large construction projects, for example a complete golf course, several consultants may be needed. Thus the responsibility for different aspects of the construction may be apportioned to different specialists (e.g. landscape or golf course architect, soil scientist, irrigation consultant, consultant agronomist) with perhaps a further consultant or organization managing the project.

The design of a scheme and its implementation are distinct stages. By whichever method it is prepared, the specification of the design and procedures to be employed should be sound, practicable and precise. Inferior designs which are doomed from the beginning continue to be produced, but more often than not it is errors and inadequacies in either the implementation of design or in subsequent maintenance which are the cause of unsatisfactory natural turf surfaces.

# Soils and site evaluation

Soil maps and soil profile examination of an area are valuable starting points for interpreting the suitability of land for amenity and recreational purposes. The greater availability of land, particularly for golf course construction work, means that it is important to assess the potential of the land in order to minimize both construction and maintenance costs. An adequate soils and site assessment is equally essential for those constructions where land availability is severely restricted (e.g. urban areas). The degree of soil assessment depends upon the extent to which the indigenous soil will be retained or used. Thus where the entire soil will be replaced it is of little importance, but knowledge of the soil on extensive areas of golf fairways is essential. Whatever the use, site features affecting drainage and the need for earthmoving for example are vital in all cases.

## *Visual examination of soil*

Test holes need to be dug to evaluate the nature of the material likely to be encountered. These should be sited in relation to differences in topography or vegetation which are likely to reflect soil variability on the site. For sportsturf use it is best to examine the soil in winter or early spring because excessive wetness is usually an overriding problem to be dealt with.

Test holes should be excavated to around 800 mm depth. Drain laying is unlikely to be necessary below this depth. The number of holes needed depends upon the variability of the site but more than four are unlikely to be needed for a cricket, soccer or rugby ground. The cut soil face (profile) should be examined visually and samples collected for key analyses. When samples are collected they should normally be taken thoughout the depth of relatively uniform (visually) layers in the soil profile. These are called horizons. Sampling depths must be recorded. In special circumstances, for example when it is intended to strip off a predetermined depth of soil, then sampling solely on depth criteria may be used.

Visual assessment should include the following:

1.  Soil depth to rock (if encountered) and stoniness. Record the depth to rock, layers where stones occur, stoniness and size and general shape (flat or round) of stones.
2.  Is the soil on made-up ground? If the soil is on an old landfill this is usually clearly evident from the material present. What is the depth of topsoil if it is a capped site? An odd combination of layers in the soil suggests it is a disturbed site.
3.  Note the type of vegetation and the general nature of the biologically active soil layers (topsoil). Does the vegetation suggest wetness or acidity or

alkalinity? Is the topsoil black (peaty), and if so how deep is this layer? How profuse and deep is rooting in the soil? Is the topsoil friable and permeable to water?

4.    Note the wetness, colour and presence of fissures in the soil. All well-drained soil will range in colour from dull brown in the topsoil to brighter brown/orange/red in the subsoil. Grey or greenish or bluish colours indicate that the subsoil is excessively wet for long periods in the year. A topsoil with rusty red mottles within it is a seasonally wet one. Is the soil severely compact (indurated) at any depth? If so, over what depth range does this occur?

5.    Note if water moves into the soil profile and at what depth the water is moving.

### Soil analyses

The visual examination of the soil should be complemented by a few analyses. Sampling depth must be recorded but there are no hard and fast rules on the depth to sample. If a soil has been ploughed recently the topsoil will be fairly uniform. This is not so in uncultivated soils. The three most useful soil analyses are:

1.    particle size distribution;
2.    pH;
3.    organic matter content.

Available P, K and Mg in the topsoil may be useful. An exception to these quite basic analyses is when it is suspected that the site has been chemically contaminated. Most toxic organic contaminants are decomposed reasonably quickly and do not normally pose a long-term threat. Contamination by heavy metals is persistent. Turfgrasses are more resistant to heavy metal toxicity than most plant species but instances of Zn toxicity have occurred. If contamination by heavy metals is suspected soil sampling has to be quite detailed because of the heterogeneity of polluted sites. Both the diagnosis and treatment of contamination are expensive.

Implicit in the nature and detail of soil examination for sportsturf is that the native soil and its drainage will need to be upgraded. Another approach is to use land for a purpose for which it is well suited. This is the type of approach used generally in the agricultural classification of land but it is only relevant to turfgrass areas where minimal change to the existing soil and site is practicable – usually on grounds of cost. The suitability criteria detailed by Palmer and Jarvis (1979) are useful as a general background especially when little change to the native soils is possible, for example with parkland and golf courses excluding greens and tees.

*Important site information*

In addition to a soil evaluation, special emphasis should also be placed on the following aspects:

**1.** Make a detailed survey of aspect, slopes and levels on the site and adjoining land.
**2.** Establish suitable outfalls for drainage noting depth below the soil surface at maximum anticipated stream or river levels.
**3.** Identify any services running under or near the site such as water, gas, electricity (permission will be needed for re-routing services).
**4.** Establish any restrictions on usage or access (e.g. residents in urban areas may oppose floodlighting).
**5.** Assess risk of flooding and susceptibility to inflow of water from off-site.
**6.** Note the presence of rivers, streams and canals adjacent to the site since any alteration of public water courses requires permission from the relevant authority (e.g. National Rivers Authority).

# Preparing and implementing the contract documents

The primary role of the consultant in large sportsturf construction developments is to provide contract documents in the form of (i) The Specification of Works; (ii) The Bill of Quantities; (iii) The Drawings.

*Specification of Works*

This document provides the contractor with all the information necessary to carry out the work including the standards of workmanship expected, the precise details of the materials to be used and the order in which the work is to be executed. In many cases the type of machinery or equipment to be used will also be stipulated. The Specification of Works includes preambles like objectives, site access and maintenance, timetable of events, inclement weather procedures, soil handling, quality control and resource allocation, in addition to the actual construction work.

Competently written documents specify acceptable or precise methods of, for example, topsoil removal, storage and redistribution. The contractor would then be left in no doubt as to what was required. Specifications would therefore include weights of machinery, modes of operation, traffic regulation, heights and position of storage, stating defined objectives so that the contractor understands the practical nature of the specifications. Traffic regulation is particularly important in the broader context of contractor discipline to avoid damage to installed drains and minimize soil compaction. Part of a typical Specification of Works embracing the above philosophy is shown in Table 4.6.

**Table 4.6.** Part of a Specification of Works for sportsground construction. (Reproduced with permission of the Commonwealth of Land Use and Land Reclamation Consultants, Devon.)

---

Topsoil reinstatement

---

**General**

Topsoil reinstatement follows the installation of cut-off drains and precedes lateral drain installation. Topsoil will have been laid out in heaps to the south of the plateau and outside positions of batters.

The depth of topsoil required on the plateau and batters is 175 mm. The manner in which this soil is distributed on the pitch area is crucial to the success of the pitch construction. The contractor shall carefully adhere to the soil-handling specifications and the order of events below.

**Procedure**

**1.** A 360° tracked excavator shall put soil over batters (from the heaps created during topsoil stripping) to a depth of 175 mm. The finished heights shall be 175 mm above subsoil heights as shown on drawing 90IWM/05. Care shall be taken to prevent soil from rolling into cut-off drains.

**2.** Acceptable grades on cut and fill batters shall be such that average heights are true to required levels ±75 mm and deviations shall not exceed 100 mm under a 5 m straight edge.

**3.** A 360° tracked excavator shall be used to load transport vehicles with topsoil from storage heaps. Transport vehicles shall not exceed 10 tonnes fully laden. Rubber tyred traffic shall never travel on reinstated topsoil. Routes to and from the topsoil heaps and the pitch shall be checked with a CLULRC representative.

**4.** Transport vehicles shall supply topsoil to an approved low ground pressure bulldozer. This unit shall have an average ground pressure of 5 psi (35 kPa). The proposed unit shall be detailed in the contractor's tender.

**5.** An evenly consolidated 175 mm layer shall be laid down over the plateau area. All soil shall have received at least four passes of the bulldozer tracks.

**6.** The final grade shall be such that average heights are true to required levels ±10 mm. Deviations under a 5 m straight edge shall not exceed 50 mm.

---

## Bill of Quantities

The Bill of Quantities accompanies the Specification of Works and itemizes every component described in the Specification of Works to enable the contractor to cost the contract.

A Bill of Quantities is typically made up of five components. These are Item Description, Quantity, Unit, Rate and Price (Table 4.7). Additionally, specifically numbered paragraphs in the Specification of Works may be referred to in the Bill of Quantities for ease of cross-referencing. It is usually the responsibility of the consultant to calculate or determine the values in the Quantity

**Table 4.7.** Part of a typical Bill of Quantities for sportsground construction. (Reproduced with permission of the Commonwealth of Land Use and Land Reclamation Consultants, Devon.)

| Bill of quantities three: drainage phase one | | | | |
|---|---|---|---|---|
| Item description | Quantity | Unit | Rate | £ p(ex VAT) |
| **Pitch main drain/cut-off drain** | | | | |
| Excavation of 225 mm trench | 126 | m | 5.40 | 680.40 |
| Supply 110 mm perforated pipe | 126 | m | 0.61 | 76.86 |
| Supply multi junctions | 11 | nr | 1.76 | 19.36 |
| Supply 110 mm end stop | 1 | nr | 0.39 | 0.39 |
| Supply 5-10 mm gravel | 45 | t | 7.15 | 321.75 |
| Supply 5 mm grit | 6 | t | 7.61 | 45.66 |
| Pipe laying/backfilling | 126 | m | 2.65 | 333.90 |
| | | | | |
| **Lateral drains** | | | | |
| Excavation of 150 mm trenches | 750 | m | 4.60 | 3450.00 |
| Supply 80 mm perforated pipe | 750 | m | 0.40 | 300.00 |
| Supply 80 mm end stops | 10 | nr | 0.28 | 2.80 |
| Supply 5-10 mm gravel | 110 | t | 7.15 | 786.50 |
| Supply 5 mm grit | 27 | t | 7.61 | 205.47 |
| Supply 80/80 mm straight connectors | 5 | nr | 1.75 | 8.75 |
| Pipe laying/backfilling | 750 | m | 2.51 | 1882.50 |
| | | | | |
| **Silt trap/inspection chamber** | | | | |
| Excavate/refill chamber | item | LS[a] | – | 57.50 |
| Supply reinforced concrete base | 1 | nr | 21.25 | 21.25 |
| Supply bricks (standard size) | 500 | nr | 0.37 | 185.00 |
| Supply Grade I cast iron cover | 1 | nr | 50.10 | 50.10 |
| Construction of chamber | item | LS[a] | – | 135.00 |
| | | | | |
| Allowance for extra handwork .............................. £ | | | | – |
| | | | | |
| Allowance for transport of machinery ................... £ | | | | 75.00 |
| | | | | |
| Allowance for site accommodation/management .............. £ | | | | 650.00 |
| | | | | |
| Allowance for (please specify) ............................ £ | | | | – |
| | | | | |
| **Bill three total** c/f £ | | | | 9288.19 |

[a] Lump Sum.

and Unit columns, leaving the contractor to fill in the Rate and Price columns. However, this procedure should not absolve contractors from agreeing that they are satisfied that, having referred to the Specification of Works, the quantities shall be sufficient to carry out the item of work without shortfall. This is a particularly difficult area because, whilst such items as earthmoving requirements can be determined reasonably accurately using computing packages, it is for example very difficult to determine precise quantities of backfill required in excavated trenches (e.g. for the trench in Fig. 4.6). It is not unusual for drain trenches to need 1.5 times the amount of material originally calculated simply because stoniness or soil instability prevent trenches being cut to the width specified. The amount of sand required over gravel in sand/gravel slit drains is 1.5 to 2 times the theoretical volume due to a combination of excessively wide slits and over-spill.

Unavoidable inaccuracies by a consultant in estimating quantities required can be compensated for by including sensible contingencies in the event of a shortfall due to unforeseen circumstances. These contingencies may be as much as 10% of the total tender price.

### The Drawings

The provision of clear, accurate scale drawings is essential to understand fully the nature of the proposed construction work. As with the written documents, there is a tendency to put too much information onto one drawing, making it difficult for the contractor to identify and interpret specific components of the construction work. Alternatively, drawings may not contain enough detail and simply be based on a few heights unrelated to existing ground levels. High quality drawings should contain enough information to enable the contractor to carry out the work accurately and in a logical order (Hawtree, 1983). Frequent reference to drawings in the Specification of Works is essential, since the information conveyed by one drawing saves many lines of potentially ambiguous writing.

The number of drawings required depends on the nature of the construction job. For sportsground constructions, up to ten drawings may be required, although for new golf courses as many as 40 or 50 drawings is not untypical since each green and tee usually requires a separate drawing. Table 4.8 summarizes typical types of drawings for sportsground and golf course constructions and the appropriate scale of the drawing. Copies of all drawings must be left with the client as a record of works done. Attempting to rediscover the location of drains and other important site features and even to find out what work was done in the first place is extremely difficult without a record of construction drawings.

### Conditions of Contract

Having prepared the Specification of Works, the Bill of Quantities and the Drawings, these three components are then tied together with a Conditions of Contract document which provides a legally binding framework containing such documentation as definitions, interpretation, general obligations,

**Table 4.8.** Typical drawings expected for major sportsground or golf course construction.

| Description | Typical scale |
|---|---|
| **Sportsground** | |
| Site survey | 1:500 to 1:2500 with existing contours at 1 m intervals and relevant site features |
| Proposed finished site plan | 1:500 to 1:2500 with details of finished sportsground layout, access, buildings, etc. |
| Site access and storage of materials | 1:200 to 1:500 with details of site accommodation, storage of vehicles, no-access areas etc. |
| Site levels and topsoil stripping | 1:200 to 1:500 with details of storage of transported topsoil |
| Sequence of levelling | 1:200 to 1:500 (for cut-and-fill excavations) |
| Subsoil levels | 1:200 to 1:500 with spot heights relative to finished heights and natural contours |
| Finished heights plan | 1:200 to 1:500 with finished spot heights |
| Field drainage plan | 1:200 to 1:500 with details of cut-off drains, main carrier drains, lateral drains, slit drains |
| Drain sections/outfalls | 1:10 to 1:50 with details of drain depths, widths, pipe sizes, depth of permeable fill, etc. |
| **Golf course** (from Hawtree, 1983) | |
| Site survey | 1:2500 with existing contours at 1 m intervals and relevant site features |
| General layout of holes | 1:2500 with contours at 1 m intervals |
| General drainage plan | 1:2500 with contours at 1 m intervals |
| Site clearance/ditch clearance | 1:2500 with contours at 1 m intervals |
| Main borrow areas/cultivation treatment | 1:2500 with contours at 1 m intervals |
| 18 green plans (existing and proposed) | 1:200 with contours at 0.25 m intervals |
| 18 tee plans (existing and proposed) | 1:200 with contours at 0.25 m intervals |
| Detailed features | 1:100 to 1:500 depending on feature (cross-section of green, layout of water features, longitudinal sections). Exceptional detail down to 1:20 |
| Tree planting and landscaping | 1:2500 |

workmanship and materials, completion certificates, alterations and additions, settlement of disputes and so on.

There are a number of different Conditions of Contract produced by professional societies for use by their members, and the most commonly used ones in the UK for sportsturf purposes are the Institute of Civil Engineers (ICE) and the Joint Council of Landscape Industries (JCLI) contracts. Modifications to these contracts are often included in the introduction of a

Specification of Works to allow for specific on-site constructional features or requirements. For example, the term 'Engineer' might be replaced by the term 'Consultant-in-Charge'.

*Tendering*

Contracts are usually let out by competitive tendering. This is the process whereby a range of suitable contractors are invited to tender for the sportsturf construction contract, to complete the Bill of Quantities and sign a free tendering declaration to indicate that they have not been privy to information which was not available to competitors. A strict closing date for submission of tenders is required.

The selection of contractors invited to tender must be chosen with caution. Their credentials must be checked and an examination of previous work is highly advisable. Geographical location and existing workloads being undertaken may also affect the choice of contractor. It is usual to send out invitations to five contractors in anticipation that at least three tenders will be returned for comparison.

Of utmost importance is the ability to recognize whether a returned tender is financially viable and whether the contractor is technically capable of doing the work. An unusually low tender must be investigated very thoroughly to ensure that the contractor appreciates the requirements in the contract. A contractor may deliberately submit a low tender to obtain the contract knowing that it will make a loss. Such a contractor may well fail to complete the contract.

## Procedural techniques and standards in construction

Employing one contractor to carry out the construction work is preferable, but often impractical because of the different specialisms required. Thus there might be an earthworks contractor, a drainage contractor and an irrigation contractor. It is the consultant's job to coordinate the activity of different contractors to make sure that each one knows what the other ones are doing. The employment of different contractors must be kept to a minimum and in situations where subcontractors are required to carry out highly specialized work like irrigation installation, it is acceptable to state a nominated subcontractor in the contract documents.

*Timing of works*

Sportsturf construction is essentially a soils technology and as such it must reflect changes in soil characteristics that occur in relation to the time of year. For this reason, timing of works is absolutely critical.

Timing depends upon the understanding of objectives and the materials involved in the construction. Thus forward ordering of materials to be used, regular contact with site supervisors and consultants, reasonable response to weather conditions and flexibility of response to site conditions are all key components of timing of works. It is insufficient to state that construction work must be carried out between March and September. Specific instructions must also be given as to the ground conditions within this period under which machinery may work and how certain operations are to be organized. An acceptable timetable for the reconstruction/upgrading of a winter games pitch in the UK with no loss of playing time in the season is outlined below.

- **October to December:** prepare contract documents and drawings.
- **January:** meet potential contractors on site, discuss construction problems, and obtain completed tenders.
- **February:** make final selection of contractor on the basis of tender price, contractor reliability, availability and track record.
- **March:** engage contractor to begin work immediately the pitch becomes available.
- **April/May:** complete work during May to allow 80–100 days between sowing grass seed and beginning of playing season.

### Quality control and contractor supervision

Standards of construction and maintenance have been receiving close attention over the last few years as the turfgrass industry gears itself to the greater standards being demanded by both the public and private sectors. The economic pressures of the 1980s and 1990s to reduce agricultural surpluses and diversify land use into amenity areas, the rationalization of existing natural turf facilities particularly within local authorities, and the advent of the 1988 Local Government Act which has seen the privatization of grounds maintenance have all served as catalysts in this respect.

It is therefore very timely that standards should be defined for the finished or maintained natural turf surface. Quality control during the process of construction should help meet these standards. Thus quality control should include such measurements as playing quality tests, water infiltration rates, general drainage capability, grass cover and purity, surface conditions (e.g. presence of stones) and a defined grade tolerance as well as more basic aspects such as checking to see if the sand delivered is of the correct specified type or whether end stops have been put on the ends of pipe drains.

Quality control and standards are closely linked to contractor supervision. For example, work deliberately covered over by contractors before it has been inspected could be assumed to be faulty, so the inspection of key stages of a construction must be carried out prior to completion. Materials used in construction work must always conform to British Standards where they exist and the Specification of Works must quote the relevant BSI numbers. Every effort should therefore be made to order, deliver, store and use materials suitable for the function intended. In recognizing this crucial aspect of construction, the Land Drainage Contractors Association (Sportsturf Section) have compiled practical guidelines which broadly represent Codes of Practice for sportsturf installation (LDCA, 1991), and the National Turfgrass Council, through its coordinating role within the turfgrass industry, serves as a forum for improving standards (e.g. Shildrick, 1989, 1991).

Although contractors may employ their own Clerk of Works to organize day-to-day operations on-site, this person is unlikely to have a sufficient breadth of knowledge and understanding of the whole construction project. Therefore the chief responsibility for supervision must come from an appointed consultant who has been involved in the design and planning stage. This person will need to make frequent visits on-site at key stages in the construction work to ensure that the design and specification components are being followed. Furthermore, for extremely large construction projects like golf courses, it is not unusual to employ a quantity surveyor on a full-time basis whose primary role is to measure the work done and provide up-to-date and accurate statements for the client with regard to the finances of the contract (Perris, 1991).

## Pitfalls in construction or upgrading operations

In the remaining sections of this chapter, pitfalls in various construction or upgrading operations are considered under the following series of task headings: pipe drainage, slit drainage, stratified constructions including soil and materials handling, and cultivation.

### Pipe drainage

Once a pipe drainage system has been covered over, it is extremely difficult to identify and locate errors. The pitfalls listed below are typical examples of inadequate installation:

1.   Failure to lay pipes to a specified grade, such that silting up, surcharging or backflow occur (occasionally no pipes installed at all!).
2.   Drain trenches not cleaned out properly resulting in uneven lie of the pipe and resultant silting up; drain trenches not backfilled immediately with permeable fill to prevent soil falling over the pipes.
3.   Damaged pipe-work not replaced and/or proper junctions and end stops not used, resulting in entry of soil into the pipe drain and reduction in efficiency of the system.

4. Failure to comply with specifications on particle size and depths of permeable fill, or permeable fill installed incorrectly causing trench wall collapse and soil contamination of gravel.

5. Inadequate filling of trenches with permeable fill (trenches *must* be over-filled in virtually all schemes to prevent soil sealing over the permeable fill).

## Slit drainage

Most of the potential pitfalls listed above also apply to slit drainage, but in this case they cause much more serious damage to the functioning of the scheme. Typical pitfalls experienced in slit drainage work are:

1. Failure to clean out bases of excavated slit drains before they are filled with permeable material.

2. Failure to comply with specifications on particle size, hydraulic conductivity and depths of permeable fill.

3. Inadequate filling of slit drains with permeable fill to the soil surface or to a permeable rootzone layer. Gibbs *et al.* (1991) reported an example where slit drains beneath a sand carpet construction were left partly filled with gravel over the winter period. This, combined with poor traffic discipline where machinery frequently ran over exposed slit drains, caused a layer of soil to cap over the gravel which was not removed prior to the application of the blinding layer. The construction lasted one year before complete re-slitting was necessary.

4. Inadequate connection of the base of the slit drain with the permeable fill over a lateral drain caused either by insufficient depth of excavation of slit drains, insufficient backfilling over lateral pipe drains or severe soil contamination of permeable fill in lateral drains. An overlap of at least 25 mm is required.

5. Poor contractor discipline with regard to failure to prevent support machinery crossing over slit drains causing compaction and capping.

6. Re-grading operations carried out after slit drainage installation causing soil to be pushed over the slit drains rendering them incapable of functioning in the desired fashion.

7. Lack of immediate aftercare particularly with respect to protective sand topdressings.

## Stratified constructions and soil/materials handling

There are a number of rules of operation which must be followed in handling soils and other material used for the construction of natural turf surfaces. These are listed below:

1. Traffic discipline is vital. The ground pressure of wheeled vehicles is high and the routing of traffic must minimize compaction and avoid rutting.

Loaded vehicles should not cross drain lines. The use of tracked vehicles for moving and distributing construction materials is preferable to rubber-tyred vehicles.

**2.** All soil handling should occur at moisture contents where the soil is at least firm and dry enough not to smear. Work involving soil handling should not occur during rain, drizzle or any other free-water conditions to minimize structural damage of the soil.

**3.** Cross contamination of materials must be avoided. This applies as much to separating topsoil from subsoil as it does to contaminating gravel and sand with topsoil.

There are some further key points which apply especially to stratified constructions whether these involve sand or sand/soil mixes over the native soil or as part of a complete profile construction:

**4.** Where a polythene liner is used, extreme care is needed to avoid puncturing. Any damage must be repaired.

**5.** Layers in a construction must be level and to the specified depth. The spreading technique used must be capable of achieving these requirements. This usually involves either small bulldozers or 360° tracked excavators. When applying the rootzone layer, it is essential that these machines always work off a 'platform' of the rootzone mix to avoid cross contamination of materials. The rootzone and blinding layers are best applied in a moist condition to prevent wind-blow and cross contamination.

**6.** When a sand-dominant rootzone is produced on-site, a fixed tine cultivator on a tracked machine should be used to lift the underlying soil into the sand top. A rotovator should not be used for this purpose, unless the pan created at its working depth is subsequently broken. The sand top must be of an even depth and the depth of working of the fixed tine cultivator must be precise and constant.

**7.** When a sand/soil rootzone is produced off-site, there must be sufficient 'clean' space for mixing and the mixing technique must produce a mix of consistent quality and uniformity.

## REFERENCES

Adams, W.A. (1981) Soils and plant nutrition for sportsturf: perspectives and prospects. In: Sheard, R.W. (ed.) *Proceedings of the 4th International Turfgrass Research Conference*. Guelph, Canada, pp. 167–179.

Adams, W.A. (1986) Practical aspects of sportsfield drainage. *Soil Use and Management* 2, 51–54.

Adams, W.A. and Gibbs, R.J. (1989) The use of polypropylene fibres (VHAF) for the stabilisation of natural turf on sports fields. In: Takatoh, H. (ed.)

*Proceedings of the 6th International Turfgrass Research Conference.* Tokyo, Japan, pp. 237–239.

Adams, W.A. and Smith, J.N.G. (1993) Chemical properties of rootzones containing a black layer and some factors affecting sulphide production. In: Carrow, R.N., Christians, N.E. and Shearman, R.C. (eds) *International Turfgrass Research Journal* 7, 540–545.

Adams, W.A., Stewart, V.I. and Thornton, D.J. (1971) The assessment of sands suitable for use in sportsfields. *Journal of the Sports Turf Research Institute* 47, 77–85.

Baker, S.W. (1982) Regional variation of design rainfall rates for slit drainage schemes in Great Britain. *Journal of the Sports Turf Research Institute* 58, 57–63.

Baker, S.W. (1985a) The selection of topsoil to be used for sand–soil rootzone mixes: a review of current procedures. *Journal of the Sports Turf Research Institute* 61, 65–70.

Baker, S.W. (1985b) Topsoil quality: relation to the performance of sand–soil mixes. In: Lemaire, F. (ed.) *Proceedings of the 5th International Turfgrass Research Conference.* Avignon, France, pp. 401–409.

Baker, S.W. (1988) The effect of rootzone composition on the performance of winter games pitches. III. Soil physical properties. *Journal of the Sports Turf Research Institute* 64, 133–143.

Baker, S.W. (1990a) *Sands for Sportsturf Construction and Maintenance.* The Sports Turf Research Institute, Bingley, 58 pp.

Baker, S.W. (1990b) STRI trials of turf reinforcement. In: Shildrick, J.P. (ed.) *Turf Reinforcement.* National Turfgrass Council Workshop Report No. 19, pp. 4–32.

Baker, S.W. (1991) The effect of a polyacrylamide co-polymer on the performance of *Lolium perenne* L. turf grown on a sand rootzone. *Journal of the Sports Turf Research Institute* 67, 66–82.

Baker, S.W. and Hacker, J.W. (1988) The use of peat in a Prunty–Mulqueen sand carpet construction: effects of application rate and depth. *Journal of the Sports Turf Research Institute* 64, 87–98.

Baker, S.W. and Isaac, S.P. (1987) The effect of rootzone composition on the performance of winter games pitches. II. Playing quality. *Journal of the Sports Turf Research Institute* 63, 67–81.

Baker, S.W. and Richards, C.W. (1991) Rootzone composition and the performance of golf greens. II. Playing quality under conditions of simulated wear. *Journal of the Sports Turf Research Institute* 67, 24–31.

Baker, S.W., Gibbs, R.J. and Taylor, R.S. (1991) Particle migration from the sand layer of slit drains into the underlying gravel. *Journal of the Sports Turf Research Institute* 67, 93–104.

Beard, J.B. and Sifers, S.I. (1989) A randomly oriented, interlocking mesh element matrices system for sport turf rootzone construction. In: Takatoh, H. (ed.) *Proceedings of the 6th International Turfgrass Research Conference.* Tokyo, Japan, pp. 253–257.

Canaway, P.M. (1992) The effects of two rootzone amendments on cover and playing quality of a sand profile construction for football. *Journal of the Sports Turf Research Institute* 68, 50–61.

Castle, D.A., McCunnall, J. and Tring, I.M. (eds) (1984) *Field Drainage: Principles and Practices.* Batsford Academic, London, 250 pp.

DES (1982) *Playing Fields and Hard Surface Areas.* Department of Education and Science Bulletin 28, HMSO, London, 70 pp.

Gibbs, R.J., Adams, W.A. and Baker, S.W. (1991) Making the most of natural turf pitches. Final results of a case studies approach: IV. Soil physical properties. *Natural Turf Pitch Prototypes Advisory Panel Report No. 9.* The Sports Council, London, 48 pp.

Gooch, R.B. and Escritt, J.R. (1975) *Sports Ground Construction–Specifications*, 2nd edn. National Playing Fields Association, London, 126 pp.

Hawtree, F.W. (1983) *The Golf Course: Planning, Design, Construction and Maintenance.* E. and F.N. Spon, London, 212 pp.

Johnson, M.S. (1984) Effect of soluble salts on water absorption by gel-forming soil conditioners. *Journal of the Science of Food and Agriculture* 35, 1063–1066.

LDCA (1991) *Guidelines for Sportsturf Drainage Installation.* The Land Drainage Contractors Association (Sportsturf Section), National Agricultural Centre, Warwickshire, 15 pp.

Luthin, J.N. (1966) *Drainage Engineering.* John Wiley and Sons, New York, 250 pp.

MAFF (1982) *The Design of Field Drainage Pipe Systems*, Ministry of Agriculture, Fisheries and Food, Reference Book 345, HMSO, London, 20 pp.

Palmer, R.C. and Jarvis, M.G. (1979) Land for winter playing fields, golf course fairways and parks. In: Jarvis, M.G. and Mackney, D. (eds) *Soil Survey Technical Monograph No. 13, Soil Survey Applications.* Soil Survey, Rothamsted, pp. 152–165.

Perris, J. (1991) How a consultant can help: Specifications, Bills of Quantity and Supervision. In: Shildrick, J.P. (ed.) *Minimum Standards for Golf Course Construction.* National Turfgrass Council Workshop Report No. 20, pp. 17–27.

Raadsma, S. (1974) Current drainage practices in flat areas of humid regions in Europe. In: van Schilfgaarde, J. (ed.) *Drainage for Agriculture.* Agronomy Series No. 17, American Society for Agronomy, Madison, Wisconsin, pp. 115–143.

Shildrick, J.P. (ed.) (1989) *Grass Surface Standards.* National Turfgrass Council Workshop Report No. 16, 73 pp.

Shildrick, J.P. (ed.) *Minimum Standards for Golf Course Construction.* National Turfgrass Council Workshop Report No. 20, 94 pp.

The Sports Council (1977) *The Grass Fleece Experiment.* Sports Council Study 12 by P. Summerside and M. Bowen. The Sports Council, London, 47 pp.

Thomasson, A.J. (1975) Soil properties affecting drainage design. In: Thomasson, A.J. (ed.) *Technical Monograph No. 7, Soils and Field Drainage.* Soil Survey, Harpenden, pp. 18–29.

Thornton, D.J. (1978) The construction and drainage of some specified sports field playing surfaces. Unpublished PhD Thesis, University College of Wales, Aberystwyth.

USGA Green Section Staff (1993) USGA recommendations for a method of putting green construction. *USGA Green Section Record* Mar/Apr 1993, 1–3.

Ward, C.J. (1983) Sportsturf drainage: a review. *Journal of the Sports Turf Research Institute* 59, 9–28.

# 5

# Principles of Turf Establishment and Maintenance

## Introduction

The scope of this chapter is turf establishment, general turf culture in terms of mowing, weed, pest and disease control, irrigation and fertilizer application and the need for soil treatments including thatch control, topdressing and aeration/decompaction. It will be evident throughout that priorities range from providing an acceptable grass cover at minimum cost to creating and maintaining very high quality natural turf playing surfaces irrespective of cost. Despite these contrasting goals and funding for provision, a maintenance regime which is sympathetic to the physiology of grasses is a common requirement.

## Establishment

In Britain the choice is between seed or turves[1] since vegetative establishment from stolons is not a practicable alternative in temperate species even when it is theoretically possible. In contrast, some tropical species including Bermuda grass (*Cynodon dactylon*) are normally established from stolons.

Several factors need to be taken into account when deciding whether to use seed or turves for establishment. There is no difference in the soil preparation required but pieces of turf[1] (or sods as they are described in the USA) provide an immediate cover and maintenance is less demanding in the

---

[1] Turf is an area of mown or grazed grass inclusive of stolons, rhizomes and shallow roots which bind it into an intact layer. A sod is a piece of turf. Turves, that is pieces of turf, are often referred to as turf.

establishment phase. The initial cost is greater for turf than for seed but, this aside, its immediate effect is an attractive proposition for the establishment of domestic lawns and amenity areas.

One potential problem with turf is poor quality in terms of species and cultivars present and the presence of weeds. There are no legally enforced standards in the UK but there is a British Standard for turf for general purposes (BS 3969: 1990) that defines the information which should be available to the purchaser.

The main recommendations on quality of turf in the standard relate to botanical composition. An increasing proportion of turf sold in the UK is produced from sown mixtures rather than from agricultural or sea marsh turf. In the case of turf from seed, the supplier should provide details on the seed mixture used. At lifting, the turf should have at least 95% ground cover by sown species and there should be less than 1% of annual meadowgrass or any other unsown species – unless declared by the seller. An alternative specification intended for turf created from agricultural pasture is a statement on all species present at more than 1% ground cover together with an estimate of their cover percentage. Some species, including Yorkshire fog (*Holcus lanatus*), should be named even if they constitute less than 1% of the sward.

Other recommendations relate to soil type (freedom from stones and not more than 40% clay), aspects of maintenance, uniformity of cut turf, freedom from disease and insect pests and a limit on the depth of thatch (less than 10 mm).

BS 3969:1990 provides recommendations on turf quality which enable a purchaser to evaluate the product for general, that is amenity or ornamental use, but criteria are inadequate for specific sportsturf uses. There are two main reasons for this. The first, which could be accommodated by agreement between grower and purchaser, is that the standard does not provide either for a turf produced from a single cultivar of one species, which is especially relevant to golf and bowling greens, or for a turf produced from several cultivars of the same species, which is often required for winter games pitches. The second difficulty, and one which has caused numerous problems, is the inadequate specification of the medium (soil) in which the turf is grown. Whichever sports use is envisaged, the precise specification of the growing medium is essential to ensure a high quality playing surface.

The ability both to choose specific species and cultivars and to ensure a uniform rootzone with no discontinuities with depth are the key reasons why establishment from seed is preferred in the UK for virtually all high specification sportsturf. These benefits greatly outweigh the advantage of immediate establishment provided by turfing.

Establishment from seed is more difficult on sand-dominant sportsturf soils than on the bulk of natural soils because of their poor water retention. Irrigation is usually necessary in the UK on sand-dominant soils when spring sowing is carried out. Establishment from seed is practicable on sand-dominant

**Table 5.1.** The relative merits of establishing grass cover from seed or turf.

| Factor | Seed | Turf |
|---|---|---|
| Cost | Low | High |
| Speed of visible establishment | Slow | Immediate |
| Tolerance of tearing type wear at 100 days | Good | Moderate to good |
| Weed invasion | Depends on quality of soil preparation | Depends on source |
| Ability to choose turfgrass species and cultivars | Possible | Sometimes possible |
| Ability to choose rootzone composition | Possible | Not normally possible |
| Seasonality of establishment | Spring or autumn without irrigation | Avoid drought and frost, otherwise year-round |

soils without irrigation when seed is sown in autumn. The relative merits and disadvantages of establishing from seed or turf are summarized in Table 5.1.

Weed invasion is potentially a greater problem when establishing from seed than when establishing from turf (provided the latter is weed free). In general, the problem increases as the proportion of 'natural' soil increases in the rootzone. The two herbicides most appropriate for the 'cleaning up' of sites prior to sowing are paraquat and glyphosate. Both are non-persistent in soil and are plant-acting rather than soil-acting, that is they are only effective when sprayed on leaves. Both are non-selective but glyphosate has a broader spectrum of effectiveness including perennials with underground storage organs including dock, thistles and dandelions. On the other hand, the latter weeds are not a problem in closely mown turf so that paraquat, which is very effective on grass weeds and is quicker acting than glyphosate, is usually more appropriate.

Whether establishing from seed or turf it is important prior to establishment to incorporate an application of phosphorus fertilizer throughout the rootzone. This opportunity should not be missed because of the immobility of P in soils.

# MOWING

A feature of all turfgrasses is their ability to tolerate regular mowing, but the closeness of mowing tolerated differs between species and between cultivars of the same species. Thus fine-leaved fescues and bentgrasses can be mown at heights of around 5 mm whereas most cultivars of perennial ryegrass and smooth stalked meadowgrass become severely stressed if mown shorter than around 20 mm. The stress caused by close mowing, which was discussed in Chapter 2, is primarily a result of removing a high proportion of the actively photosynthesizing leaf tissue. Species with a prostrate habit generally tolerate closer mowing than those with an erect habit. In erect growing species such as perennial ryegrass where turfgrass breeding and selection have produced cultivars which are dense tillering and produce expanded leaves close to the ground, tolerance of close mowing has been improved.

Mowing stress, which is expressed through poor tillering, weak, disease susceptible topgrowth and shallow rooting, can be minimized by mowing at a height appropriate for the cultivars in the turf and at a frequency which entails removing less than 50% of the standing height of the sward. The stress caused by close mowing is influenced by light intensity so that close mowing under high light intensity conditions is less stressful than under low light (Troughton, 1957). Severe mowing, that is a cut at less than 50% of the mean grass height, is always stressful but is especially damaging in autumn when it usually results in invasion by moss.

In sportsturf the nature of the game determines to a substantial extent the height of cut. This accepted, occasions have arisen when soccer pitches have been mown more closely than is either required by the game or justified on grounds of playing quality. In contrast, it is not unusual to see the grass on rugby pitches allowed to grow to heights in excess of 100 mm. Such tall grass causes severe shading and prevents the establishment of new tillers which is essential if a dense turf is to be sustained (see Chapter 2).

An appropriate height of cut and cutting frequency does not guarantee an even, uniform turf. All turfgrasses produce lateral growth whether these be stolons, aerial tillers or trailing stems. Lateral growth covers bare soil and shades developing seedlings and new tillers but it gives the illusion of complete turf cover. An essential complement to mowing is a maintenance treatment which lifts trailing growth so that it can be mown at the general mowing height. A variety of equipment is available to achieve this, ranging from hand rakes through mechanical rotorakes to vertical mowing solid tines which may be set to cut above or into the soil surface. The frequency of use of these implements varies but ranges from twice to eight times a year for those which rip out or cut out debris and plant tissue, up to once per week during the growing season for those which rake, brush or groom the turf but cause negligible physical damage.

Grass can be mown with different types of cutter. The traditional turfgrass machine is a cylinder mower. Considerable development has taken place since the first commercial machine was produced in 1832 by Ransomes of Ipswich. Technological developments have improved evenness of cut, closeness of cut and, most obviously, width of cut and mechanical assistance in operation.

A cylinder mower incorporates blades which rotate on a horizontal axis. These move across a fixed lower blade set at a predetermined height. The action is similar to that of scissors and a clean, even and very close cut is possible. The cut grass is not brushed into any particular direction and a roller behind the cutting blade can be used to bend the grass over, giving the groundstaff and greenkeepers the opportunity for artistic design. The light or dark stripes result from differences in the reflectivity between upper and lower surfaces of grass leaves which is especially marked in perennial ryegrass.

Rotary mowers utilize a disc or blade rotating in a vertical plane. There is no bottom blade so that the cut is less clean than with a cylinder mower but some of the vulnerability of the latter to metal and stone objects in the grass is avoided. The cleanness of cut of a rotary mower improves as the standing height relative to the cut height increases.

Flail mowers have cutters which rotate on a horizontal axis like a cylinder mower. There is no bottom blade and the cutters may be strips of metal or chains. Flail mowers are used for mowing roadside verges, hedges and in other situations where a clean cut is not required. Their major advantage is tolerance of 'foreign bodies' in their path which would destroy or severely damage a cylinder mower or indeed most rotary mowers.

It is important to select the correct mower for the particular situation. For fine, closely mown turf a cylinder mower is essential to create a clean, even finish. The evenness of cut depends upon the number of cutting blades on the cylinder. The best machines have around ten but mowers for domestic lawns have only five. When the mower height of cut is raised above about 50 mm, cylinder mowers tend to leave both trailing stems and flowering shoots uncut, whereas rotary mowers do not. A cylinder mower is essential on a golf or bowling green but a rotary mower is generally preferable for rugby pitches where a very close cut is not required and the risk of a stone or other hard objects on the pitch cannot be discounted. Another important point is that cylinder mowers tend to pull the grass so that a rotary mower is most suitable for the first cut of a newly seeded area. The main types of mower and their advantages and disadvantages are summarized in Table 5.2.

## WEEDS AND THEIR CONTROL

Weeds are species growing 'in the wrong place' and in the different classes and uses for turf there may be different reasons why particular species are unwanted

**Table 5.2.** Advantages and disadvantages of three types of mower.

| Cylinder | | Rotary | | Flail | |
|---|---|---|---|---|---|
| Advantages | Disadvantages | Advantages | Disadvantages | Advantages | Disadvantages |
| Capable of close even cut | Does not lift prostrate stems | Versatile in cutting height | Cut is not clean at low heights of cut | Suitable for tall rough grass and scrub | Cut is not clean |
| Can produce an attractive striped surface | 'Pulls' very young grass | Can cut tall seeding stems and lift trailing ones | Collection of clippings possible but more complex than cylinder | Least susceptible to damage by debris in the sward | Clippings are not normally collected |
| Design favours collections of clippings | Cut is not reliable if set higher than 40 mm | Damage by 'foreign objects' less severe than cylinder | | Tough | Close cutting is not practicable |
| Multiples have high capacity (ha h$^{-1}$) | Lower blade susceptible to damage by stones | | | | Relatively slow compared with cylinder multiples |
| | Will not cut tall grass | | | | |

or undesirable. Weeds are not restricted to non-grass species; for example annual meadowgrass and Yorkshire fog are both considered weed species in some classes of turf. Weed grasses are more difficult to control chemically than non-grass weeds and considerable effort has been put into the chemical control of annual meadowgrass with limited success (Turgeon, 1974; Goss *et al.*, 1977; Lewis and Dipaola, 1989). Cultural control of this species is often practicable (see Chapter 2). A new approach to the chemical control of all weeds in turf is to breed turfgrass cultivars which are tolerant to non-selective herbicides, such as amino triazole (Johnston, 1988). The problems of annual meadowgrass and other grass weeds are dealt with elsewhere so the following section focusses on non-grass and primarily broadleaved weeds.

## Broadleaved weeds in sportsturf

In turf used for sport the appreciation of which species constitute weeds has not changed to any great extent in the last century. The chief reason is that recognized weed species have characteristics which are detrimental to the playing quality of the surface. Whilst an effect on the visual uniformity of the surface may be most apparent to the television viewer, there are factors of more practical significance. Firstly, weeds, through the haphazard way in which they invade, are not uniform in their distribution within the grass sward. Their morphology and texture are usually different from the grasses with which they compete so that they create a non-uniform surface in terms of ball reaction. A second factor is that many broadleaved weeds, but white clover (*Trifolium repens*) and plantain (*Plantago major*) in particular, die back in winter. This senescence reduces ground cover, which for winter games pitches increases their susceptibility to damage by wear and for fine turf areas increases the risk of further weed invasion. Another but equally important factor is that weed regrowth in the spring on winter games pitches competes strongly with grasses and may result in complete weed dominance. Both plantain and knotgrass (*Polygonum aviculare*) are capable of suppressing the recovery of turfgrasses.

## Broadleaved weeds in lawns

In contrast to the objective categorization of weeds in sportsturf, assessment is subjective in domestic and ornamental lawns. Visual uniformity has been and to a substantial extent still is the main criterion for assessing the quality of ornamental lawns but a preferred appearance is subject to changes in fashion. An increasing number of people appreciate the presence of species including daisy (*Bellis perennis*), creeping buttercup (*Ranunculus repens*), white clover and the speedwells (*Veronica* spp). Whilst there may be a desire for species diversity in some ornamental lawns the maintenance of a reasonable balance between grasses and broadleaved species is difficult. In lawns on very acidic soils (below pH 4.8) fine-leaved bents and fescues can dominate because of their high tolerance of acidity compared with most broadleaved weeds. In soils with a pH greater than 5.0, several broadleaved weeds are capable of becoming dominant when mowing is frequent and close. White clover, plantian, daisy, creeping buttercup and pearlwort (*Sagina procumbens*) are the most important in the UK with their relative success dependent upon soil fertility, soil wetness and closeness of mowing.

## Broadleaved weeds in amenity turf

The perception of weeds in amenity turf has changed over the last two decades. Whilst the description 'amenity turf' covers a range of situations, distinction should be made between areas where maintenance is virtually the same as ornamental lawns in terms of closeness and frequency of cutting, to areas which are cut no more than twice each year. It is within this latter category that the perception of weed species has changed most profoundly. Mowing only once or twice per year results in a very broad diversity of species when fertilizer input is low and the soil pH is above 5.5. This is largely because tall-growing species which depend upon seeding and re-establishment for survival can compete on reasonably equal terms with tall and low-growing vegetative perennials. Management which approximates to mowing for hay followed by autumn grazing fosters the widest species diversity of grasses and herbs (see Chapter 10).

A list for the identification of the main weeds of turfgrasses is given in the Appendix to this chapter.

## Control of broadleaved weeds

Regular close mowing, which is an essential part of routine maintenance, is the main factor responsible for the invasion and success of the important weed species of turf. The only cultural control of most broadleaved weeds in mown turf is to maintain a soil in so acidic a condition that they cannot survive. A pH below around 4.8 will also result in the elimination of earthworms which themselves favour weed invasion (Table 1.3). This strategy is not advisable with sportsturf because of its effect on turf performance and soil microbial activity but it is practicable on ornamental lawns in the drier parts of the UK. In the wetter west of the UK very acidic conditions lead to the development of a peaty soil surface.

Since the discovery and development of the phenoxyalkanoic acid selective herbicides (e.g. 2,4-D) an increasing range of herbicides has become available for use on turf. Selective herbicides are commercially available for use on turf which are effective against all the broadleaved weed species. These herbicides are degraded within one or two months in soil (Upchurch, 1972) and present no environmental hazard if used correctly. Selective herbicides have a low mammalian toxicity and the only environmental dangers of the diluted products are through carelessness in application resulting in spray drift onto other sensitive plants and the use of clippings from treated areas as a mulch around sensitive plants.

The mode of action of selective herbicides used in turf involves a derangement of growth regulation (Upchurch, 1972) and maximum effectiveness is achieved when weeds are growing rapidly. Taking into account both the conditions for maximum kill and the ability of turfgrasses to colonize sites vacated by weeds, the best time for the chemical control of broadleaved weeds in turf in the UK is April or May.

The conditions required for maximum efficiency of selective herbicides create difficulties for groundstaff maintaining winter games pitches. The most serious weeds of soccer and rugby pitches (which are plantain and knotgrass) begin to grow vigorously at the end of the season when overseeding of bare areas is usually required. The problem is that all effective selective herbicides used on turf are liable to damage seedling grass. The key question is when to spray. The best solution depends on circumstances, but a standard blueprint for the most serious weed infestation is to apply fertilizer in late March or early April and to spray for weeds in late April a few days prior to overseeding. Most commercially available selective herbicides contain a 'cocktail' of ingredients which provides control over a broad spectrum of weed species.

## Moss and algae

Moss is a major invader of the sheep-grazed semi-natural pastures of the uplands, downlands and other areas of turf in the UK which were and are appreciated as having the desirable characteristics of fine turf. The many species of moss are tolerant of acidity, favoured by a cool, humid environment and have no need of soil for their mineral nutrition. Mosses grow slowly and are tolerant of shade. Close mowing (or grazing) favours them as do any treatments which cause physiological stress to turfgrasses. The exception is wear, for mosses are less tolerant of wear than turfgrasses and because of this moss is a more serious problem in fine ornamental lawns than on most sportsturf. Scarification can be considered to be a form of wear and moss can be pulled out of turf by the procedure.

Moss can be controlled by cultural measures in most circumstances. Key points are as follows:

1. Maximize air circulation and minimize water retention in the soil surface.
2. Impose 'wear' either by usage or through scarification.
3. Avoid scalping when mowing and neither mow severely at any time nor very closely at the end of the growing season.
4. Reduce shading where practicable.

Chemical control of moss is possible using ferrous sulphate or other proprietary materials such as dichlorophen but in contrast to broadleaved weed control this should be considered to be a temporary expedient rather than eliminating an inevitable problem.

Algae can cause a glutinous slime (squidge) which often constitutes a weed problem in turf, especially fine turf in spring when the surface is waterlogged. Algae are susceptible to drying out and since they have no 'roots' they require a wet soil and a humid environment. Some algae are able to fix their own nitrogen from the atmosphere but depend upon a ready supply of phosphate and other nutrients. Typically algae, which create a sticky slimy surface, succeed where soils are sealed by compaction and when fertilizer input, rain and atmospheric humidity are high. Cricket squares, especially wicket ends, the outer 2 m of bowling greens and shaded and overfertilized ornamental lawns are problem sites. Cultural control is easier than with moss because, in the UK, conditions are usually dry enough to desiccate algae over much or all of the growing season provided the general environment is managed to allow this.

Proprietary algicides are available (e.g. dichlorophen) but their use should be unnecessary.

## PESTS AND DISEASES OF TURF

### Pests

There are few animals (vertebrates or invertebrates) which cause significant damage to turf in the UK. Some animals are deemed pests because, on balance, their activities are detrimental (e.g. earthworms) whereas others are 'all bad' (e.g. leatherjackets).

The important pests of turf in the UK can be divided into two types, those which cause damage by eating live plant tissue (mainly roots) and those which mix or disturb soil and/or turf.

#### *Tissue-eating pests*

The key pests in this category are leatherjackets which are the immature stage of the daddy longlegs or cranefly (*Tipula paludosa*). Leatherjackets, which are

20–30 mm long and look like a dark greenish-grey bullet, can be a serious pest of all types of turf but especially cricket squares, golf fairways and golf and bowling greens. The turf is weakened severely by their eating roots and other live plant material below ground. Damage is most severe in spring when grasses produce and become dependent upon new roots. Invasion by this pest usually follows a moist late summer when craneflies are active. Damage by leather-jackets themselves is compounded by rooks or starlings tearing the turf to get at them. Bird activity can 'give away' the presence of an infestation but other evidence in addition to poor turf performance is easily obtained. One symptom is the occurrence of what look like small hollow tine holes. A test, which is foolproof, is to place a damp hessian sack or carpet on the soil overnight. If leatherjackets are present they will be under the cover in large numbers in the morning.

Also within this category are other insect larvae which eat roots but which are really of little or no significance as pests. Grubs of chafer beetles sometimes cause minor damage and caterpillars of the turnip moth and garden dart moth are of negligible importance.

Leatherjackets cannot be controlled culturally in fine turf but chemical control can be achieved with gamma-HCH or chlorpyrifos.

## Soil and turf disturbing pests

There are contrasting types of animal in this category. Bird damage by rooks and starlings has been referred to already. Seed-eating birds hardly fall within this category; nevertheless house sparrows and homing pigeons are often a serious problem on newly seeded areas.

Earthworms are soil movers whose beneficial and detrimental consequences were examined in Chapter 1. They are pests of golf and bowling greens in particular but can be controlled with carbaryl or gamma-HCH. Chlordane has been used for many years as a worm killer but is now banned. Its use should have been banned many years ago because it is a very persistent and hazardous chlorinated hydrocarbon.

The remaining pests in this category are mammals. The most important are moles and rabbits, with dogs and foxes less important. Moles are potential pests of all turf where earthworms are present. The damage caused by moles can be severe not only through creating 'heaps' but also by tunnel excavation underground. Moles can be trapped, gassed or poisoned. The latter two methods should only be used as a last resort.

In areas where rabbits are numerous they can cause problems through scratching holes in sand slits and in general in sand-dominant sportsturf soils. The holes created can be the cause of leg and ankle injuries to players. Dogs and foxes create similar problems although it is only in cities that foxes are likely to cause damage. Rabbits are best controlled by shooting at night.

Finally, despite the attention given to wild animals, *Homo sapiens* on a trail or scrambling motorcycle is frequently the most damaging pest in the soil disturbing category. An effective deterrent is awaited.

## Diseases

It was pointed out in Chapter 2 that pathogenic diseases are rarely if ever seriously damaging in agricultural grasslands ranging from semi-natural chalk grasslands or acidic rough grazings to the sown short-term lowlands pasture leys. Since diseases are far more prevalent in mown turf and especially closely mown turf it is reasonable to conclude that regular close mowing together with conditions of the micro-environment within turf and soil predispose turfgrasses to pathogenic attack. This conclusion is supported by the fact that the most damaging fungal pathogens including *Gaeumannomyces graminis* causing take-all patch and *Microdochium nivale* responsible for fusarium patch exist normally in soils in a non-pathogenic mode decomposing dead rather than living organic matter.

In addition to close mowing, which is inevitable on much fine turf, there are two factors which generally encourage fungal attack. One is surface compaction resulting in poor aeration, the other is the accumulation of thatch which facilitates the lateral spread of fungal hyphae. Avoidance of both of these soil properties reduces greatly the risk of fungal attack and is an essential component of an integrated programme for the control of fungal pathogens. Fungicides are available for the control of virtually all fungal diseases of turf except take-all but they should be considered only as an unavoidable complement to cultural control not a replacement.

Disease problems can be categorized into those which affect grass establishment and those which damage mature turf. With regard to the former there are many species which attack at stages ranging from pre-germination, through post-germination but pre-emergence, to the immediate post-emergence stage. Species include: *Fusarium culmorum*, *Microdochium nivale*, *Pythium* spp., *Rhizoctonia* spp. and others. The key factor in avoiding these diseases is to ensure rapid germination. Three factors are involved: (i) adequately warm soil conditions for germination (above $12°C$); (ii) moist but not waterlogged conditions; and (iii) not too deep sowing of seed. Fungicide treatment of seed can improve establishment under inadequate conditions but should not be necessary (see Chapter 2).

The important diseases of turfgrasses in the UK are identified in Table 5.3 together with visual symptoms and factors affecting their epidemiology. For a more comprehensive coverage of turfgrass diseases including the visual appearance of disease, the texts by Smith *et al.* (1989) and Baldwin (1990) are recommended.

**Table 5.3.** Major diseases of established turf in the UK.

| Common name | Systematic name | Symptoms | Species affected | Seasonality and potential damage | Conditions favouring damage | Cultural control |
|---|---|---|---|---|---|---|
| Fusarium patch | *Microdochium nivale* | Orange/brown circular spots increasing in size | Annual meadowgrass, 'Penncross' bent | Late autumn to spring (severe) | High N, humid environment, thatch | Good air flow and soil drainage eliminate thatch |
| Take-all patch | *Gaeumannomyces graminis* | Small patches of failing grass with rotted roots | Bentgrass, perennial ryegrass, smooth stalked meadowgrass | Mid-autumn (mild–serious) | Newly established sand constructions, rapid pH increase | Maintain favourable growing conditions |
| Dollar spot | *Sclerotinia homeocarpa* | Small white to yellow (dry) spots 10–20 mm | Slender creeping red fescue | Warm, moist summer/autumn (mild–moderate) | Low fertility especially N, sea marsh turf | Do not grow susceptible species, maintain fertility |
| Anthracnose | *Colletotrichum graminicola* | Yellowing plants with rotten base | Annual meadowgrass | All moist conditions (mild) | Compaction and surface wetness | Improve soil physical conditions |
| Leaf spot/melting out | *Drechslera* spp. | Oval or long bleached-brown leaf spots | Perennial ryegrass, smooth stalked meadowgrass | Warm–moist in growing season (mild–moderate) | Mowing stress, excessive or untimely irrigation | Avoid physiological stress |
| Rust | *Puccinia* spp. | Orange or brown pustules on leaves | Smooth stalked meadowgrass (SSMG), perennial ryegrass | Autumn (mild to serious) | Susceptible cultivars of SSMG, unmown or irregularly mown grass | Avoid SSMG in western UK |
| Red thread (also pink patch with different pathogen) | *Laetisaria fuciformis* | Patches in turf 50–200 mm in diameter with shrivelled leaves and pink/red appearance | Red fescue, perennial ryegrass | Mid to late summer (mild) | 'Normal' on low maintenance turf | Light N fertilizer |

Up until recent years the identification of turfgrass diseases depended upon visual diagnosis, backed up if necessary by confirmation using conventional laboratory techniques which typically take several days to complete. Technology now exists which may revolutionize the identification and management of turfgrass diseases. Well-tried serological techniques are involved and identification and assessment of the level of infection are based on the reaction between the antigen of the pathogen and specific antibodies produced in animals to the pathogen's antigen. Antigens of pathogens in infected turf can be extracted easily and quickly from small samples of leaf clippings.

Already in the USA diagnostic kits are available for several of the important turfgrass diseases including brown patch (*Rhizoctonia*) and *Pythium* blight. A positive test results in a clear colour change which can be quantified using a simple portable meter. Guidelines are provided for the interpretation of meter readings both for fungicide application and on testing frequency in relation to weather conditions. The basic kit format will probably be developed for use in the UK. Specific antibodies to some of the important turfgrass diseases in the UK are not yet available and guidelines for action in relation to weather will require modification.


## IRRIGATION


Irrigation is the input of water to supplement the natural rainfall and is used to replenish soil water reserves. The reasons or justifications for irrigation are diverse and can be summarized as follows:

1.   To ensure water supply from the soil does not delay seed germination or restrict turf growth.
2.   To prevent death of drought susceptible species (e.g. annual meadowgrass).
3.   To maintain a good verdure or turf colour.
4.   To prevent the development of 'dry patch' disease.
5.   To wash out accumulated salts.

The purely cosmetic use of irrigation water on turf to improve appearance should be resisted because it favours invasion by annual meadowgrass and is liable to result in overwatering leading to waterlogging and anaerobic soil conditions. In general, a moderate soil moisture stress for fine turf improves playing conditions and helps maintain a desirable species composition. When usage of fine turf is intensive however greenkeepers often feel the need to irrigate liberally to improve the rate of grass growth and thus recovery from wear.

The sand-dominant soils used for most sportsturf have a large bulk density and a small total porosity. Some macroporosity should be present to ensure adequate aeration so that the water-retaining porosity is small and typically around 25% of the soil volume; the USGA specifications for golf green root zones quote a lower limit of 15% (USGA Green Section Staff, 1993). In addition to the poor water retentivity of sportsfield soils they are usually quite shallow over a drainage layer or compacted subsoil. This, together with the fact that the roots of closely mown turf are shallow, means that the effective soil depth for the retention of useful water is often no more than 150 mm. With an available water capacity of approximately 20% and an effective soil depth of 150 mm a soil could hold a maximum of 30 mm of water. This would be exhausted in under one week of dry summer weather in the UK. Looked at another way the maximum potential soil moisture deficit (water depletion below the field capacity) in summer over more than three quarters of England and Wales is in excess of 100 mm which is much greater than an average retention of 30 mm of available water. Inevitably, therefore, drought stress will occur on all sand-dominant sportsfield soils, designed to have free drainage, on several occasions every year, unless they receive irrigation.

There are a few key points affecting methodology, timing and rate of irrigation. Probably the most important fact is that soils cannot be moistened to a uniform water content when input does not achieve field capacity. This means that small inputs of water to a dry soil result in uneven wetting and, although turf growth will be maintained, shallow rooting may be encouraged. A sensible strategy is to irrigate to saturation at infrequent intervals (say 20 days) during a long dry spell but to complement this by light irrigation of 1–4 mm daily or every second day. To ensure regular flushing of the soil, a strategy similar to the above must be followed in arid areas if the quality of irrigation water is poor and salt accumulation is to be prevented.

Sand-dominant rootzones release virtually all of their available water at tensions less than 200 kPa. That is, the available water is easily available with little reserve of water held at high tensions. As a result, roots remove water easily from moist sandy soils but a point is reached where the supply falls off very abruptly compared with loamy soils. This sensitivity to sudden drying out must be recognized by groundstaff and greenkeepers so that they can avoid 'dry patch' developing.

Irrigation systems are available where watering is triggered by soil moisture sensors. Field capacity approximates to −4 kPa in turfgrass soils and soils should be allowed to dry out to not less than around −15 kPa before irrigation is begun to avoid excess wetness and poor aeration (Marsh, 1969).

A wide range of systems is available for the delivery of overhead irrigation water. Simple travelling sprinklers are satisfactory when the need for irrigation is infrequent and where need is restricted to sections of the playing area. Pop-up systems are in common use and are concealed below the surface when not in use. As a general rule pop-up sprinklers should not be installed on the playing area. Achieving this presents no problems with golf and bowling greens but it is more difficult with soccer and rugby pitches. The main reason for avoiding the playing area is that the moving parts are liable to become jammed by eroding rootzone material, especially on winter game pitches. Also players may trip up over irrigation heads, or maintenance operations (e.g. spiking) may cause damage to the irrigation heads.

Some constructions have a facility for sub-irrigation (e.g. Fig. 3.12). The control of drainage outflow can improve water-use efficiency and reduce the risk of pollution of drainage water by nitrate and pesticides. It was explained in Chapter 1 that it is not good practice to maintain a constant watertable depth in sand profile constructions and so the subsoil drainage layer should be emptied from time to time. If the drainage layer is drained, much of the potential economy of water use in sub-irrigation is lost. For example, the refilling of a 150 mm depth drainage layer would require around 60 mm of water which is around twice the total available water capacity of sand-dominant rootzones.

It is evident from present trends that water for irrigation of turf of all kinds will become more expensive and less available. Planners and designers of golf courses and other sports facilities are already implementing schemes to store and re-use drainage water for irrigation. Improved economy of water use is vital.

## FERTILIZERS AND THEIR USE

Fertilizer use is an established practice in agriculture and horticulture. The main reason why an input of plant nutrients to soils is required for crops is because the nutrients are removed in crops more rapidly than they can be replenished by the weathering of soil minerals, biological fixation of nitrogen and deposition from the air. An exception to this rule is in the industrialized countries where sulphur deposition from the air, whence it arises through the burning of fossil fuels, currently exceeds plant demand.

If grass clippings are allowed to fly, as they are on large areas of amenity turf, the loss of plant nutrients is small and the requirement for fertilizers is minimal. In contrast, when clippings are removed the offtake of plant nutrients can be very large and fertilizer input is essential. Thus in turf areas in general, the fate of clippings is a dominant factor in determining fertilizer requirement. An additional factor in intensively used sportsturf is the destruction of turf by wear. Wear reduces the the uptake of nutrients and thus their recycling back to the soil in plant residues. It also tends to increase leaching losses. The uptake of mineral nitrogen and its return in organic forms in plant residues is particularly important. Therefore areas of sportfields which become severely worn require an extra input of fertilizer, especially nitrogen.

Although the supply of one of the recognized fertilizer nutrients, N, P and K, is most likely to control turfgrass growth, any one of the essential plant nutrients may be limiting. In practice few are likely to be in critically short supply. Nutrients in addition to N, P and K which may become limiting are

Mg, Cu and B. Magnesium is required in moderate amounts by turfgrasses and levels can become very low in some sandy soils. For this reason many compound turf fertilizers contain Mg. Although both Cu and B are required in very small amounts, reserves of both nutrients are usually very low in sandy soils and the retention of both is improved by association with organic matter which is present in small amounts in some turfgrass soils. Rootzones most susceptible to deficiencies of trace elements are those comprised largely of clean quartz sand with a low organic matter content.

## Nutrient balance in fertilizers

It was explained in Chapter 2 that the ratio of N:P:K removed in the clippings of turfgrasses remains reasonably constant so that if the main function of fertilizers is to replenish these then an appropriate overall ratio is around 4:1:3 for $N:P_2O_5:K_2O$. Deviation from this ratio might result from factors such as a greater loss in drainage water of one or more nutrients or soil processes which rendered a nutrient in the fertilizer unavailable. Nitrogen as nitrate is very susceptible to loss through leaching but provided there is complete grass cover and individual applications of fertilizer N are not greater than around $6 \, g \, m^{-2}$ ($60 \, kg \, ha^{-1}$) little loss can be expected. Phosphate is 'immobilized' even in sandy soils (see Chapter 1) but in the long term this does not affect the fertilizer nutrient ratio provided a large input of fertilizer P is incorporated into the rootzone prior to establishment. Simplistically, therefore, the use of a 4:1:3 fertilizer when conditions are favourable for growth will maintain a supply to turf of the major nutrients in the ratio required. This prescription has become a recommendation for many turfgrass situations in the USA and the main manufacturers in the UK produce fertilizers with this formulation. This recommendation is simpler and more soundly based than the autumn and spring/summer formulations still promoted in the UK (Adams, 1983).

An alternative strategy of fertilizer use which has a good scientific basis is the occasional use of a high P fertilizer, say once per year or less often, in conjunction with more frequent applications of N:K fertilizer when environmental conditions favour growth. There are two factors which support this approach. Firstly, P is retained strongly in soils and P fertilizers have a high residual value (persistence of effect). Both N and K are removed in large amounts in clippings and both are mobile and readily leached from sand-dominant soils. Also in contrast to clay-rich soils, K is not released in significant amounts by either weathering or from 'fixed' reserves.

As a general rule it is unwise to run the risk of allowing a major nutrient other than N to restrict turfgrass growth. The best strategy for fertilizer use on turf is to ensure that all essential nutrients other than N are non-limiting and to use N input as the growth controller. Soil analysis on a regular basis is essential if this plan is to be followed.

## Chemical sources of fertilizer nutrients

A detailed description of all sources of fertilizer nutrients is given by Tisdale *et al.* (1984). The more important fertilizer sources for turf are summarized in Table 5.4. Not included in the table are fertilizers marketed as 'organic based' or 'organic' which utilize seaweed or other natural organic materials including lignite, sludges, manures or slurry products. These can be useful sources of trace elements although, until legislation on pollution is improved, sewage sludges will continue to be contaminated by heavy metals. In order to make most of these organic sources useful as fertilizers, they must be 'reinforced' with the chemicals in Table 5.4. Two factors should be borne in mind when using organic fertilizer sources. The first is that in many turfgrass situations a major difficulty is in reducing soil organic matter levels. The second is that the addition of a readily decomposable organic material to a compacted soil may increase the microbial demand for oxygen to such an extent that it transforms marginally adequate aeration to distinctly anaerobic conditions.

Some additional points can be made concerning particular nutrient sources. As a general rule nitrogen fertilizers are acidifying in the long term. This is basically because most are sources of $NH_4^+$ and the oxidation of this to nitrate ($NO_3^-$) is an acidifying reaction. Ammonium sulphate is especially important because nitrification results in the formation of two strong acids.

$$(NH_4)_2 SO_4 + 4O_2 \rightarrow \underset{(acid)}{H_2 SO_4} + \underset{(acid)}{2HNO_3} + 2H_2 O$$

In contrast to this general statement, the hydrolysis of urea fertilizer which occurs within two to five days of application causes an alkaline reaction through the production of ammonium hydroxide.

$$(NH_2)_2 CO + 2H_2 O \rightarrow (NH_4)_2 CO_3$$

$$(NH_4)_2 CO_3 + 2H_2 O \rightarrow 2NH_4 OH + H_2 CO_3$$

If the pH rises to over 7.5 much of the ammonium will exist as dissolved ammonia which is subject to loss by volatilization. Clearly if the soil is already alkaline a significant proportion of urea-nitrogen will be lost from the soil. Normal applications of urea fertilizers to acidic soils will not result in ammonia loss.

The oxidation of $NH_4^+$ can be inhibited and fertilizers containing Didin (dicyandiamide) are marketed. Didin is not as specific an inhibitor of nitrification as N-serve (Nitrapyrin) but it is claimed to improve the efficiency of nitrogen fertilizer use. The $NH_4^+$ cation behaves in a very similar way to $K^+$

**Table 5.4.** Major sources of fertilizer nutrients.

### *Nitrogen*

*Water soluble forms*:

Urea $(NH_2)_2CO$ (46% N)
Ammonium nitrate $NH_4NO_3$ (35% N)
Ammonium sulphate $(NH_4)_2SO_4$ (21% N)

*Natural organic*:

Dried blood (10-14% N)
Hoof and horn (13% N)

*Synthetic organic*:

Nitroform (38% N)
Isobutylidene diurea (IBDU 31% N)
Crotonylidene diurea (CDU 32% N)

### *Magnesium*

*Water soluble forms*:

Magnesium sulphate sources (10-16% Mg)

*Insoluble forms*:

Dolomitic limestone $Ca/MgCO_3$ (more than 3% Mg)

### *Potassium*

*Water soluble forms*:

Potassium chloride KCl (60% $K_2O$)
Potassium sulphate $K_2SO_4$ (50% $K_2O$)

### *Phosphorus*

*Water soluble forms*:

Superphosphate (19% $P_2O_5$)
Triple superphosphate (47% $P_2O_5$)
Mono-ammonium phosphate $NH_4H_2PO_4$ (62% $P_2O_5$, 12% N)
Di-ammonium phosphate $(NH_4)_2HPO_4$ (54% $P_2O_5$, 21% N)

*Sparingly soluble forms*:

Magnesium ammonium phosphate $MgNH_4PO_4$ (52% $P_2O_5$, 18% Mg, 10% N)

*Insoluble forms*:

Ground rock phosphate (25-35% $P_2O_5$)
Bone meal (22% $P_2O_5$, 4% N)

Delayed release fertilizers are also produced by coating water soluble fertilizers with materials to impede dissolution. Coatings may be sulphur, waxes or a range of types of polymer.

and, since K is leached quite readily from very sandy soils (Chapter 1), the benefit of a nitrification inhibitor is potentially much greater in loamy soils with a moderate cation exchange capacity than in sand-dominant soils.

Dried blood is degraded so rapidly that it is equivalent to a water soluble source of N. Hoof and horn are degraded slowly but the synthetic organic nitrogen sources have advantages in release characteristics over this natural product. They are hydrolysed to $NH_4^+$ over time and currently IBDU is perceived to have the best slow release characteristics. Hydrolysis or mineralization of IBDU is related most closely to the dissolution of particles, the rate of which increases with decrease in particle size. Mineralization is temperature and moisture dependent and conditions favourable for mineralization are those which favour plant growth.

All the important fertilizer sources of K are water soluble. In most natural soils this presents no problem but in very sandy soils K is easily leached because of the low cation exchange capacity.

Trace element fertilizers are not covered in Table 5.4. When a trace element problem is suspected the best approach is to use a trace element cocktail. These are available as frits, which are in effect glass particles which release trace elements slowly. Organic fertilizers, notably those based on cattle slurry or seaweed, are also important sources of trace elements.

There is no need to use insoluble sources of P to gain the benefit of slow release because the P in water soluble materials is rapidly immobilized in soil. Nevertheless, because of its cheapness, ground rock phosphate is a good long-term P source on extensive areas of amenity turf.

## Timing and rate of fertilizer applications

It is a simple rule that if fertilizers are applied when weather conditions are not conducive to turfgrass growth their use will be inefficient. Essentially nitrogenous fertilizers are likely to be leached from the soil whilst phosphatic fertilizers will become immobilized. The application of so-called Autumn/Winter fertilizers is a case in point where justification seems to be based on the philosophy that if you are determined to apply fertilizer then one high in P and low in N is safest.

High levels of available N result in large, thin-walled leaf cells. It is reasonable to assert therefore that high N fertilizer inputs in autumn, when fungal attacks are most prevalent, increase the risk of disease. Not only this but, if surface wetness is a problem, late applications of N and P will favour algae.

Nitrogen as $NO_3^-$ is the most mobile major nutrient and also N is required in the largest amounts. Thus it is in relation to N that recommendations on the frequency of fertilizer applications are based. Mature perennial ryegrass turf can utilize about 2 kg ha$^{-1}$ d$^{-1}$ of N in good growing conditions so the question arises as to how long available N will remain within the rootzone, accessible to turfgrass roots. The answer depends upon the nature of the soil and the amount of precipitation in relation to evapotranspiration. There is also the point that it takes over 2 weeks for fertilizer $NH_4^+$ to be oxidized to mobile $NO_3^-$. From May to September the nitrogen in water soluble fertilizer would be expected to stay within the rootzone for at least two months provided irrigation was not excessive. Nevertheless a substantial proportion of water soluble nitrogen fertilizer is likely to be leached from very sandy rootzones in a much shorter time than this if it is applied in early spring.

In summary, it is advisable to apply water soluble nitrogen fertilizers at intervals not greater than two to three months during the growing season on very sandy rootzones and at rates of N not exceeding $6 \, \text{g m}^{-2}$ ($60 \, \text{kg ha}^{-1}$). The potential benefits of slow release sources of N are evident. Other situations where materials such as IBDU are particularly valuable are establishment from seed on very sandy rootzones and prior to overseeding severely worn areas on sand-dominant soils.

# Calculating the rate of fertilizer application and the relative cost of fertilizers

The concentration of each nutrient contained in a fertilizer is stated as a percentage in catalogues and on the bag of fertilizer. The total concentration of one nutrient may be divided into different forms, for example water soluble N (ammonium plus nitrate) and IBDU – slow release N. It is current practice in the UK to quote concentrations of the three main fertilizer nutrients (N, P, K) as N, $P_2O_5$ and $K_2O$. The expression of P and K as oxides should be discontinued. Other nutrients, e.g. Mg and Fe, are stated as concentrations of the element.

Fertilizer recommendations are given as amounts of plant nutrient per unit area, not the weight of fertilizer, because fertilizers vary widely in the concentration of nutrients they contain. Units for fertilizer recommendations are usually either $\text{g m}^{-2}$ or $\text{kg ha}^{-1}$. Conversion between the two is straightforward because $1 \, \text{g m}^{-2}$ is equal to $10 \, \text{kg ha}^{-1}$. Whichever unit you are dealing with, the conversion of recommendation into a weight of fertilizer per unit area simply involves multiplying the recommendation by 100 divided by the percentage concentration of the nutrient in the fertilizer. The following example is used as an illustration:

- **Recommendation**: Application of $20 \, \text{g m}^{-2}$ of N and $15 \, \text{g m}^{-2}$ of $K_2O$ over the growing season in four equal applications.
- **Implementation**: Firstly identify a fertilizer with a nutrient ratio of $20:0:15$, i.e. $4:0:3$ of N, $P_2O_5$ and $K_2O$. Suppose the closest to this is a fertilizer containing 12% N and 8% $K_2O$.

Having selected the best fit, calculate on the basis of the N recommendation. The total amount of fertilizer required is:

$$20 \times 100/12 = 167\,\mathrm{g\,m}^{-2}$$

This would be applied in four equal applications of $42\,\mathrm{g\,m}^{-2}$. Had the same recommendation been given in $\mathrm{kg\,ha}^{-1}$ it would have been for $200\,\mathrm{kg\,ha}^{-1}$ of N and $150\,\mathrm{kg\,ha}^{-1}$ of $K_2O$ and the total fertilizer application would have been $1667\,\mathrm{kg\,ha}^{-1}$ amounting to $420\,\mathrm{kg\,ha}^{-1}$ on each of four occasions. A groundsman/greenkeeper must know the area of land he is responsible for so that he can order fertilizer (and topdressing) appropriately. Often, parts of the total area are subjected to different types of use and management and have different fertilizer requirements, for example, greens and fairways on a golf course. This situation demands separate calculations on the amounts of fertilizer needed.

Since the fertilizers on the market contain different concentrations of plant nutrients, assessment of their value must be based on the cost per unit amount of nutrient contained. Thus a fertilizer containing twice the concentration of the same nutrients as those present in another fertilizer would be worth buying if it was anything less than twice the cost per bag. Assessment of value must also take into account the chemical form of the nutrient in the fertilizer. Thus IBDU-N, because of its slow release character, is worth more per unit of N than water soluble N. The problem is in assessing how much more. The N in IBDU may in some situations be up to twice as efficient as water soluble N so that if the cost exceeded twice that of a water soluble source, its purchase would need to be justified on additional benefits, for example evenness of growth or impracticability of making more than one or two fertilizer applications per year.

## Assessing soil nutrient status

Typical fertilizer recommendations are given in chapters devoted to particular sports but there are general guidelines. Soil analysis for major plant nutrients and also pH is essential to ensure an efficient and successful fertilizer programme. Analysis for available nutrients in soils is not absolutely precise but this is unimportant because no one should be maintaining nutrient status on the absolute verge of deficiency. Routine soil extraction can be carried out for available P, K and Mg and the amount extracted converted into an Index of Availability. This is illustrated in Table 5.5 for extraction with $0.5\,\mathrm{M}$ ammonium acetate/acetic acid which can be used for all three nutrients. The preferred Index is 2 or 3. If the Index is less than 2, fertilizer input should be increased, if the Index is 4 or greater it should be decreased. Alternative extractants may be used to determine available nutrients. The most widely used

**Table 5.5.** Relationships between amounts of potassium, phosphorous and magnesium extracted from soil, and availability index. The extractant used was 0.5 M ammonium acetate-0.5 M acetic acid at a soil:extractant ratio of 1:5 (w/v).

| Index | K (mg kg$^{-1}$) | P (mg kg$^{-1}$) | Mg (mg kg$^{-1}$) |
|---|---|---|---|
| 0 | 0-50 | 0-2.0 | 0-25 |
| 1 | 51-100 | 2.1-5.0 | 26-50 |
| 2 | 101-200 | 5.1-10.0 | 51-100 |
| 3 | 205-350 | 11-20 | 101-175 |
| 4 | 355-500 | 21-40 | 176-250 |
| 5 | 505-700 | 41-70 | 255-350 |
| 6 | 705-1200 | 71-125 | 355-600 |
| 7 | 1210-2000 | 126-200 | 605-1000 |
| 8 | 2010-3000 | 201-300 | 1010-1500 |
| 9 | over 3000 | over 300 | over 1500 |

extractants in the UK are 0.5 M sodium bicarbonate for P and 1 M ammonium nitrate for K and Mg (MAFF, 1991). Different extractants are used for trace elements, for example boiling water is used to extract available B and the complexing agent ethylenediaminetetracetic acid (EDTA) for Cu and Zn.

Immediately available N in the forms of $NH_4^+$ and $NO_3^-$ can be measured but there is no satisfactory method to determine available N relevant to a total growing season. The main reason is that the availability of N in soils is under the control of microorganisms and the whole system is dynamic. It is very difficult to simulate or predict what will happen to the available N status over a growing season.

## Soil sampling for available nutrients

The errors incurred in assessing the availability of a nutrient using an extractant are small compared with differences due to sampling over different depths in turfgrass soils. Because turfgrass soils are 'uncultivated', marked gradients can develop in the concentration of available nutrients with depth. This is especially the case with P where the amount extracted may differ by more than two-fold over 20 mm depth. Thus sampling of turfgrass soils must not only be representative of the area of interest but also of a precise and constant depth range. A 0-100 mm depth range is recommended for most situations but a 0-50 mm depth range is appropriate for golf and bowling greens because of the shallow root development. In addition to sampling to a constant depth it is vital that all of each core sample is rubbed down for analysis after cutting the grass to ground level.

In order to obtain a soil sample representative of an area of turf, several samples must be taken and combined because the risk of a single sample being non-representative is high. Important questions are how many samples? and to what sampling pattern? The first point, however, is that samples should only be combined from areas which are apparently uniform. It is not sensible, for example, to sample a cricket square in general when it is known that pitches have been constructed at different times. Given that an area which may be a green, a pitch or a field is apparently uniform then the sampling should be to a formal pattern or grid and the number of samples sufficient to ensure that a single errant sample does not have an unduly large effect on the analysis of the bulked sample. The minimum number of samples is five taken from the points of an imaginary W set on the area. The sample number can be much greater and different types of grid are acceptable. The samples should be bulked, rubbed down and a subsample of appropriate amount used for analysis.

The strategy for sampling described may be appropriate for other soil analyses in addition to available nutrients. Nevertheless, a very different strategy may be necessary. The most common alternative sampling approach is when the variation in a soil property with depth is critically important. Quite often, changes with depth in, for example, organic matter content or particle size distribution exert major effects on playing quality and in these cases sampling in relation to depth is vital.

## TOPDRESSING AND THATCH CONTROL

### Definition and purposes of topdressing

In agriculture the term topdressing is used almost exclusively to describe the practice of applying fertilizer to a standing crop whereas in turfgrass science the term is largely restricted to the application of a sand, soil, loam or other non-fertilizer dressing to the turf/soil system. The latter usage is adopted here.

There are three main reasons why topdressings are applied:

1. To restore a level soil surface.
2. To create or maintain a desirable particle size distribution in the surface soil.
3. To control (usually reduce) the organic matter content of the surface soil.

A particular topdressing programme may be intended to achieve one or more effect, although, in all types of sportsturf, because of the demand for surface trueness, all topdressing schemes have this as a function.

# Topdressing winter games pitches

Because of the turf damage inflicted by play in winter, topdressings are not usually required to control a build-up of organic matter on high wear areas of a pitch. The main purposes are to restore trueness and to maintain the required soil particle distribution of the rootzone. The former is especially necessary and demanding on maintenance time when sand-dominant root-zones lose turf cover and become physically unstable. The amounts of sand topdressing required are variable but are generally related to the damage to surface trueness. Topdressings to maintain surface trueness are applied mainly during the playing season. Topdressings designed to maintain a sufficiently sand-dominant rootzone are especially important where sand carpet or sand/soil rootzone schemes are used (Chapter 4). In this case topdressings of sand should be applied in the close season. The most appropriate time is late May to mid July and applications should form part of an integrated programme of decompaction plus topdressing to gain maximum benefit.

The amount of topdressing required to maintain an appropriately sand-dominant rootzone must, in the long term, be assessed with reference to soil analysis. Figure 5.1 suggests that there is a tendency in routine maintenance to underestimate the amount of topdressing required but it is not possible to give a general prescription. There are several reasons but the key ones are that earthworm activity, which is responsible for bringing up underlying soil to the surface, is variable, the other is that the underlying soil differs from site to site. Possibly the best estimate of the amount of soil brought to the surface by casting earthworms in the UK is $25 \, t \, ha^{-1} \, y^{-1}$ (Adams *et al.*, 1992). Using this it is possible to calculate the amount of sand topdressing required to maintain a particular rootzone composition provided the particle size distribution of the soil deeper than 100 mm is known. Thus:

$$Q = 100(1 - 1.25x)$$

where  $Q = t \, ha^{-1}$ of sand topdressing required to maintain a $D_{20}$ of 125 $\mu$m using a topdressing sand with no particles less than 125 $\mu$m; $x$ = proportion in excess of 125 $\mu$m in the underlying soil expressed as a decimal.

Thus for a subsoil containing 50% in excess of 125 $\mu$m the required topdressing would be $37.5 \, t \, ha^{-1}$ or a little over 25 t of sand per pitch per year.

A requirement for sand topdressing on total profile constructions with a sand rootzone, where earthworms are absent, is only evident in areas of the pitch which are not severely worn. In these cases topdressing is mainly geared to the control of thatch. The amount of topdressing is similar to that for bowling greens and golf greens, and can be applied in a single dressing in the close season.

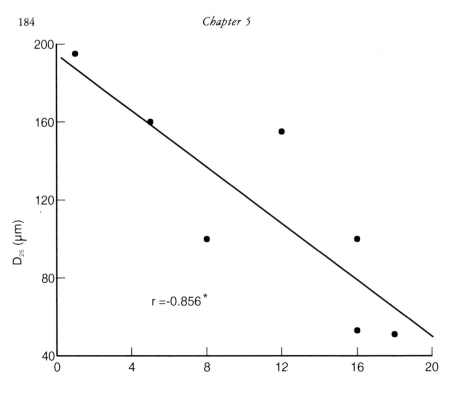

Time since construction (years)

**Fig. 5.1.** The relationship between the $D_{25}$ of seven winter games pitches constructed using a sand/soil rootzone and time since construction (from Adams *et al.*, 1992).

## Topdressing golf greens and bowling greens

Earthworms are normally absent from these sportsturf areas and there are two consequences. One is that topdressings are not needed to maintain the particle size distribution of the mineral framework, the other is that they are required as a component in thatch control.

Surface accumulation of organic matter is, as was explained in Chapter 1, a natural process in uncultivated soils which are devoid of earthworms. Thatch is in effect an incipient organic horizon on the mineral soil which contains plant-derived organic matter ranging from live roots to recently dead debris to colloidal humus. If thatch is allowed to develop uncontrolled, it eventually dominates soil behaviour in many aspects affecting turfgrass performance and playing quality. The nature of thatch, its occurrence and control are reviewed by Shildrick (1985).

Thatch is a consequence of the lack of comminution and incorporation into the mineral soil of plant residues. The build-up of thatch depends upon the level of plant residue input in relation to decomposition. In essence this means that provided surface organic matter accumulation does not itself affect its own rate of decomposition (for example by creating long-term waterlogging or severe acidification) an equilibrium depth of thatch will be attained. In sports-turf this is irrelevant since an equilibrium depth is much deeper than is acceptable.

From a practical point of view, since thatch causes a soft, water-retentive surface and adversely affects rooting, aeration and drainage, it has to be con-trolled. There are two components to this control. One is the removal by comb-ing, raking, scarifying or vertical mowing of both living and dead plant debris to reduce input. The second is topdressing with an appropriate material of low organic matter content to 'dilute' the residual organic matter. The aim is to prevent the organic matter content of the surface soil exceeding around 10% when it begins to control soil physical behaviour.

Since the main aim of topdressing in thatch control is to 'dilute' added organic matter it would seem that an estimate of the amount of topdressing required should be related both to its own organic matter content and to the organic matter added to the soil in plant residues. Adams and Saxon (1979) used this approach to develop the following predictive equation:

$$T = \frac{I\,(D - 100)}{P - D}$$

where  $T =$ annual topdressing (kg m$^{-2}$); I $=$ input of humified plant residues (kg m$^{-2}$) (plant residue $\times$ 0.25); $D =$ required soil organic matter con-tent (%); $P =$ organic matter content of the topdressing (%).

For a typical input of turfgrass residues of 6000 kg ha$^{-1}$ (0.6 kg m$^{-2}$) and an organic matter content of 1.5% in the topdressing, the anticipated annual topdressing required to maintain a 10% organic matter content in the topsoil would be 1.6 kg m$^{-2}$.

The annual topdressing required on golf and bowling greens amounts to 1.0–1.5 mm depth. If this is applied in a single dressing it is difficult both to integrate it with the rootzone surface and to avoid smothering grass. Also heavy topdressings tend to remain as discrete thin layers as dressings build up. This can cause shallow rooting. Madison *et al.* (1974) developed a system of frequent topdressings for golf greens. Topdressing on four to six occasions during the growing season can be carried out with minimal disruption to play and topdressings can be integrated much more easily. The practice of applying split multiple topdressings is increasing on bowling greens and it has largely replaced the traditional single topdressing in autumn on golf greens.

Whatever system of topdressing is used, a vital factor is consistency in par-ticle size composition. Changing between sand grades is a common cause of root breaks in rootzones.

It is worth noting that whilst clean sand is often used as topdressing on

golf and bowling greens, shallow rooting has often been a long-term consequence. Sand-dominant soils with no more than 15% less than 125 $\mu$m and
organic matter contents of 1–2% appear to reduce the problem.

## Topdressing cricket squares and tennis courts

Thatch development on cricket squares where earthworms are controlled
chemically is as inevitable as it is on golf and bowling greens unless control
measures are implemented. Indeed the problems created by thatch on cricket
squares and tennis courts are much more serious because of its effect on ball
bounce and susceptibility of the surface to disintegration. Uncontrolled thatch
development on cricket squares is rarely seen because of the obvious
problems it creates. Nevertheless, equally serious problems are created when
topdressings are applied which bury even imperceptibly thin layers of thatch.
The compressed layers of organic material, separating mineral soil topdressings,
act as lines of weakness preventing vertical integration. The total lack of
integration caused by burying thatch is an extreme situation but, more commonly, layering, through poor integration, is a result of inadequate preparation prior to topdressing.

A fundamental principle in applying topdressings to cricket squares and
tennis courts is that they must knit with the existing mineral soil. Two factors
are essential:

1. The existing turf must be scarified and mown so closely that ground cover
is reduced to not more than 20%. The soil surface must be scored.
2. The topdressing (usually called loam) must be compatible with the existing
soil in terms of particle size distribution and particularly swell/shrink
characteristics. Plate 2 shows the type of layering which used to be present on
most county cricket squares. Cricket square topdressings should only be applied
in autumn otherwise the soil does not have an opportunity to integrate. Topdressing during the playing season is legitimately considered as 'doctoring' (see
Chapter 8).

## COMPACTION AND POOR AERATION

### The nature of the problem

It is clear from Chapter 2 that turfgrass root growth and function are affected
by several factors. Roots depend upon a supply of photosynthate from leaves
and nutrients from the soil, but even when the supply of both of these is adequate the performance of roots may be very poor due to adverse soil physical

conditions. There are two potential problems which are widespread on all types of sportsturf and which are a result of soil compaction.

Soils not subjected to either foot or vehicular traffic have a moderate volume of macropores (pores greater than 75 $\mu$m in diameter). These pores, which are usually produced by aggregates of clay and other small particles, drain under gravity and provide open space for the growth of turfgrass roots. When solid particles of sand size create large pores rather than aggregates, this is achieved at moderate sand contents in uncompacted soils because of the heterogeneous organization of particles of different size. Aggregates are fragile, especially when wet, so that compaction results in the destruction of the large pores which they create. Thus total porosity decreases much less rapidly than macroporosity. In single-grain structured sandy soils, compaction results in small particles being pressed within the framework of larger particles. This phenomenon of interpacking occurs not only in sportsturf but in agricultural soils compacted by intensive cattle grazing (Davies *et al.*, 1989). When sandy soils are compacted it is only if the fine material present is insufficient to occlude the large pores created by the sand that macropore space can be guaranteed (Adams, 1982).

In all but the sandiest soils, compaction can eliminate all free-draining macropore space. As a consequence, compacted soils which become saturated through rain or irrigation remain saturated until water is lost by evaporation and transpiration. Aeration is inadequate and the sequence of anaerobic processes described in Chapter 1 is set in train.

Another consequence of compaction is that the pores within the soil matrix are too small for turfgrass roots to extend into. No turfgrass roots are smaller than around 60 $\mu$m in diameter so extension into pores of this size or smaller requires the mechanical creation of the necessary space. Their ability to do this depends upon the pore sizes in the compacted zone, the mechanical strength of the compacted zone and the ability of roots to gain leverage from adjacent soil (Russell, 1977).

Thus soil compaction has a detrimental effect on roots, not only by causing anaerobic conditions which through anoxia (lack of oxygen) or the presence of toxic products affects root development and function, but also by creating mechanical impedance to rooting.

## Depth of compacted or anaerobic zone

When soils are compacted the problems created occur over different depths depending upon the cause of the compaction. This must be recognized and identified in order to prescribe the most appropriate technique for ameliorating the situation. The most straightforward example is on a cricket square where pitch preparation is designed to cause reasonably uniform compaction to

100 mm depth (Chapter 8). Turf subjected to horse traffic, whether a show-ground or racetrack, is a different situation where compaction increases with depth to a maximum in the 80–120 mm depth range (Chapter 9). There are two distinct types of problem on winter games pitches. One is compaction due to treading which has a maximum effect at 30–60 mm depth. The other, which is much more important, is a smearing and kneading of the top 30 mm which is most damaging when play occurs during rain and when ground cover is reduced (less than 20%). On golf greens, compaction is due to compression only (no shear) by foot traffic and vehicles. Whilst significant compaction is caused down to around 80 mm, macroporosity in the top 20 mm of sod or turf is maintained, when turf cover is intact, by the presence and resilience of stolons, rhizomes and stem bases.

Another factor comes into play which determines the zone of maximum anaerobiosis. This is one of variation in oxygen demand with depth. The demand for oxygen by the soil microbial biomass is positively related to soil organic matter content and so decreases with depth. Given that compaction forces by players are expressed down to around 80 mm and the demand for oxygen decreases with depth, severely anaerobic conditions are unlikely to occur below this depth. If grass cover is lost and the soil surface smeared, the most severe conditions occur close to the surface. With grass cover some air entry usually occurs close to the surface and the zone most likely to suffer severely anaerobic conditions is the 15–60 mm depth range. When metal sulphides are produced on golf greens under severe reducing conditions it is usually at a maximum within this depth range (Plate 3).

## Principles of decompaction and aeration improvement

New macropore space can only be created in a compacted soil by two means. One is by raising the soil surface so that the same mass of soil occupies a greater volume, the other is by removing cores or segments of soil so that a smaller mass of soil occupies the same field volume. Any other method of macropore creation achieves it by decreasing the existing micropore space.

Solid tines and to a lesser extent slit tines create macropores by compressing the soil below and around the tines. Nevertheless they usually improve water infiltration and soil aeration. The reason why they are often beneficial is because although the macroporosity in the immediate surface may be virtually zero, it increases with depth. Thus gas diffusion into the rootzone and the drainage of water from it can often be improved by a shallow piercing of a thin saturated layer. Therefore winter games pitches with smeared surfaces usually benefit from solid tine spiking. Notwithstanding the widespread efficacy of the procedure, traditional solid tine spiking is an aeration procedure which does not relieve compaction.

**Table 5.6.** Examples of techniques for relieving compaction based on soil fracturing or loosening.

| | |
|---|---|
| Soil fracturing by injection of air or water | *Water*: Hydroject (The Toro Company, Minnesota, USA) |
| | *Air*: Terralift (Zink-Motorentechnik GmbH, Germany) Robin Dagger (Sachs-Dolmar (UK), Manchester) |
| Mechanical fracturing or loosening | *Tines*: Verti-drain (Edexim BV, The Netherlands) |
| | *Subsoil mole and shoe*: Twose turf conditioner (Twose, Devon, UK) |
| | *Tines rotating on horizontal axis*: Surface slotter (Cambridge Associates, Cambridge, UK) |

# Methodology of decompaction

Coring or hollow tining is now a universally accepted technique not only to relieve compaction but also as a component in thatch control. It can also be used in conjunction with topdressings as a soil replacement technique (Baker, 1983) and for correcting minor errors in surface levels.

Decompaction is achieved by soil removal, but nevertheless there is evidence that the penetration of hollow tines causes some compaction of adjacent soil (Rieke and Murphy, 1989). It is not clear whether the creation of a void in the soil facilitates the recovery of this effect. Certainly roots develop strongly in hollow tine holes in compacted soils (Plate 4).

There are several methods or techniques to relieve compaction by soil fracturing or loosening. Examples are given in Table 5.6. Equipment for the injection of air or water under pressure to relieve compaction is less mobile than that used for mechanical fracturing methods and is thus less appropriate for treating extensive areas. The technique is effective, however, for shattering subsoil compaction. The Verti-drain and systems based on subsoil tines or the mole plough are also effective in relieving subsoil compaction. Added to this the Verti-drain leaves tine holes connecting with the surface. Whilst subsoil compaction is often a problem, the summary analysis of situations presented earlier in this section indicated that the most serious problems are usually confined to the top 120 mm. The surface slotter causes shattering and loosening of this surface layer through blades set to a scroll pattern rotating on a horizontal shaft. The implement is a very effective machine especially in the 'restructuring' of rootzones of winter games pitches. One shortcoming is that it cannot be used until a reasonably intact turf cover has been restored.

**Table 5.7.** Air-filled porosities (% v/v) in the field condition on the high wear (HW) and low wear (LW) areas of 22 winter games pitches constructed to different designs (0-20 mm depth). (After Adams *et al.*, 1992.)

| | Sand/soil rootzone | | Sand carpet | | Sand profile | |
|---|---|---|---|---|---|---|
| | HW | LW | HW | LW | HW | LW |
| Mean | 3.3 | 7.1 | 3.6 | 4.6 | 5.9 | 9.9 |
| Standard error | 0.9 | 1.0 | 0.6 | 0.7 | 1.4 | 1.8 |

Finally, it has been assumed by many that the use of the most appropriate rootzone materials will avoid the problems of poor aeration inherent in less sandy media. In the case of winter games pitches this is incorrect both theoretically (see Chapter 1) and practically when usage is intensive. A national survey of winter games pitches in the UK with sand-dominant rootzones indicated that in order to ensure that air-filled porosities greater than 5% are regained in areas of high wear, soil loosening must be carried out at the end of the playing season (Table 5.7). Pitches were examined towards the end of the season and in the field condition only sand profile designs had mean air-filled porosities in high wear areas in excess of 5%. Even in this category of design a majority of pitches had air-filled porosities less than 5% (Adams *et al.*, 1992).

## REFERENCES

Adams, W.A. (1982) Soils under physical stress. In: Johnson, D.B. (ed.) *Welsh Soils Discussion Group Annual Report No. 23*. University College of North Wales, Bangor, pp. 41–51.

Adams, W.A. (1983) Feeding turf on sand and soil. In: Shildrick, J.P. (ed.) *Proceedings of the 2nd National Turfgrass Conference*. National Turfgrass Council, pp. 87–93.

Adams, W.A. and Saxon, C. (1979) The occurrence and control of thatch in sportsturf. *Rasen Turf Gazon* 3, 76–83.

Adams, W.A., Gibbs, R.J., Baker, S.W. and Lance C.D. (1992) Making the most of natural turf pitches. A national survey of winter games pitches with high quality drainage design. *Natural Turf Pitch Prototypes Advisory Panel Report No. 10*. The Sports Council, London, 20 pp.

Baker, S.W. (1983) Rate of soil replacement by a combined hollow tining and top dressing programme. *Journal of the Sports Turf Research Institute* 59, 146–147.

Baldwin, N.A. (1990) *Turfgrass Pests and Diseases*, 3rd edn. The Sports Turf Research Institute, Bingley.

BS3969 (1990) *Recommendations for Turf for General Purposes*. British Standards Institution, Milton Keynes.

Davies, A., Adams, W.A. and Wilman, D. (1989) Soil compaction in permanent pasture and its amelioration by slitting. *Journal of the Agricultural Society Cambridge* 113, 189–197.

Goss, R.L., Cook, T.W., Brauen, S.E. and Orton, S.P. (1977) Effect of repeated applications of Bensulide and tricalcium arsenate on the control of annual bluegrass and on the quality of highland colonial bentgrass putting green turf. In: Beard, J.B. (ed.) *Proceedings of the 3rd International Turfgrass Research Conference*. American Society of Agronomy, Madison, Wisconsin, pp. 247–256.

Johnston, D.T. (1988) Establishment of aminotriazole varieties on golf greens. In: Gibbs, R.J. and Adams, W.A. (eds) *Proceedings of the 6th Discussion Meeting of Amenity Grass Research*. University College of Wales, Aberystwyth, pp. 148–154.

Lewis, W.M. and Dipaola, J.M. (1989) Ethofumesate for *Poa annua* control of bentgrass. In: Takatoh, H. (ed.) *Proceedings of the 6th International Turfgrass Research Conference*. Tokyo, Japan, pp. 303–306.

Madison, J.H., Davies, W.B. and Paul, J.L. (1974) *An Alternative Method of Greens Management*. California Golf Course Superintendent Institute.

MAFF (1991) *Fertilizer Recommendations*, 2nd impression. Ministry of Agriculture, Fisheries and Food; Agricultural Development and Advisory Service, Bulletin 209. HMSO, London.

Marsh, A.W. (1969) Soil water – irrigation and drainage. In: Hanson, A.A. and Juska, F.V. (eds) *Turfgrass Science*. American Society of Agronomy Monograph No. 14, Madison, Winconsin, pp. 151–186.

Rieke, P.E. and Murphy, J.A. (1989) Advances in turf cultivation. In: Takatoh, H. (ed.) *Proceedings of the 6th International Turfgrass Research Conference*. Tokyo, Japan, pp. 49–54.

Russell, R. Scott (1977) *Plant Root Systems*. McGraw-Hill (UK) Ltd, Maidenhead, Berks, 298 pp.

Shildrick, J.P. (1985) Thatch: A review with special reference to UK golf courses. *Journal of the Sports Turf Research Institute* 61, 8–25.

Smith, J.D., Jackson, N. and Woolhouse, A.R. (1989) *Fungal Diseases of Amenity Turf Grasses*, 3rd edn. E. and F.N. Spon, London, 401 pp.

Stewart, V.I. and Adams, W.A. (eds) (1970) *Lectures on Sportsfield Construction and Management*, 1st edn. Sisis Equipment, Macclesfield.

Tisdale, S.L., Nelson, W.L. and Beaton, J.D. (1984) *Soil Fertility and Fertilizers*, 4th edn. Millan Publishing Company, New York, 754 pp.

Troughton, A. (1957) *The Underground Organs of Herbage Grasses*. Bulletin, Commonwealth Bureau of Pastures and Field Crops No. 44, Hurley.

Turgeon, A.J. (1974) Annual bluegrass control with herbicides in cool season turfgrasses. In: Roberts, E.C. (ed.) *Proceedings of the 2nd International Turfgrass Research Conference*. American Society of Agronomy, Madison, Wisconsin, pp. 382–389.

Upchurch, R.P. (1972) Herbicides and plant growth regulators. In: Goring, C.A.I. and Hamaker, J.W. (eds) *Organic Chemicals in the Soil Environment*, Vol. 2. Marcel Dekker Inc., New York, pp. 443–512.

USGA Green Section Staff (1993) USGA recommendations for a method of putting green construction. *USGA Green Section Record* Mar/Apr, 1–3.

**BULBOUS BUTTERCUP**

**CREEPING BUTTERCUP**

**COMMON CROWFOOT**

**LESSER CELANDINE**

**PEARLWORT**

**MOUSE-EAR CHICKWEED**

# APPENDIX: WEEDS AND HERBS IN MOWN TURF

## Ranunculaceae

1. **BULBOUS BUTTERCUP** (*Ranunculus bulbosus* L.)
   PERENNIAL. Long-stalked, rosette-like hairy plant,
   without stolons. Leaves made up of three lobes,
   each divided into three segments; lower lobes close
   together, upper one stalked. Forms corm-like growth
   at base of stem. Earliest flowering of the three
   common buttercups. Favours dry soils.

2. **CREEPING BUTTERCUP** (*Ranunculus repens* L.)
   PERENNIAL. Stoloniferous creeping plant forming
   self-rooted rosettes. Long, hairy leaf stalks. Leaves
   triangular in outline, slightly hairy; made up of three
   lobes, each acutely divided into three segments,
   central lobe long-stalked projecting beyond others.
   Long stout roots. Favours moist situations and heavy
   soils.

3. **COMMON CROWFOOT** (*Ranunculus acris* L.)
   PERENNIAL. Long-stalked, semi-creeping, softly
   hairy plant without stolons. Leaves mostly stalked,
   divided into two to seven lobes and again into three
   toothed rather pointed segments. Short rootstock.
   Favours moist situations.

4. **LESSER CELANDINE** (*Ranunculus ficaria* L.)
   PERENNIAL. Long-stalked, rosette-like plant with
   fibrous roots producing oblong or cylindrical annually
   renewed tubers. Fleshy leaves bluntly angled;
   hairless, dark green and shiny. Flowers very early
   spring. Favours moist or shaded situations.

## Caryophyllaceae

5. **PEARLWORT** (*Sagina procumbens* L.)
   PERENNIAL. Small tufted plant of stoloniferous habit,
   forming dense colonies of connected rosettes. Very
   narrow leaves. Seeds prolifically. No particular soil
   preference. Often associated with too close mowing.

6. **MOUSE-EAR CHICKWEED** (*Cerastium* spp.)
   ANNUALS OR PERENNIALS. A variable group of
   species forming dense mats of stems which lie close
   to the ground. The stalkless, oval, somewhat hairy
   leaves are arranged in pairs on the stems which in
   some forms root at the nodes. Common on most
   types of soil. May be reduced by dressings of
   proprietary lawn sands, and soil improvement by
   fertilizers.

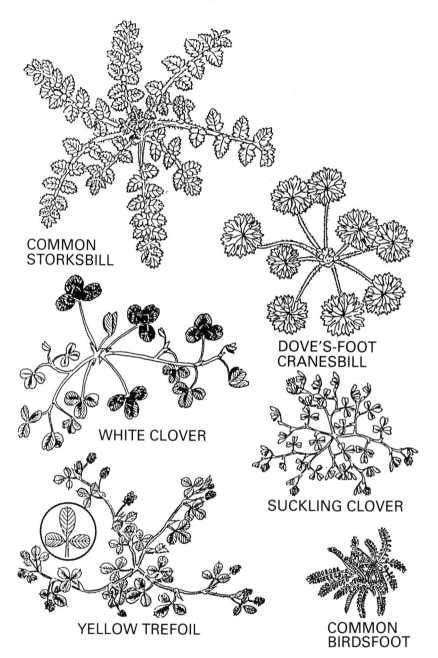

COMMON
STORKSBILL

DOVE'S-FOOT
CRANESBILL

WHITE CLOVER

SUCKLING CLOVER

YELLOW TREFOIL

COMMON
BIRDSFOOT

## Geraniaceae

7. **DOVE'S-FOOT CRANESBILL** (*Geranium molle* L.)
ANNUAL. A long-stalked rosette-like plant with tap
root. Stalks hairy and of a reddish tinge at base.
Leaves rounded and divided into seven to 11
wedge-shaped lobes, each having three to five
segments. Flowers purple. Seed pods beak-shaped.
Common on light dry soils.

8. **COMMON STORKSBILL** (*Erodium cicutarium* L'hér.)
ANNUAL OR BIENNIAL. A hairy plant with rather
fern-like leaves prostrate on the surface. Numerous
short-spreading stems produced. Thick tap root.
Flowers pink or purple. Seed pods beak-shaped.
Common on light sandy soils.

## Papilionaceae

9. **WHITE CLOVER** (*Trifolium repens* L.)
PERENNIAL. A hairless creeping plant rooting at the
nodes. Leaves made up of three toothed leaflets
often with pale inverted V-shaped marking on each.
Stipules at leaf bases small and oblong. Flowers
white or pinkish. Common on most soils.

10. **SUCKLING CLOVER** (*Trifolium dubium* Sibth.)
ANNUAL. Slender procumbent sometimes hairy
plant with short branched stems. Leaves small, often
folded, made up of three leaflets with central one on
longer stem. Stipules broad and pointed. Small heads
of minute yellow flowers; prolific seeder. Common,
especially on dry soils.

11. **YELLOW TREFOIL** (*Medicago lupulina* L.)
ANNUAL. A procumbent downy plant of spreading
habit, similar to suckling clover, though often has
small projection at apex of leaflets. Small heads of
bright yellow flowers. Seed pods black when ripe.
Common on most soils.

12. **COMMON BIRDSFOOT** (*Ornithopus perpusillus* L.)
ANNUAL. Small hairy prostrate plant, the leaves of
which consist of five to ten pairs of oblong or oval
leaflets. Minute white flowers, veined with red.
Common on dry sandy soils.

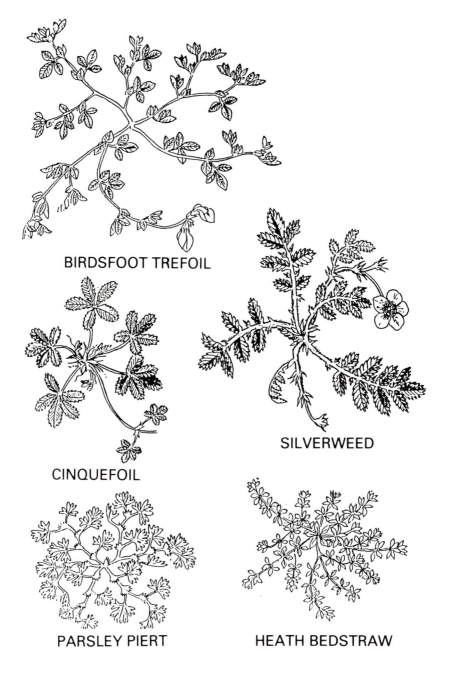

BIRDSFOOT TREFOIL

SILVERWEED

CINQUEFOIL

PARSLEY PIERT

HEATH BEDSTRAW

**Papilionaceae**–*continued*

13. **BIRDSFOOT TREFOIL** (*Lotus corniculatus* L.)
    PERENNIAL. A tap-rooted hairless plant producing
    trailing stems on the surface. Leaves consist of five
    leaflets, the lower two of which are 'stipule leaves';
    this distinguishes it from the true trefoils. Flowers
    bright yellow often tinged reddish brown. Common
    on most soils.

## Rosaceae

14. **CINQUEFOIL** (*Potentilla reptans* L.)
    PERENNIAL. A strawberry-like plant with slender
    prostrate stems often rooting at the nodes. Leaves
    made up of five oval or wedge-shaped, toothed
    leaflets. A pair of stipules is present at base of each
    stalk. Parent plant has long woody tap root. Common
    on many soil types and situations.

15. **SILVERWEED** (*Potentilla anserina* L.)
    PERENNIAL. A silky plant with fern-like leaves, the
    under-sides being silvery white. Produces long,
    creeping, rooting and flowering stolons. Leaves have
    seven to twelve pairs of main leaflets alternating with
    smaller ones, all oval and deeply toothed. Flowers
    produced singly on long stalks. Favours damp heavy
    soils.

16. **PARSLEY PIERT** (*Alchemilla arvensis* Scop.)
    ANNUAL. Small pale green much-branched plant.
    Leaves fan shaped, divided into three lobes each
    with two or more teeth. Flowers minute, greenish
    colour. Common on dry light soils. Prolific seeder.
    May be reduced by lawn sand treatment and general
    soil improvement by fertilizers.

## Rubiaceae

17. **HEATH BEDSTRAW** (*Galium saxatile* L.)
    PERENNIAL. Small much-branched plant with
    prostrate shoots. Stems quadrangular, leaves in
    whorls of four to six, lower ones oval, terminating in
    a point, upper ones rather narrow. Flowers small and
    white. Favours heaths, moors and woods, especially
    where soil is acid.

DANDELION

SMOOTH HAWK'S-BEARD

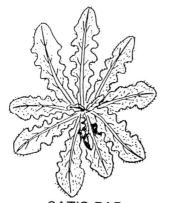

CAT'S-EAR

COMMON HAWKBIT

AUTUMN HAWKBIT

MOUSE-EAR HAWKWEED

## Compositae

18. **DANDELION** (*Taraxacum officinale* Weber.)
PERENNIAL. A tufted plant varying considerably in size, usually smooth but occasionally downy or even hairy. Leaves deeply toothed with triangular lobes pointing downwards. Strong tap root. Common on all soils.

19. **SMOOTH HAWK'S-BEARD** (*Crepis capillaris* Wallr.)
ANNUAL OR BIENNIAL. An erect branched hairless plant. Leaves narrow, toothed with triangular or narrow short lobes pointing up or down. Tap root. Prefers dry situations.

20. **CAT'S-EAR** (*Hypochaeris radicata* L.)
PERENNIAL. An erect branching plant forming a rosette of broad oblong leaves, narrowing gradually to broad stalk-like base. Leaves have rounded lobes at right angles to mid-rib. Whole plant covered with short stiff hairs. Common on most soils.

21. **COMMON HAWKBIT** (*Leontodon hispidus* L.)
PERENNIAL. A long narrow-leaved plant covered with short stiff hairs (hispid). Leaves toothed but not deeply so, lobes pointing downwards. Common, but favours chalk soils.

22. **AUTUMN HAWKBIT** (*Leontodon autumnalis* L.)
PERENNIAL. A long narrow-leaved plant usually without hairs or only slightly hairy. Leaves toothed with narrow lobes pointing upwards towards leaf apex. Common on most soils.

23. **MOUSE-EAR HAWKWEED** (*Hieraceum pilosella* L.)
PERENNIAL. A creeping plant with slender rhizomes and numerous long stolons which produce rosettes of oblong or oval leaves. Leaves with long hairs on upper surfaces and white down on undersides. Common on most soils, favouring dry situations.

RAGWORT

DAISY

YARROW

LESSER KNAPWEED

COMMON CHAMOMILE

## Compositae — *continued*

24. **RAGWORT** (*Senecio jacoboea* L.)
PERENNIAL. A short, erect, tufted plant forming a
rosette. Leaves coarsely toothed with up to six pairs
of lobes; terminal lobe pronounced and blunt; the
whole of dark green colour and firm crisp texture.
Roots short, thick and white. Common on most soils.

25. **DAISY** (*Bellis perennis* L.)
PERENNIAL. A short, erect plant forming rosettes
of oval leaves, which are notched at the edges.
Leaves narrow abruptly into short broad stalk. Roots
stoutly fibrous. Common on most soils. Lawn sands
are usually effective but more than one application
may be necessary.

26. **LESSER KNAPWEED** (*Centaurea nigra* L.)
PERENNIAL. A branched plant with tufted clusters
of oblong rather narrow leaves. The lower leaves
almost entire, but may have few teeth at base. Plant
roughly hairy to the touch. Common in pastures and
waste places. Close mowing will usually keep it in
check.

27. **YARROW** (*Achillea millefolium* L.)
PERENNIAL. A strongly creeping plant, spreading
by well-developed underground stems. Leaves fern-
like, finely divided, sometimes hairy. Common on
most soils and extremely drought resistant.

28. **COMMON CHAMOMILE** (*Anthemis nobilis* L.)
PERENNIAL. Pleasantly scented, with short,
much-branched creeping stock. Stems hairy. Leaves
fern-like with short narrow segments, not so finely
divided as yarrow; sparsely hairy. Favours sandy
soils. It is sometimes used in ornamental lawns but is
objected to in putting and bowling greens.

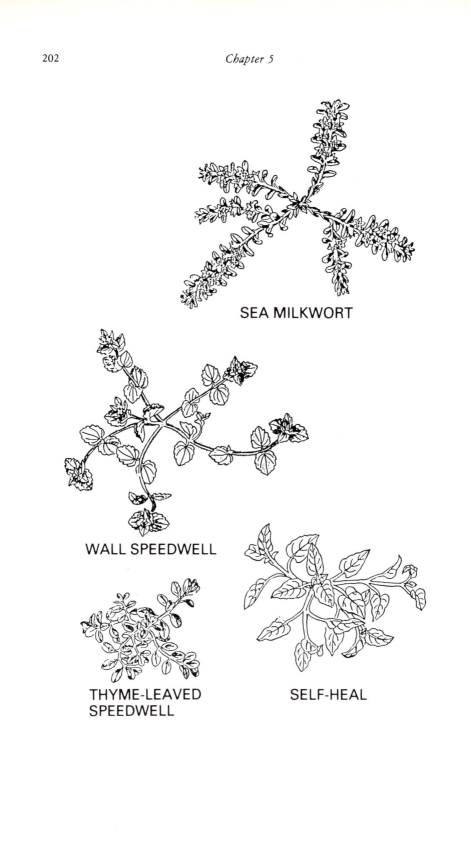

SEA MILKWORT

WALL SPEEDWELL

THYME-LEAVED
SPEEDWELL

SELF-HEAL

## Primulaceae

**29.** **SEA MILKWORT** (*Glaux maritima* L.)
PERENNIAL.   A creeping prostrate plant with small opposite oval leaves and tiny pinkish flowers. The stems die back during the winter and fresh growth is made in the spring. Common on sea sands and salt marshes. Occurs in bowling greens which have been laid with sea-washed turf.

## Schrophulariaceae

**30.** **WALL SPEEDWELL** (*Veronica arvensis* L.)
ANNUAL.   A normally erect, hairy plant of variable size, branched at base and becoming prostrate under close mowing. Leaves almost stalkless, of triangular/oval shape, coarsely toothed. Flowers blue or whitish. Favours dry soils.

**31.** **THYME-LEAVED SPEEDWELL** (*Veronica serpyllifolia* L.)
PERENNIAL.   A short creeping plant rooting at the nodes. Lower leaves oval or oblong, entire; upper slightly toothed; all hairless and of pale green colour. Flowers pale blue. Prefers moist situations.

(The *Veronica* spp. encountered in turf are all difficult to eliminate. Lawn sands may give some measure of control.)

## Labiatae

**32.** **SELF-HEAL** (*Prunella vulgaris* L.)
PERENNIAL.   A slightly hairy plant with short rhizomes. Stems procumbent, rooting at nodes. Leaves stalked, oval or spade-shaped and deeply veined. Flowers purple. Favours heavy soils and moist situations.

HOARY PLANTAIN

RIBWORT or NARROW-LEAVED
PLANTAIN

STARWEED or
BUCK'S-HORN
PLANTAIN

GREATER or BROAD-LEAVED
PLANTAIN

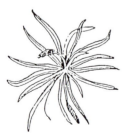

SEA PLANTAIN                    SEA PINK or THRIFT

## Plantaginaceae

33. **RIBWORT or NARROW-LEAVED PLANTAIN**
(*Plantago lanceolata* L.)
PERENNIAL. Forms rosettes of narrow, slightly notched leaves having from three to five well-defined veins. A tuft of white woolly hair is present in the crown of the plant. Common on most soils.

34. **HOARY PLANTAIN** (*Plantago media* L.)
PERENNIAL. Frequently mistaken for *Plantago lanceolata*, but can be distinguished by its rounded leaves, which are covered with short down giving the plant a dusty appearance. Common on chalk and alkaline soils.

35. **GREATER or BROAD-LEAVED PLANTAIN** (*Plantago major* L.)
PERENNIAL. Has broad oval slightly toothed leaves with five to nine veins carried on long stalks. The leaves die off during winter. Common, especially on gravelly and heavy soils.

36. **STARWEED or BUCK'S-HORN PLANTAIN** (*Plantago coronopus* L.)
ANNUAL or PERENNIAL. The deeply divided narrow leaves, which are supposed to resemble the antlers of a deer, should prove sufficient to identify this plantain. It is common on sandy soils and is frequently found on golf courses and bowling greens, especially near the sea.

37. **SEA PLANTAIN** (*Plantago maritima* L.)
PERENNIAL. The narrow strap-shaped leaves are similar to those of the sea pink, but are rather more fleshy. Common on seashores and salt marshes. Frequently occurs with sea pink on bowling greens made from sea-washed turf.

## Plumbaginaceae

38. **SEA PINK or THRIFT** (*Armeria maritima* Willd.)
PERENNIAL. Forms tufts with narrow leaves, similar to those of *Plantago maritima*. The globular heads of pinkish flowers are seldom produced where plants are growing in cultivated turf. Common on sandy seashores and salt marshes.

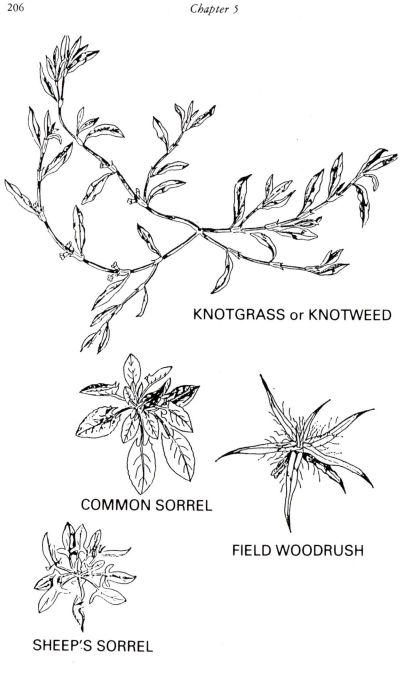

KNOTGRASS or KNOTWEED

COMMON SORREL

FIELD WOODRUSH

SHEEP'S SORREL

## Polygonaceae

39. **KNOTGRASS or KNOTWEED** (*Polygonium aviculare* L.)
ANNUAL. This plant produces an abundance of long wiry prostrate stems with clusters of small reddish flowers. The white-toothed stipules at the base of the leaves are a prominent feature. Common on most soils. Not essentially a grassland weed, but quickly colonizes bare areas in cultivated turf, especially the goal-mouths of football pitches during the off-season.

40. **COMMON SORREL** (*Rumex acetosa* L.)
PERENNIAL. A slightly creeping plant having arrowhead-shaped leaves with pointed lobes which tend to turn in towards the stems. Common on moist, acid soils.

41. **SHEEP'S SORREL** (*Rumex acetosella* L.)
PERENNIAL. A frequent plant on poor, light, dry soils, and a fairly reliable indicator of acid conditions. The leaves differ from common sorrel in having a well-defined waist and prominent lobes which stand out at right angles from the leaf. Numerous underground creeping stems are produced. The plant often assumes a reddish tinge, especially when in flower.

(Sorrels are difficult to control because of their association with acidity of soil.)

## Juncaceae

42. **FIELD WOODRUSH** (*Luzula campestris* Br.)
PERENNIAL. Much like a grass in general appearance. The leaves are fringed with long white hairs, which are also conspicuous in the crown of the plants. Produces short underground creeping stems. Distinguished from the grasses by its leaves being grouped in rows of three on the stem, instead of two.

# SOCCER AND RUGBY GROUNDS

## INTRODUCTION

Winter games pitches are used primarily for sports involving kicking, throwing or carrying a ball into an opposing team's territory. In Britain, the principle of this type of sport is best reflected in the games of soccer (Association Football) and rugby football (both League and Union). However there are also a variety of other sports played around the world in addition to soccer and rugby which have similar playing characteristics, notably American football, Gaelic football and Australian Rules football. All these sports share a common development which goes back as far as the Ancient Greeks and Chinese who were both believed to have independently developed some form of game involving kicking or carrying a ball into an opponent's territory.

Under the same general category of winter games are the sports of hockey, lacrosse and the Celtic game of hurling, the last-mentioned being closely associated with early forms of hockey. This second group of sports are all stick and ball games which, like football, have descended from the earliest civilizations, except for lacrosse which is thought to have been developed by the North American Indians.

The common feature of all winter sports on natural turf is that the games are played at a time of year when the soil is often at or wetter than field capacity and when grass growth is minimal. In the absence of good surface drainage, games affected by wet weather either have to be postponed or are played when the surface is excessively soft or wet. All too often grass cover is seriously eroded and quagmire conditions develop in the high wear areas of a pitch. Overall this action places extreme pressure on the turfgrass plants to provide a satisfactory quality of play. Indeed, for some sports, notably hockey, the difficulties in maintaining a high quality natural turf surface in winter have resulted in the adoption of synthetic turf surfaces for a large part of club hockey and for all major International Hockey Federation tournaments. In the early 1980s it looked as though artificial surfaces would make significant inroads into the traditional surfaces favoured by soccer players, but the final removal of the remaining three synthetic turf surfaces from professional soccer clubs (Luton Town, Oldham Athletic and Preston North End) in Britain is a credit to the faith of football clubs and the English Football League in modern turf culture. Similarly the widespread use of synthetic turf surfaces for American football is being challenged in favour of a return to natural turf, primarily as a result of a large number of injuries sustained by players.

The purpose of this chapter is to examine the relationship between two of the most popular winter sports, soccer and rugby football, and the turfgrass medium. It is with these two sports, particularly soccer, that British turfgrass scientists and others overseas have been active since the late 1960s in developing sportsturf construction and management systems that are able to provide high quality surfaces sustained under intensive usage. Although this chapter concentrates primarily on only two winter sports, the principles set out have relevance for all types of sport played in winter on natural turf.

## NATURE OF THE GAMES

## Historical background

Although it is thought that the Romans were responsible for bringing the game of football in its broadest sense to Britain, it was not until the 11th or 12th century that its popularity spread. It was then only to be banned by various monarchal proclamations in the 13th century, firstly because of the 'evil' associated with the game, and later because the sport tended to rival archery, a sport with considerable military value, unlike football. For the next 600 years or so, football developed very slowly and continued to be a brutal game of little order and unwritten rules, involving competition between mobs of opponents rather than teams of players attempting to move a 'roundish' object forward by kicking.

It was not until the Industrial Revolution of the late 18th and 19th centuries that football took on a shape that would be recognized today as a sort of cross between soccer and rugby. This was a development that occurred principally because of the mass movement of people from rural areas to high density urban housing. At the same time, the growth of private education for the priveleged few, often for those with money from the Industrial Revolution, resulted in the concurrent development of the game.

However, although these 'public' schools all played football, the game varied according to the tradition and custom of the school, which meant that the game continued to develop in a rather haphazard way. A significant development was the birth of rugby football at Rugby school, which allowed the ball to be picked up and run with, unlike other schools such as Eton, Charterhouse and Westminster which concentrated on the 'dribbling' game.

The difference between these two dominant types of the game caused a clash of rules when the sports were taken up by undergraduates at Oxford and Cambridge. Hockey was then the most popular winter sport at the universities. It was not until 1863 that a clear distinction between soccer and rugby became evident by the setting down of the Cambridge University Football Rules and the formation of the Football Association. Within the space of a few years the 'rugby' element of the game disappeared, and the game of soccer has remained basically the same ever since.

The now separated game of rugby football obtained its own governing body, the Rugby Football Union, in 1871 to coordinate the variety of rules still being used for the game and to compete against the popularity of soccer. Unlike soccer, the game of rugby football remained a strictly amateur game which caused particular problems amongst certain clubs who remained exasperated at repeated refusals by the ruling body to allow them to compensate players for time off playing the game. In August 1895, 21 clubs in northern Britain decided to break away and form their own Rugby Union, later to be called the Rugby Football League in 1922. This game broke away from the traditional Rugby Union concept by setting out to cater more for spectators, with the introduction of rules that promoted a better continuity of play.

Soccer and Rugby Union football are now international games. To a lesser extent, the same is true of Rugby League football although this game is played mainly in the UK, France, Australia and New Zealand.

## Pattern of play

Rugby Union and Rugby League football are games which have a similar appearance to American football. This is not surprising considering that American football dates back to 1609 when English colonists developed the game and later adapted it from traditional English Rugby Union rules in the late 19th century. Players of all the three games adopt 'front line' positions prior to commencement of each play, use throwing and running as the primary methods of moving the ball, only kick the ball on a few specific occasions, employ body tackling as a major skill and are involved in stop and start passages of play lasting between 5 and 90 seconds (Douge, 1988). These games take place on areas of turfgrass of typically 0.5 to 1 hectare in size (Table 6.1).

In contrast, the appearance of soccer is very similar to Australian Rules

**Table 6.1.** Size of pitch, length of matches and average ground space per player for different types of winter games played on natural turf.

| Sport | Number of players per game | Typical area of full-size pitch (m$^2$) | Length of one match (min) | Area of turf per player (m$^2$) |
|---|---|---|---|---|
| Rugby Union | 30 | 9936[a] | 80 | 230[b] |
| Rugby League | 26 | 8296[a] | 80 | 262[b] |
| American football | 22 | 5348[a] | 60 | 203[b] |
| Soccer | 22 | 7700 | 90 | 350 |
| Gaelic football | 30 | 11234 | 60 | 374 |
| Australian Rules football | 36 | 25000 | 100 | 694 |
| Hockey | 22 | 5027 | 70 | 229 |
| Lacrosse (mens) | 20 | 6400 | 90 | 320 |
| Hurling | 30 | 11234 | 60-80 | 374 |

[a] Including dead ball area.
[b] Excluding dead ball area.

football and Gaelic football. Players of these games adopt positions which are dispersed over the playing surface, use kicking as the primary method for moving the ball, use limited body tackling and are involved in relatively continuous play (Douge, 1988). With the exception of Australian Rules football, both soccer and Gaelic football, the latter of which is essentially a compromise between soccer and Rugby Union football, need similar sized playing areas to rugby (Table 6.1). Australian Rules football, which was developed by two keen cricketers in 1858, is played on large oval-shaped grounds which are typically controlled by cricket clubs (e.g. the Melbourne Cricket Ground). Although popular in Australia, the game has never seriously spread beyond that country. The dimensions and line markings on soccer and Rugby Union pitches are illustrated in Fig. 6.1.

## Type and location of wear

The total number of players in football-related games varies from 22 for soccer (i.e. 11 players per team) to 36 for Australian Rules football (Table 6.1). In terms of the intensity of players per square metre of turf, it would initially appear that there is plenty of space per player (Table 6.1). However, the nature of soccer is such that about 70% of the game takes place on about 30% of the pitch, leading to a characteristic 'diamond' pattern of wear from goal to

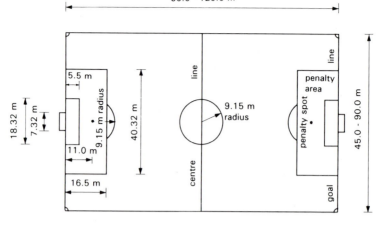

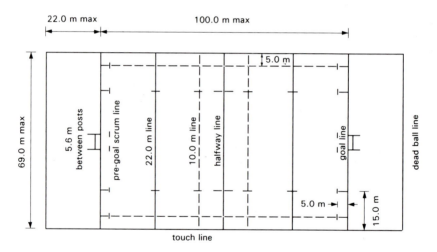

**Fig. 6.1.** Pitch dimensions and layout for soccer (above) and Rugby Union football (below).

goal (Plate 5). Extreme wear takes place in the goal areas which often become devoid of grass before the rest of the pitch, due to the intense concentration of activity around them, and it is not unusual for the entire centre strip of a pitch to be completely devoid of grass by the end of a season. The effect of this intense wear on soil physical properties is illustrated later in this chapter.

Rugby-type wear contrasts with the game of soccer. In rugby, the most intensive wear occurs along the length of the pitch 5–15 m in from the touchlines (particularly those in front of grandstands) because of lineouts and scrummages. Wear is more spread out than for soccer as greater utilization is made of the pitch during the course of a game.

It has been estimated that in the game of soccer, a player will typically travel 10 km (Asami *et al.*, 1988). Baker (1991a) took this a stage further and estimated that if the average stride is assumed to be 0.85 m, this results in about 12,000 foot imprints into the soil per player per game. With the trampling of all the players in both teams added together, the total number of foot imprints during a match comes to approximately 250,000 (including the referee), excluding the additional effect of two goalkeepers who cause immense damage to goal areas by their continuous, start/stop movement. On average, this number of foot imprints is equivalent to 42 per m$^2$ for a modest 6000 m$^2$ soccer pitch which, if multiplied by the number of games per season (typically 100), is an average of 4200 foot imprints per m$^2$ of turf over a period of approximately 35 weeks. It is therefore not surprising that a major component of wear associated with winter sports is that of soil compaction.

In addition to compaction, acceleration, deceleration and sliding cause tearing of the turf and smearing of soil. Direct effects of wear result in the gradual loss of grass through compression and tearing of the turf by the player's studded footwear (Chapter 2). There is also a direct effect of wear on soil, causing compaction and smearing. As the soil surface becomes more exposed, particle migration is likely to occur, particularly in wet weather. The net effect is the creation of a thin crust of fine particles at the immediate soil surface which reduces water infiltration rate making the soil even more susceptible to compaction and smearing as the season progresses.

Increased compaction leads to poorer soil physical conditions for turfgrass growth, particularly in terms of poor aeration and drainage (see The potential for poor drainage in amenity soils – Chapter 3). Ironically, in Britain and elsewhere in the world (e.g. New Zealand, France), the major sports of soccer and rugby are played at a time of year when rainfall usually exceeds the amount of water which is lost by evapotranspiration. Thus play generally takes place in conditions under which maximum soil damage will occur.

Research to address the demand for better quality surfaces brought about not least by the introduction of colour television, has resulted in an extensive range of construction and maintenance options now being available.

# Grass species

The main criterion for selecting turfgrasses for soccer and rugby grounds is that of wear tolerance followed by other desirable attributes such as good winter colour (growth at low temperatures), compactness and recovery after wear. The need for the selection of cultivars that establish quickly after sowing is also a key requirement given the long playing seasons that now apply in soccer and rugby, which result in minimal time for renovation (often no more than 100 days).

Bryan and Adams (1971) carried out a detailed assessment of grass species persisting at the end of the playing season on eight First and Second Division English League soccer grounds. The only species making significant contributions to the sward were, in order of quantitative importance, annual meadowgrass (*Poa annua*), perennial ryegrass (*Lolium perenne*), common bentgrass (*Agrostis capillaris*) and rough stalked meadowgrass (*Poa trivialis*). Fine-leaved fescues (*Festuca* spp.) and timothy (*Phleum pratense*) were included in most seed mixtures used for overseeding but none survived in other than trace amounts. More recently, comprehensive studies have been carried out on wear tolerance by Canaway and his colleagues at the Sports Turf Research Institute (STRI) (Canaway, 1983). The wear tolerance of annual meadowgrass (or at least its potential to survive) was confirmed, but it had a susceptibility to being kicked out and did not produce the best quality of playing surface (Table 6.2). Perennial ryegrass and smooth stalked meadowgrass (*Poa pratensis*) showed very good wear tolerance and produced the best quality of playing surface. Despite the favourable attributes of the latter species (Shildrick, 1984), it is too slow to establish (Chapter 2), it is disease susceptible (Adams and Bryan, 1974) and does not persist on winter games pitches in the UK.

Of the fine grasses, there is little doubt about the inability of fine fescue and bent species (e.g. *Festuca rubra* subsp. *rubra*, *F. rubra* subsp. *commutata*, *Agrostis castellana* and *A. capillaris*) to tolerate football-type wear (Table 6.2). Despite this, fine fescue and bent species are still frequently included in seed mixtures for winter games pitches. There is some evidence to support the use of bentgrasses in these mixtures because of their good playing quality characteristics with limited wear. This is of no practical value in the case of intensively used winter games pitches, but is appropriate for hockey pitches or where less intensively used winter games areas also serve as cricket outfields during the summer months.

Turf-type tall fescue (*F. arundinacea*) cultivars which have been bred primarily in the USA, have little potential for winter games pitches in the UK. The difficulty with this species is that once worn it cannot, like smooth stalked meadowgrass, compete with perennial ryegrass in re-establishment. Large-leaved timothy (*Phleum pratense*) is included in some seed mixtures for winter games pitches, but it is less tolerant of wear and slower to recover than perennial ryegrass.

Although new and established cultivars of turfgrass species like perennial ryegrass are evaluated annually by STRI, an equally important consideration for the ability of turfgrasses to tolerate football-type wear is the effect of rootzone composition. There is little merit in sowing expensive turf-type perennial ryegrass into poorly drained and compacted soils, since the lack of adequate soil physical conditions will completely override the small

**Table 6.2.** Suitability of turfgrass species for football, based on wear tolerance, shear strength and ball rebound resilience (from Canaway, 1983).

| | Wear tolerance | Shear strength [a] | | Ball rebound [a] | | Overall suitability for football |
| --- | --- | --- | --- | --- | --- | --- |
| | | Before wear | After wear | Before wear | After wear | |
| *Agrostis castellana* | x | x | x | (_) | x | Totally unsuitable |
| *Festuca arundinacea* | (_) | _ | _ | _ | _ | Worth investigating in mixtures |
| *F. rubra* subsp. *commutata* | x | _ | _ | _ | (_) | Too little wear tolerance |
| *F. rubra* subsp. *rubra* | x | _ | _ | _ | (_) | Too little wear tolerance |
| *Lolium perenne* | _ | _ | _ | _ | _ | Suitable |
| *Phleum pratense* | (_) | _ | x | _ | _ | Worth investigating in mixtures |
| *Poa annua* | _ | x | x | x | x | Too little shear strength, low ball rebound |
| *Poa pratensis* | _ | _ | _ | (_) | _ | Suitable but lacking persistence in practice |

Ratings _ good, (_) intermediate, x poor. [a] Ratings for shear strength and ball rebound resilience after wear were derived mainly from sand data, as all species species were unsatisfactory on soil.

additional benefits in wear tolerance conferred by selecting the top-of-the-range cultivars.

The introduction of sand-dominant rootzones in the 1970s posed the question as to whether the same wear tolerance ranking of different turfgrass species applied equally to sand-based constructions as it did to soil-based ones. The experiments by Canaway (1983) established that the same ranking did apply, and all the species tested, survived. This result contrasted with the soil-based constructions in which only the more wear-tolerant species survived (annual meadowgrass, smooth stalked meadowgrass and perennial ryegrass).

In summary, in Britain when the use of winter games pitches is quite

intensive, say three games per week on average or more, no benefit is gained
by using any other species than perennial ryegrass. This also applies to overseeding during renovation. A tighter, more even turf is possible with bentgrasses but to retain them wear must be low. Fine fescues are least tolerant
of wear.


## PERFORMANCE CRITERIA


Wear tolerance of the turf is one amongst several factors which influence
the quality of the playing surface.

Rugby Union was the first winter game to make reference to performance
criteria in its rules. The laws of the game state that 'the field must be covered
with grass or where this is not available, with clay or sand, provided the surface
is not of a dangerous hardness'. In Rugby, Warwickshire, where the game
originated, the soil-based fields were often softened by rain so that they never
became particularly hard. Moreover it was accepted that rugby fields would
become slippery and muddy later on in the season. However, with the international spread of the game to countries including Australia, South Africa and
France, the fields would often be much drier than in Britain, and therefore
harder. In particular in Australia and New Zealand, parts of the field can
be very hard indeed because rugby is often played on cricket grounds where
play in the vicinity of the wicket ends produces a surface markedly different
from the rest of the field. Thus the range of surface conditions for which
the game was originally intended has changed considerably as a result of its
international popularity.

There are at least three general factors which should be considered in
relation to the performance criteria of natural turf winter games pitches,
particularly those used for soccer. These are:

1. Firmness of the surface influencing grip/traction and ball bounce.
2. Evenness of the surface affecting the run of the ball.
3. Percentage grass cover.

The last factor has a major aesthetic or cosmetic component, but turf also
has an important cushioning effect and turfgrass roots are the main stabilizers
of the surface of sand-dominant soils (Gibbs *et al.*, 1989).

The need for a series of objective measurements that provided a quantitative assessment of different sports surfaces was recognized some time ago
(e.g. Adams and Jones, 1979; Ward, 1983). Prior to this, sportsturf research
was dominated by agronomic measurements on turfgrasses and physical
measurements on rootzone characteristics with few assessments of factors
directly concerning the player. An exception occurred with the study of shear
strength which had been used for many years as an agronomist's tool for

measuring sod strength and which was easily adapted as an objective measurement of traction or grip.

It is perhaps ironical that it was the introduction of artificial surfaces by the English Football League in 1980 that was one of the catalysts for prompting research into the playing quality characteristics of natural turf. Player comments on synthetic soccer surfaces around this time were frequently derogatory because the surfaces were characterized by excessive ball bounce, ball roll and hardness ('ping-pong surfaces'). It quickly became apparent that a set of comparative measurements was needed to devise standards for a playing surface more or less matching an acceptable natural one. The response to this dilemma was met by the Sports Council and Football Association and resulted in a substantial report being produced in the mid 1980s on the playing quality characteristics of natural and synthetic soccer pitches (Winterbottom, 1985). However, at this stage of development of performance criteria, more was known about the playing quality of synthetic turf than of natural turf. Consequently in 1983 the Sports Council commissioned a four-year project at STRI to develop playing quality test methods and standards for natural turf, concentrating primarily on soccer pitches, but also on flat bowling greens (Bell and Holmes, 1988).

The removal of synthetic turf from the professional soccer scene in the UK has not diminished the need for standards for natural turf winter games pitches. Dramatic changes in local government organization and schools in the late 1980s which resulted in the introduction of Compulsory Competitive Tendering (CCT) for grounds maintenance and Local Management of Schools (LMS), have furthered the need for objective standards of performance in contract documents relating to all aspects of winter games pitch management.

The consequence of all this change is that the provision of existing, new or upgraded natural turf pitches requires very careful planning with an increased emphasis on obtaining 'value-for-money', i.e. providing a specified sustainable standard of surface most economically. Moreover those responsible for providing and maintaining these facilities now demand standards of performance that can be set and measured in order to justify payment for work carried out on a contract basis.

STRI have developed four main playing quality components for characterizing the game of soccer (Canaway *et al.*, 1990). These components can be split into the ball/surface characteristics of ball rebound resilience and ball roll, and the player/surface characteristics of surface traction/grip and hardness (Fig. 6.2). These playing quality measurements are now used in their own right in the evaluation of different turfgrass species and cultivars and of different construction systems. The component variables are not all independent and simplification would seem to be practicable. For example ball rebound resilience and surface hardness are interrelated, with a soft muddy surface giving low rebound resilience and a hard surface giving a high bounce.

Performance standards for the playing quality of natural turf soccer pitches

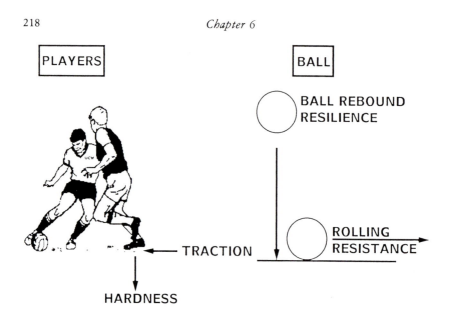

**Fig. 6.2.** Components of playing quality for soccer (reproduced with permission of STRI, Bingley).

currently being recommended have been derived by analysing player responses shortly after they had experienced playing on the surface that was measured. These standards which are inevitably somewhat subjective and liable to revision are shown in Table 6.3.

Although the above four components of playing quality enable assessment of the surface, there are other properties which can also be used as performance criteria. These are: surface evenness, for which standards of acceptability also exist (Table 6.3); percentage ground cover and species composition; and water infiltration rate which gives an indication of the likelihood of water ponding on the surface. In this respect, a poor draining surface can be considered to have an infiltration rate of less than $5 \, mm \, h^{-1}$, with good drainage being $10 \, mm \, h^{-1}$ or greater. For comparison, the International Hockey Federation recommendations for synthetic hockey fields stipulate an infiltration rate of more than $600 \, mm \, h^{-1}$.

In a project financed by the Sports Council and the Department for Education between 1986 and 1991, case studies over four years were used to evaluate the performance of natural turf winter games pitches constructed to different designs. One outcome of this project was mechanisms by which the inevitable variations in playing quality over different parts of a pitch could be accommodated within an assessment of the overall playing quality of a pitch (Table 6.4). In the system, Grade A pitches would meet the preferred standards for all tests in all sampling areas, Grade B pitches would meet the acceptable range for all tests and Grade C pitches would be allowed certain permitted

**Table 6.3.** Playing quality standards for soccer (from Canaway *et al.*, 1990).

| Quality | Minima | Maxima |
|---|---|---|
| Rebound resilience | | |
| Preferred | 20% | 50% |
| Acceptable | 15% | 55% |
| Distance rolled | | |
| Preferred | 3 m | 12 m |
| Acceptable | 2 m | 14 m |
| Traction | | |
| Preferred | 25 N m | – |
| Acceptable | 20 N m | – |
| Surface hardness | | |
| Preferred | 20 gravities | 80 gravities |
| Acceptable | 10 gravities | 100 gravities |
| Surface evenness | | |
| Preferred | – | 8 mm standard deviation |
| Acceptable | – | 10 mm standard deviation |

**Table 6.4.** Practical application of standards for football fields (from Baker and Gibbs, 1989).

| Grade | Standard |
|---|---|
| A | Preferred range for all tests |
| B | Acceptable range for all tests |
| C | Certain permitted deviations[a] from acceptable range |
| F | Fail |

[a] Field edges may be below standards for: rebound resilience, distance rolled, surface hardness. Other parts may exceed standards for: rebound resilience, distance rolled, surface hardness.

deviations from the standards. Those pitches unable to make even Grade C where considerable deviations were permitted, would not be up to standard for a playing surface for soccer.

Temporal variation in playing quality also has to be considered if pitches are to be monitored against specified performance standards. This is because moisture content (and hence antecedent rainfall) strongly influences certain playing quality characteristics like surface hardness and ball rebound resilience. Not only does the playing quality of an individual pitch vary on a temporal

**Table 6.5.** Maximum permitted rainfalls for test results to be valid (from Baker, 1991b).

| Grade of pitch | Maximum permitted rainfall in the time period before testing (mm) | | | |
|---|---|---|---|---|
| | 1 hour | 6 hours | 12 hours | 24 hours |
| Basic | 2 | 4 | 6 | 10 |
| Standard | 5 | 8 | 12 | 20 |
| High | 10 | 15 | 20 | 30 |

basis, for example, following a storm, but the persistence of poor playing conditions will vary between pitches of different inherent quality. Thus a well-drained pitch will recover more quickly than a badly drained one. It will also perform better under adverse weather conditions.

In order to overcome the difficulty, Baker (1991b) suggested that pitches should be placed in one of three primary grades based on inherent quality. These grades then determine the permitted antecedent conditions for quality assessments to be valid (Table 6.5). The two components, primary grades and performance tests provide a system for playing quality assessment. The practicability of using these in construction design or maintenance contracts still has to be tested fully. It is possible for example to stipulate that a pitch in the basic primary grade should satisfy performance grade A under specified weather conditions but in order to check this not only does the pitch have to be tested, but the antecedent conditions need to be monitored. Furthermore, stipulations on playing quality cannot be made without a statement on the usage sustained when these criteria must be met.

In summary, therefore, playing quality criteria for soccer pitches on natural turf have been established and they can be used to assess the success of construction or maintenance operations as well as helping to plan day-to-day usage. Using the two components of classification, pitches of the different primary grades (Basic, Standard and High) can be specified to satisfy the playing quality criteria (i.e. secondary grade A, B, C or F) in the pitch positions specified by Baker and Gibbs (1989) under prescribed weather conditions. Contract documents can be drawn up incorporating performance criteria but they have to include specifications on usage. Furthermore, the testing for compliance may be complex.

Although soccer is the winter game for which performance criteria have been developed, the desirable playing quality attributes for rugby and hockey have features in common with it. Surface traction is essential for hockey but, in common with a cricket outfield (see Chapter 8), smoothness of ball roll is the most important playing quality characteristic and this accounts for the widespread adoption of synthetic surfaces. In rugby, both surface traction and hardness are key factors but because of frequent body impact with the surface, hardness is more important with rugby than with soccer.

## OPTIONS FOR UPGRADING OR RECONSTRUCTION

Since the early 1970s considerable strides have been made in improving the drainage of winter games pitches using different types of construction and upgrading techniques varying in cost and complexity. Even simple operations such as regular topdressing with sand, which normally fall into the category of routine maintenance procedures, can cause substantial upgrading over a period of years (see Maintenance regimes).

The main improvement in the playing quality of winter games pitches has been gained by recognizing that the low permeability of the surface soil caused by compaction is usually the critical problem and that improved drainage requires designs and methodologies to ensure the rapid flow of surface water to drainage systems.

Improved playing quality through better drainage has meant fewer games postponed or cancelled and an increase in use by satisfying a latent demand. Many soccer and rugby clubs in Britain have an increasing membership and this requires either more pitches in total or upgraded pitches able to withstand a higher intensity of use. In the public sector, Local Authorities have been encouraged to get rid of 'surplus' land and the area perceived to be necessary for sportsfields has diminished as improvements in wear tolerance and playing quality have been demonstrated. Despite these improvements, increases in carrying capacities conferred by improved designs have not met the speculative predictions for natural turf made in the early 1980s (see Cost of provision).

## Construction types

There are several options for reconstruction or upgrading which are summarized in Table 6.6 in order of increasing cost (acceptable surface levels are assumed at this stage). Details of each design have been discussed in detail in Chapter 3. Specifications are available on drain and slit spacing and acceptable particle size distribution of sands, sand-dominant soils and permeable fill materials (Chapter 4). Some options can be attained in a step-wise order over time, but advice should be sought on the correct sequence to adopt. As a rough guideline, undrained and pipe drained constructions can be classified as 'Basic' grade pitches in the system proposed by Baker (1991b). Slit drained and sand/soil rootzone pitches can be classified as 'Standard' grade pitches, with pure sand rootzone constructions and sand/soil rootzone constructions with complementary slit drains falling into the 'High' grade category.

**Table 6.6.** Types of winter games pitch construction or upgrading.

| System description | Abbreviation |
| --- | --- |
| Pipe drains with permeable fill to, or close to, the surface | PD |
| Pipe drains as above with a shallow surface layer of sand (30-50 mm) | PDS |
| Sand/gravel slit drains with pipe drainage as above | SD |
| Sand/soil (sand-dominant) rootzone with pipe drainage as above or plus sand/gravel slit drains | SSR |
| Sand carpet (100 mm) with pipe drainage as above and linking sand/gravel slits | SC |
| Total profile construction with a sand or sand-dominant rootzone (e.g. suspended watertable pitch or enclosed system) | SWT |

## Implications of choice for reconstruction or upgrading

Three main benefits are gained by an increased investment in reconstruction or upgrading. These are:

1. A reduced risk of postponement.
2. Better playing conditions in wet weather.
3. Less damage to turf by play in adverse weather.

The minimum upgrading, which is pipe drainage, should not be expected to give significant improvement in any of these areas over an undrained pitch because the soil physical performance of this design is still dependent upon native soil and local site conditions (Baker and Canaway, 1991; Gibbs *et al.*, 1993a). Topsoils other than loamy sands and some sandy loams will fail to allow water to be transmitted to the pipe drains fast enough to prevent waterlogging (see Symptoms, causes and characteristics of deterioration). Pipe drainage must therefore be seen as a method of controlling the watertable in the subsoil (Chapter 3), and possibly as a precursor to a more elaborate drainage system (e.g. slit drainage) or physical soil conditioning treatment (e.g. mini-mole ploughing).

Supplementation with bypass drainage (slit drains) produces a major upgrading, because the slit drains facilitate rapid removal of water by typically producing a ten-fold increase in infiltration rate compared with undrained or pipe drained constructions (Gibbs, 1988). However, the improvement in soil physical conditions conferred by slit drainage is mainly for the benefit of the player rather than the turfgrass plant which still has to grow in what might be unfavourable native soil between the slit drains. Moreover, there is considerable variation in the amount of drainage improvement associated with slit drainage because of the variety of installation procedures and differing on-site conditions especially concerning topsoil and subsoil quality.

Despite the improvements gained by installing a slit drainage system, this type of drainage creates three new problems: firstly, surface unevenness always develops in clayey soils through their swell/shrink characteristics, and untrueness is likely to develop in any soil because of erosion of slits unprotected by grass. The second factor is that because slit function depends upon contact with the surface, frequent and substantial sand topdressings are required to protect them. Thirdly, an aspect which has been highlighted by a recent sequence of dry summers over much of England and Wales is the difficulty in establishing, re-establishing and maintaining grass over the lines of slit drains in the absence of irrigation.

Sand-dominant rootzones, sand carpet and total profile constructions do not have the inherent surface heterogeneity of slit drained pitches because their installation places progressively less dependence on native or indigenous soil conditions. These designs can therefore provide free-draining, true pitches of high quality providing, in many cases, play irrespective of the weather. Here again however there are important maintenance implications. All require regular sand topdressings to control surface organic matter accumulation and prevent deterioration in water infiltration rates. In sand carpet and sand/soil rootzone designs where the native soil is retained within the profile, sand topdressings are required to prevent surface sealing by silt and clay brought to the surface by casting earthworms (Fig. 6.3). In addition, with sand-dominant rootzones, sand topdressings help counteract the ill-effects of particle segregation which occurs near the surface when play takes place in severely wet weather. Growing good quality turf on sand-dominant soils also requires a much more refined approach to fertilizer use than with loamy soils. Indeed, for this reason alone, it is unwise to consider installing a total profile construction unless it can receive high quality maintenance.

Whilst pitches in the sand or sand-dominant category will retain grass with higher levels of use than softer and less well-drained pitches, surface stability can be lost especially towards the end of the season on areas of high wear if grass cover falls to less than about 15%. Indeed preventing loss of surface stability on sand or sand-dominant constructions is a major management consideration for these designs (see Symptoms, causes and characteristics of deterioration). It is important therefore not to let high quality drainage systems become a victim of their own success because there is a biological limit to the wear tolerance of turf for all natural turf surfaces. A further consideration, with sand carpet and total profile constructions, is that an irrigation system is vital for establishment, reliable recovery at the end of the season and for sustaining growth in the summer. For sand/soil rootzones an irrigation system is advisable but not always essential.

**Fig. 6.3.** Resin-impregnated section of a 22-year-old sand carpet construction at Queen's University, Belfast. As a result of intense earthworm activity and absence of sand topdressings, the original topsoil underneath the sand carpet has been deposited on the surface creating an 80 mm-deep layer of soft, easily poached soil. The scale is in units of millimetres.

## Typical problems of pitch playing quality

Poor quality may be due to inherent features such as excessive slope or surface unevenness or such inadequate drainage as to prevent virtually any play in the winter months. These apart, it is a consequence at least to some extent of the

**Fig. 6.4.** Quagmire conditions developed on this soil-based soccer pitch when its usage intensity was doubled to take on an extra team.

intensity of use. Intensity of use or level of demand is particularly important because many minor soccer and rugby club pitches have natural drainage which provides acceptable playing conditions only for the equivalent of a one-team club. The same pitches become a quagmire when the club expands to running two teams (Fig. 6.4). At the other end of the scale, the most sophisticated and well-maintained constructions will retain an acceptable grass cover up to quite high intensities of use.

## Slope

A slope aids surface drainage and makes the laying of pipe drains to a consistent gradient simple to achieve. Constructions should never be regraded to a flat surface unless watertable control is prescribed in the specifications. When a natural slope exists, the pitch should be orientated so that the maximum slope is across the pitch.

A slope much in excess of 1 in 50 is unlikely to be acceptable to a soccer or rugby club, but if regrading is decided upon, two factors need to be recognized. The first is that in order to reduce a slope which is severe enough to warrant alteration, the topsoil has to be stripped off, stored and replaced after regrading. Space must be available to stockpile topsoil and around 1200 tonnes of topsoil has to be moved twice for a typical soccer or rugby pitch in addition to the cut-and-fill of underlying soil. The second and most important factor is that, however carefully the work is carried out, natural drainage will be impaired. Special attention has to be paid to infield drainage and, in addition, the 'cut' of the regrading usually requires a cut-off drain to intercept seepage water from further up the slope.

## A *locally uneven surface*

When an agricultural field is taken over it is quite likely to be too uneven. In addition to this possibility, unevenness can develop for several reasons on a pitch which initially had a true surface. There are three main causes: subsidence, erosion or settling of slit drains and erosion through excess wear.

Subsidence is a common feature of pitches built on landfills of domestic waste and settlement can continue for decades. The built-up material is usually permeable to water but, as depressions develop, surface water runs into them and ponding occurs. Depressions are usually shallow but extend over several metres.

Sand or sand/gravel slit drains which are around 50 mm wide or wider are susceptible to erosion both before grass is established and when grass is destroyed by wear. Erosion is worst where the soil dries out approaching the end of the season. Clayey topsoils whose drainage has been improved by slit drains will also develop undulations with time because of the swell/shrink character of the soil.

When grass cover is destroyed, play causes erosion of two main types. One type of erosion is the creation of distinct hollows, typical of goal-mouths on soccer pitches and the centre spot on rugby pitches. The other type occurs on areas of high wear when turf is not completely destroyed. In this latter case, erosion leads to a knobbly surface where eroded soil is held within the residue of turf.

Hollows caused by subsidence or excessive wear should never be trued up simply by filling with sand or suitable soil because the soil within the depression will be compact and act as a drainage barrier. Compaction should always be relieved to at least 150 mm depth. Shallow depressions can be built up over a period of time without destroying the turf through applications of sand or sand-dominant soil at a maximum rate of about 10–15 mm per application. Applications can be made quite frequently (1–2 monthly intervals) during the growing season (April to September). If the depressions are too deep to raise by topdressings. there are two approaches. One is to strip off the turf, cultivate the underlying soil and raise it to a true level with sand-dominant soil before replacing the turf. Often depressions have no turf worth saving. In this case the soil can be broken up to relieve compaction and the level raised by mixing in appropriate sand or sandy soil. A rotovator must not be used. Establishment from seed should follow, and if problem areas are few, germination sheets or plastic covers will help retain moisture and speed up establishment.

Sand or sand/gravel slit drains of 50 mm width or greater should not be installed as the only means of surface drainage improvement on soils containing more than about 20% clay if a true surface is essential. Creating a sand-dominant rootzone to complement slit drainage avoids the progressive development of undulations.

When slit drainage is installed, it should be done very early in the close season to give the maximum period for grass establishment. Permeable fill within slit drains must be compacted. When use has been excessive and the turf over slit drains has been destroyed, the only practical means to reduce erosion is to ensure that sand is moist and firm when play takes place.

The reinstatement of soccer goal-mouths is often exacerbated by players and children 'kicking in' during the close season, so goal posts should be removed. When space is available the rotation of a pitch through 90 degrees is advantageous. In the last few years synthetic materials to reinforce natural turf have become available (see Chapter 4). The use of random fibres for reinforcement is expanding rapidly. The use of VHAF is best restricted to small areas of extreme wear such as soccer goal-mouths (Fig. 6.5).

The knobbly type of unevenness caused by excessive wear is the most difficult to rectify without cultivation. Cultivation of sand-dominant rootzones may be practical when the period for recovery is adequate, but it is not an option on pitches with slit drainage. When cultivation is not possible, some removal of local undulations can be achieved by the use of specialist equipment (e.g. a surface slotter) capable of lifting the rootzone without destroying the turf, in combination with rolling and sand topdressing.

### *Unsatisfactory surface conditions at a fixed level of use or deterioration through increasing usage*

The intensity of winter games pitch usage in the UK ranges from a one-team club when the pitch may be used on average for no more than one game per week, to sports centres and dual/joint use systems where the demand for use may exceed six games per week. Irrespective of the level of demand, the first requirement of a pitch is that games can be played when scheduled, on acceptable surface conditions in all but the most severe weather. For this to be achieved natural drainage must be good or it must be upgraded.

Increasing usage expresses itself in two main ways. The first is that it causes more soil compaction and often poaching in wet weather and the second is a greater wear of turf at a time when grass is growing slowly, if at all. Sand-dominant sportsturf soils remain firm when wet and do not poach easily. They also drain more quickly. Few pitches of this type occur naturally in the UK

*Chapter 6*

**Fig. 6.5.** Installation of the soil stabilizing material 'VHAF' in a soccer goal-mouth.

but there are some, for example Blackpool AFC and the St Helens ground at Swansea. Other soils by no means as sandy as these will tolerate a low level of use with no input of artificial drainage provided there is a clear policy to postpone in adverse weather.

Turf is less susceptible to being torn out of sandy soils, so these soils benefit in firmness, drainage and turf persistence. Grass is nevertheless susceptible to damage and an intensity of use is reached where, irrespective of the quality of turf maintenance, grass is lost, creating a whole new series of problems of pitch unevenness and the need for reseeding and possible recultivation. Winter games pitches constructed and maintained to the highest specification will not tolerate more than an average of around five games per week from the beginning of September until the end of April without serious loss of grass cover in areas of high wear. Use in excess of this amount creates a loose surface requiring frequent restoration and is little better than playing on a beach.

To summarize the complex considerations for upgrading and reconstruction, the following key points should be noted:

1. Many undrained natural pitches will tolerate a low level of use and remain in reasonable condition with a low maintenance input (one-team pitches).

2. When pipe drainage is needed to remove water from the subsoil this will usually have little if any effect on the wear tolerance or playing quality of the pitch. This applies when drains are as close as 4–5 m spacings with permeable fill close to the surface. Nevertheless, the installation of drains with permeable fill is normally necessary as a basic component of pitch improvement.

3. Sand/gravel slit drains facilitate the rapid removal of surface water to a compatible underdrainage system. They increase a soil's ability to accept rain and shorten the time needed to return pitches to good playing conditions following heavy rain. Continuity of slits to the surface is vital and this requires a large sand topdressing at the time of installation and at regular intervals subsequently. Slit drains can be as beneficial on a deteriorated sand ameliorated or sand topdressed soil as they are on unimproved soils. Slit drainage systems without topsoil upgrading should not be chosen as the sole improvement system if a major increase in demand for use is envisaged.

A decrease in efficiency of slit drainage systems is inevitable (see Symptoms, causes and characteristics of deterioration) and the rate of deterioration depends on usage level and quality of maintenance. Slit drainage systems on heavy soils lead to the development of corrugations which may not be acceptable on a soccer pitch.

4. Improvement schemes which involve the creation of a sand rootzone or a sand-dominant soil provide firmer playing conditions, improve the wear tolerance of turf and increase in water infiltration rate and the flow of surface water to permeable fill and thereafter to underdrains. All require regular topdressings of sand to sustain their quality.

**5.** The most efficient pitches in terms of drainage are those with a virtually pure sand rootzone. These designs demand highly qualified groundstaff and timely maintenance. When used close to their potential maximum, extra problems of turf culture and soil maintenance occur which are costly to contain.
**6.** If the projected intensity of use of a natural turf pitch is five or more games per week in winter and a reliable, good quality of surface is required, providers are faced with installing a sophisticated design of pitch with high construction costs as well as high maintenance costs. Providers must be absolutely certain that the demand for use exists if cost-effectiveness is of primary importance. There are a number of factors to consider: (i) it is difficult to generate usage levels greater than eight hours per week unless floodlighting and adequate changing facilities are available; (ii) adult demand is often concentrated at weekends when a single surface cannot meet peak demand; (iii) repeated training on selected areas of sand-based pitches is particularly damaging to the surface and small erosion holes can seriously reduce the overall quality of the pitch (if this type of play represents the majority of demand, a synthetic surface is a better option); and (iv) the moderate upgrading of two natural turf pitches to create a lower use per pitch is preferable to a single *'premier'* pitch receiving an inordinately high demand for use.

## Cost of provision

Prior to 1986, usage of up to 20 hours team use per week was thought possible for high quality natural turf in winter (Hayes, 1987) and up to 30 hours use per week for schools' use (DES, 1982). Research carried out in the UK since 1986 at the University College of Wales, Aberystwyth, and the Sports Turf Research Institute, Bingley, has shown that such high levels of use cannot be sustained in winter whilst retaining what could be described as a turf surface. The research also assessed the cost-effectiveness of a range of pitch construction schemes.

Table 6.7 summarizes the overall cost of a range of different natural turf surfaces examined in the study where the cost has been divided into the cost of the initial construction plus drainage, and the cost of maintenance (materials and labour). Assessment of maintenance is based on that needed to sustain the quality of the surface close to that gained by the initial improvement at a near-maximum acceptable intensity of use. No attempt has been made to incorporate land purchase, administrative costs, rent or lease or any notional appreciation or depreciation of the value of the land or the facility. Also the prices given assume that no significant levelling or earthworks are required. The construction and maintenance costs will increase with inflation, but the relative difference is likely to remain. Moreover, at any one time, tenders for the installation of the same construction specifications usually vary by 30% or more.

**Table 6.7.** Installation costs, maintenance costs and estimated carrying capacities for a range of natural turf winter games pitches in the UK (1991 prices).

| System [a] | Capital cost (£) | Inclusive of irrigation (£) | Maintenance cost (per annum) (£) | Maximum winter usage (hours per week) | Games per season [b] |
|---|---|---|---|---|---|
| PD | 9000 | – | 1000 [c] | 1-3 | 25-75 |
| PDS | 20000 | – | 2000 | 2-4 | 50-100 |
| SD | 23000 | – | 2500 | 3-5 | 75-125 |
| SSR | 33000 | (48000) | 3000 | 3-5 | 75-125 |
| SC | 45000 | 60000 | 4500 | 4-6 | 100-150 |
| SWT | 85000 | 100000 | 6000 | 5-7 | 125-175 |

[a] For explanation of abbreviations, see Table 6.6.
[b] Usage level when end of season recovery requires overseeding or reseeding on less than 20% of the surface.
[c] Sand topdressing (approx. £1000 p.a.) not included.

The study showed that the maintenance input invariably increased with increasing usage, and that the proportion of time spent on each operation also varied considerably with construction design. For example, a poorly drained pipe drained pitch in a high rainfall area gave exceptionally few hours of use per hour of maintenance mainly because mowing and marking had to be carried out even though very often games were postponed because of water-logging. Not surprisingly, this pitch was the least cost-effective of all options studied (see also Baker and Canaway, 1991).

In contrast, an undrained pitch in the same study which was located on a naturally permeable soil in a low rainfall area, allowed much greater usage and therefore a more efficient use/maintenance ratio and a far better cost-effectiveness, partly as a result of minimal construction costs (Gibbs *et al.*, 1992). But although good natural drainage can provide a pitch which is cheap to run, particularly when winters are mild, there is still the risk that the quality of the playing surface will deteriorate drastically in wet weather unless play is postponed. Protecting the surface from damage in adverse weather must have a high priority in these low input schemes.

Slit drained pitches, pitches with sand-dominant rootzones and sand carpet pitches have all been shown to have similar ratios of maintenance hours or costs to potential usage. In 1991, the basic maintenance costs of labour and materials for these types of pitch were estimated to be in the region of £25-35 per game (Fig. 6.6). However, similar maintenance costs are only achieved when these construction types are used close to or at their maximum sustainable carrying capacity.

The increase in potential usage gained by total profile constructions with pure sand rootzones is modest over less complex systems (e.g. slit drained,

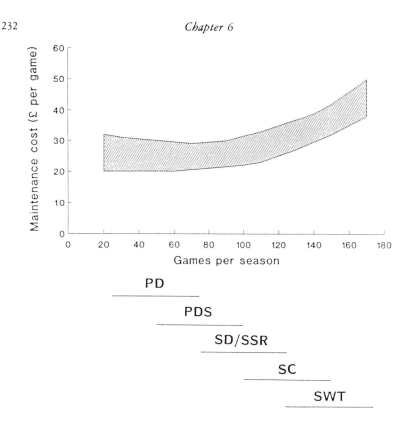

PD

PDS

SD/SSR

SC

SWT

**Fig. 6.6.** The relationship between the maintenance cost per game and the number of games played per season for different designs of winter games pitches used at an intensity within their design capability. Bars indicate the range in usage which the designs are capable of sustaining. See Table 6.6 for explanation of abbreviations.

sand/soil rootzone and sand carpet systems) and it is clear that the increase in usage is not proportional to the increase in capital costs (Table 6.7). Moreover, they cannot maintain grass cover at intensities of use which would make them economical. For the above reason, typical pitch maintenance costs per hour of use increase as usage of the pitch rises. Thus, for example, a scheme of improvement in pitch playing quality enabling two or three games a week to be played with little risk of postponement would have a maintenance cost of about £25 per game. In contrast a total profile construction able to receive and receiving over five games a week would have a maintenance cost of the equivalent of around £40 per game of adult use.

If resources are unable to cope with the maintenance requirements which high intensities of use cause, then lowering the intensity of use on a total profile construction makes this type of pitch an expensive drainage option, even though it will provide an excellent playing surface with minimum risk of postponement. When the high cost of construction is added on to high maintenance costs, it is never as cost-effective as alternatives with a similar wear tolerance, for example a sand/soil rootzone complemented with slit drains. Therefore the installation of total profile constructions can only be justified if the highest quality natural turf surface is demanded and if play has to be guaranteed under all weather conditions except snow and frost. These circumstances apply to professional sports clubs and national stadia (Gibbs *et al.*, 1993b).

## Pitch longevity

The question of pitch cost-effectiveness cannot really be considered without some idea of how long the different drainage systems will last. A properly installed pipe drainage system can be expected to last in excess of 40 years. Slit drainage systems have a useful life span which is modified by the intensity of use, quality of maintenance and nature of the native soil. Well-installed and well-maintained slit drainage systems should continue to give some benefit for at least five years, but deterioration is inevitably progressive and they are unlikely to last more than ten years without major renovation. In this respect, the long-term cost-effectiveness of slit drained pitches has yet to be evaluated against sand/soil, sand carpet and total profile constructions which have a much greater life span than slit drained pitches.

Even if the sand-dominant or pure sand constructions were to last less than 15 years, simple cheap renovation could still be carried out using traditional agricultural implements (e.g. power harrow), but these same implements could not be used on slit drained pitches because the integrity of the slit drains would be completely destroyed. Instead, at the very least, specialized 'sand grooving' implements would be required to reconnect the slit drains with the surface, although some pitches would undoubtedly need complete re-slitting if the original slit drains had become badly capped with soil.

Another point to note is that once the construction costs for the more expensive sand/soil rootzone, sand carpet and total profile constructions have been paid off (say after 10 years), they become more financially attractive. However it is precisely at this point that slit drained pitches are likely to need a large injection of money for renovation purposes. Thus it must be stated that, although the installation of slit drainage brings about an immediate benefit at relatively low cost the benefit may be short-term and the long-term financial comparison with alternatives is less favourable.

Despite the above conclusion, if a 20 year life span is taken as the maximum life span of sand carpet or total profile constructions without major renovation, a slit drained pitch can be re-slitted twice during the 20 year period and still be the most cost-effective option (Gibbs *et al.*, 1992) although the quality of playing surface may be inferior. However, if re-slitting is needed more than twice in the 20 year period, a slit drained pitch loses its advantage over alternatives which involve alteration or replacement of the total rootzone.

## SYMPTOMS, CAUSES AND CHARACTERISTICS OF DETERIORATION

Deterioration of natural turf surfaces used for winter games can be categorized into two components, that of a deterioration in playing quality associated principally with a loss of ground cover and change in soil moisture content, and that of a deterioration in soil physical conditions. Loss of ground cover is a relatively short-term problem in that ground cover can usually be regained in the close season. However a deterioration in soil physical conditions tends to be more long term and can be progressive throughout the life of a pitch. It is also more difficult and expensive to restore once substantial deterioration has occurred. The two types of deterioration are linked since a progressive deterioration in soil physical conditions makes it more difficult to establish well-rooted grass for the next season, which then becomes more quickly worn out than in the previous season. This in turn offers less protection for the soil surface and therefore there is more potential for compaction and smearing. The cyclical nature of this deterioration applies to all winter games pitches. Most causes of deterioration are man-made (i.e. through use), but some components can be natural (e.g. accumulation of organic matter in low wear areas and contamination of sand-dominant rootzones by earthworm activity).

Various attempts have been made to quantify the nature of deterioration of winter games pitches used for soccer in the UK to assist in management planning. These attempts have been complicated by the variety of interacting factors influencing the rate of deterioration including construction type, age of the player, intensity of use, soil moisture content and presence or absence of earthworms.

Loss of ground cover is most people's first impression of deterioration of a natural turf surface. The rate of loss of ground cover is influenced generally by the intensity of play but is more rapid on poorly drained surfaces.

Loss of ground cover is only one aspect of a deterioration of pitch quality and performance. Another is the development of a soft, slippery, poached surface often with ponded water causing low ball bounce and unacceptably variable ball roll. Since these conditions are caused by inadequacies of the drainage rates of the surface layer, it also follows that the playing performance of undrained and pipe drained constructions is very dependent upon the weather conditions. In the worst case, these types of pitches can become waterlogged for much of the season, making play virtually impossible. Towards the end of the season, the surface may dry out, producing exactly the opposite undesirable playing characteristics of a very hard surface with excessive ball bounce.

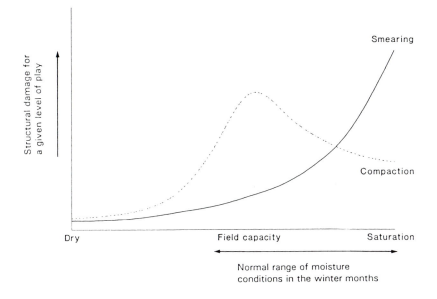

**Fig. 6.7.** The influence of moisture content on soil damage by compaction and smearing (from Baker, 1991a).

The effect of soil moisture state on the susceptibility to physical deterioration of aggregated native soil on pitches is illustrated in Fig. 6.7. The two main components of physical deterioration (i.e. compaction and smearing) respond differently to changes in soil moisture conditions. When the soil is dry, physical deterioration is negligible because aggregates are mechanically stable. At field capacity and somewhat wetter, aggregates are saturated which makes them fragile but there is still air-filled pore space which will be eliminated by compaction. When the soil becomes completely saturated further compression is resisted, but any local surface loading effectively kneads the soil into a plastic mass. This smearing process at or near saturation caused by the horizontal and tangential forces of sliding has a more dramatic effect than local loading on sealing or capping the surface especially if worm casts are present.

Between November and February soils in the UK are for the most part at or wetter than field capacity. Games played on virtually all native soils are therefore likely to cause severe deterioration in soil physical conditions. Even a single game played in extremely wet conditions can cause damage which is irreparable that season.

Soil aeration is inadequate for grass when air-filled porosity in the rootzone falls to below about 5% by volume for prolonged periods (Chapter 5). For soccer there have been several detailed studies to establish the spatial variability

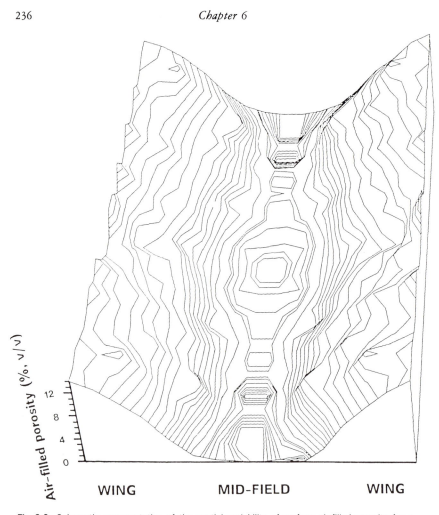

**Fig. 6.8.** Schematic representation of the spatial variability of surface air-filled porosity for a soccer pitch at the end of a season. Intense activity down the centre of the field between the two goals frequently causes air-filled porosity to fall below 5% (v/v) regardless of construction design (adapted from Gibbs *et al.*, 1991).

of soil physical properties across a pitch in relation to the nature of the game (e.g. van Wijk, 1980; Karlsson, 1988; Gibbs *et al.*, 1991; Adams *et al.*, 1993). Gibbs *et al.* (1991) showed that the upper rootzone of the central strip from goal to goal was frequently very close to zero air-filled porosity by the end of a season, irrespective of construction type (Fig. 6.8). Even sand-dominant or all-sand constructions whose rootzones have desirable physical properties in terms of drainage and stability, still provide inadequate aeration for roots when high wear areas are compacted (see also Table 5.7 in Chapter 5). The relief of compaction and creation of air-filled porosity is therefore an essential component of the close-season maintenance of winter games pitches.

Slit draining a pitch can reduce the risk of smearing provided an adequate sand topdressing layer is in place. However, a deterioration in drainage can be rapid even when an adequate sand topdressing layer is present. For example, Gibbs *et al.* (1993a) recorded a decrease in infiltration rate from approximately 50 mm h$^{-1}$ to below 5 mm h$^{-1}$ after just three seasons of use in both intensively and non-intensively used areas of a slit drained soccer pitch receiving moderate to high levels of use. The pattern of deterioration was cyclical in that during each summer period of monitoring, there was a temporary improvement in infiltration rate as the soil dried out and the slit drains opened up. In fact the exceptionally dry summer of 1989 restored drainage rates to those immediately after construction, but overall it was clear that the slit drains were nearing the end of their useful life after four seasons of use.

Earthworm activity may increase the rate of deterioration in slit drainage efficiency, but this aspect has yet to be quantified. There is no doubt that fine soil moved by earthworm activity contaminates the sand topdressing layer of slit drains, but the relative importance of earthworm activity and smearing by play has not been assessed.

Some of the most serious physical deterioration associated with intensive winter games use occurs towards the end of a season on sand-dominant rootzones because of a loss of surface stability associated with a loss of ground cover. Not surprisingly there have been several studies illustrating the close relationship between the percentage ground cover of sand-dominant or pure sand rootzones and stability, as measured by surface traction (Baker, 1989, 1991b; Gibbs *et al.*, 1989; Adams *et al.*, 1993).

Retention of ground cover is therefore very important in maintaining stability and good traction (i.e. values of 25 N m or more). However, it is not the ground cover itself that confers stability in sand-dominant rootzones, but the root biomass associated with ground cover (Adams *et al.*, 1985). In a detailed study of factors affecting the surface stability of a sand rootzone, Gibbs *et al.* (1989) found a highly significant linear relationship between surface traction and ash-free root organic matter in the 0–20 mm depth, which essentially existed whether the grass leaves remained intact or whether they were shaved off to the surface.

There is experimental data suggesting that a level of traction currently acceptable (20–25 N m) can be achieved in a sand rootzone without the presence of roots or ground cover. The problem in practice is that the stability of moist compacted sand is lost once it is disturbed (Adams and Gibbs, 1989). Thus the retention of ground cover on sand rootzone pitches is essential to maintain the quality of playing surface of which they are capable. Whilst it is not possible to be precise, ground cover should not be allowed to fall below about 15% to ensure traction remains in the recommended range. For a sand-based suspended watertable soccer pitch, Gibbs *et al.* (1989) found

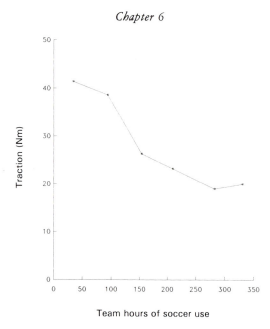

**Fig. 6.9.** Changes in traction characteristics in the centre circle of a sand rootzone suspended watertable soccer pitch with cumulative adult-equivalent team hours of soccer (from Gibbs *et al.*, 1989).

that traction fell below 25 N m after approximately 150 hours of team use or 100 games (Fig. 6.9).

Finally, there are two other important factors relating to the nature and intensity of deterioration of winter games pitches. The first is that, in the absence of earthworm activity and sand topdressings, organic matter accumulates on less intensively used areas of a pitch. The extent of organic matter accumulation on winter games pitches is not usually as severe as with fine turfgrass systems (e.g. golf greens) because of more intensive turf damage inflicted by players (see Chapter 5). Nevertheless, wear is very uneven on a soccer pitch and this is reflected in a much greater accumulation of organic matter in areas of low wear compared with high wear areas (Adams *et al.*, 1992)

The second factor relates to the age of the player. There is no doubt that children cause less surface damage than adults, but a means of apportioning relative damage to different age groups is exceedingly difficult. In Britain, the usual scaling factor assumes that children under 15 years of age cause half the wear of adult users, thus effectively doubling the carrying capacity. The Netherlands Sport Foundation uses a similar system with more categories of age groups. Gibbs and Adams (1990) studied the above problem by recording the damage caused to two soil-based soccer pitches by children of two different age groups (under 12 years of age and 12–16 years of age). Their

**Table 6.8.** Effect of intensity of use of high quality natural turf winter games pitches on percentage ground cover in high wear areas measured towards the end of the season (from Adams *et al.*, 1993).

| Number of games per week | Number of hours play per season | Mean percentage ground cover | Percentage of pitches with no ground cover |
|---|---|---|---|
| Two or less | 100 or less | 40 | 0 |
| Two or three | 100-150 | 32 | 0 |
| Three of four | 150-200 | 21 | 43 |
| More than four | over 200 | 0 | 100 |

results demonstrated that, in general, the relationship between age and extent of wear is reasonable. However, different aspects of wear were not affected equally and that soil moisture conditions affected the relative wear inflicted by different age groups. Thus experiments with players of a single age group (old or young) cannot be used to predict wear by players of a different age group.

## MAINTENANCE REGIMES

The basic consideration for the maintenance of winter games pitches is that high intensities of use cannot be sustained without appropriate maintenance, irrespective of the quality of the installation. Moreover there is no prescriptive frequency of operations to follow for any particular pitch because ultimately it is the drainage capability, local climate and intensity of use which govern the maintenance input. An illustration of this point relates to the amount of reseeding required in the high wear areas of soccer and rugby pitches used to different intensities over the season (Table 6.8). Furthermore, maintenance schedules should only provide guidelines on the frequency of operation, but specify the site conditions necessary for a successful operation to be carried out (e.g. level of ground cover, soil wetness and weather).

## Presentational maintenance

Routine maintenance of soccer and rugby pitches is directed mainly towards visual appearance (i.e. weed control, mowing and marking). Most operations also have some effect on playing quality; for example, if the sward is allowed to grow too long for soccer, this will have a detrimental effect on ball roll and ball bounce. During the playing season, soccer pitches should be cut at a height of 20–30 mm and rugby pitches at a height of around 75 mm. The direction of mowing should be alternated to control growth and keep an even surface. If mowing height is not dictated by usage in summer (e.g. cricket outfield) it is most important that mowing in the close season is frequent and at a height not exceeding around 50 mm irrespective of the type of winter game. Frequency of cut and height of cut are examined in detail in Chapter 2. In practice, during the growing season, mowing is carried out at a frequency of at least once a week. The fate of the clippings is an important consideration for presentational reasons as well as for determining fertilizer requirement (see Chapter 5). Most pitches are mown with a gang mower and clippings are allowed to fly. It is vital that mowing is sufficiently frequent to prevent clippings 'clumping' on the surface where they ferment and smother the underlying grass. There should be enough flexibility in any mobile mowing maintenance schedule to avoid this being an inevitable problem should a few consecutive days of wet weather occur. Clearly the risk of this happening increases with increase in annual rainfall. It is advisable to remove clippings for at least some of the growing season on sand-dominant or pure sand constructions with minimal earthworm activity to avoid humified clippings contributing to surface organic matter accumulation. For pitches in the public eye, appearance alone may dictate that clippings are removed. Types of mower have been discussed in Chapter 5.

Apart from cosmetic striping of the grass surface by directionally brushing or rolling the grass leaves, straight uniform marking is essential for a good quality of appearance. For soccer, pitch line markings need to be 100 mm wide to achieve maximum visual effect (particularly for television), but a general rule of thumb is that markings should be clearly visible from a distance of 30 m for both soccer and rugby, straight or curved as appropriate to the required measurements for the game. Because of the longer cutting height for rugby, it is common practice to cut a mower width to 25 mm over the lines prior to marking.

Some groundstaff still tend to use paraquat, diesel or wood preserver instead of white marking compounds, particularly on rugby pitches. Although this method reduces the frequency of marking required, it is not a particularly satisfactory technique as the lines eventually erode. Moreover if markings are followed along string lines, great care must be taken to avoid killing off other areas of grass as the string lines are moved across a pitch.

## Renovation and close-season maintenance

The aim is to return grass cover to near 100% in desirable species and to restore soil composition and physical properties to those most favourable for turf growth and wear tolerance. There are seven recommended components although not all of these may be practicable:

- overseeding (or reseeding);
- fertilizer input;
- aeration/decompaction;
- weed control;
- sand topdressing;
- mowing;
- irrigation.

The soccer and rugby seasons now extend to late April or early May which is in the middle of the spring flush of grass growth. The grass which has withstood the winter's wear will benefit from fertilizer (N in particular) from mid March and conditions for rapid germination of seed will occur well before the end of the season. Thus a fertilizer application applying about 30–40 kg N ha$^{-1}$ (P and K are less important) should be made before the end of March. About 30 g m$^{-2}$ of seed can be scattered on the bare areas of the pitch around this time to be kicked in by play. This is an extravagant use of seed but greatly increases the rate of grass restoration.

Renovation should begin immediately after the last match. The first priority should be given to breaking up soil compaction and reinstating true levels in small areas of extreme wear, for example the goal-mouth on soccer pitches. Other bare areas should be brought up to true level before overseeding with perennial ryegrass. Restoring grass cover is the next priority and casual use of the most severely worn areas must be prevented. Overseeding can be carried out by hand at a rate of about 30 g m$^{-2}$ or by an implement such as a contravator when not less than two passes should be made.

Frequent mowing is an essential complement to overseeding. Failure to mow the residual grass creates excessive competition with establishing seedlings and results in an open, patchy turf.

It is during the re-establishment of grass on bare or thin areas of turf that irrigation is most valuable. Often only quite small areas are involved and installing a sophisticated irrigation system may not be justified.

Weed control on winter games pitches was discussed in Chapter 5. This is a vital component of close-season maintenance because the most serious broadleaved weeds not only compete with turfgrasses during the close season but die back in winter leaving unprotected soil. Fertilizer input over the growing season, sand topdressings and aeration/decompaction are the other aspects of close-season maintenance.

## Fertilizer

Principles of fertilizer use and fertilizer strategy have been covered in Chapter 5. Table 6.9 summarizes guidelines on annual fertilizer input for winter games pitches. The recommendations for school and Local Authority pitches with no

**Table 6.9.** Guide to annual fertilizer requirements for winter games pitches.

| Type of pitch | Fertilizer requirements (kg ha$^{-1}$ y$^{-1}$) | | | Source |
|---|---|---|---|---|
| | N | P$_2$O$_5$ | K$_2$O | |
| Local Authority pitches or school or club pitches (clippings not removed, earthworms present, loamy soils) | 35-45[a] | 5-15 | 10-20 | Adams, 1977 |
| | | P$_2$O$_5$ and K$_2$O can be applied every third year | | |
| As above, but on loamy sand or sandy loam soil or having been slit drained | 45-65[b] | 15-25 | 20-40 | Adams, 1977 |
| Major sports stadia (free-draining, intensively used, clippings removed, irrigated) | 160-240[c] | 60-90 | 140-200 | Adams, 1977 |
| Soil rootzone (clippings returned) | 80-100 | 20-50[d] | 20-50[d] | Lawson, 1989 |
| Soil rootzone (clippings removed) | 160-200 | 80-100[d] | 80-100[d] | Lawson, 1989 |
| Sand rootzone (clippings removed) | 250 | 80-100[d] | 80-100[d] | Lawson, 1989 |

[a] Annual application in March or split March/April.

[b] Application split March/April.

[c] Minimum of four applications; insurance trace element application every five years; controlled-release fertilizers and occasional Mg fertilizer recommended.

[d] Requirement for potassium and phosphorus application depends on soil test results.

clippings removed were based on four years of trials on school playing fields in Breconshire with the cooperation of Richard Bowering, the County Playing Fields Officer (Adams, 1977). Recommendations for sand-dominant rootzone systems with clippings removed were initially based on the amount of nutrients removed in clippings in field experiments (Adams, 1977). Subsequent research by Canaway (1985) and Canaway and Hacker (1988a, b) relating fertilizer input to percentage ground cover and various playing quality components has resulted in similar recommendations. The natural growth pattern of grasses when they are allowed to flower and set seed is modified by frequent mowing (Chapter 2). Thus whilst for a hay meadow fertilizer input (especially N) should be strongly biased in favour of the first half of the growing season, with sportsturf a more modest bias is appropriate.

Benefits in fertilizer use efficiency and evenness of grass growth are gained on sand-dominant rootzones by the use of slow release sources of N such as IBDU. Their high cost can be justified in the reinstatement of bare areas.

## Aeration / decompaction

It is likely that on at least the high wear areas of any winter games pitch air-filled porosity will be reduced to less than 5% by the end of the season. Whilst summer drying and worm action may increase macroporosity to some extent, to all intents and purposes mechanical decompaction is essential to restore adequate physical conditions. It is not necessary to carry out decompaction immediately the season ends, indeed some types of equipment (e.g. surface slotter) require an intact turf cover. Serious compaction does not normally extend below about 150 mm but fracturing or loosening to this depth is necessary. There is still little reliable data on the comparative performance of equipment designed to relieve compaction; however, it is clear that a solid tine slitting or spiking is of little if any benefit in relieving compaction. In essence the soil must be moved to create fissures and several types of equipment are available to achieve this (Chapter 5). The soil must be relatively dry and the work should be carried out before the end of June.

## Sand topdressing

Although sand is frequently applied to winter games pitches during the season to help firm up a plastic surface, sand topdressings should be applied during the close season.

Sand topdressings are an integral part of maintenance for all types of winter games pitch. The actual amount needed varies typically between 20 and 60 tonnes per pitch per year, although it can be calculated more precisely (see Chapter 5). A dressing of approximately 30 tonnes per pitch per year is a good guideline for pitches with a native subsoil. This topdressing can be applied in one application annually or twice the application biennially. It is important that the right type of sand is used for topdressings. Considerable benefit can also be gained by combining sand topdressings with decompaction treatments such as Verti-draining or surface slotting as the sand helps to keep the holes and fissures open.

Baker and Canaway (1990) studied the effect of sand topdressings on the performance of winter games pitches of different construction types. The results of their experiments are summarized in Fig. 6.10. In these trials sand topdressings of 0, 4, 8 and $16 \, \text{kg m}^{-2}$ per year were used (equivalent to 0, 25, 50 and 100 tonnes per year for a $6250 \, \text{m}^2$ pitch). A total of 80% of the sand was applied during the summer in two applications with two further

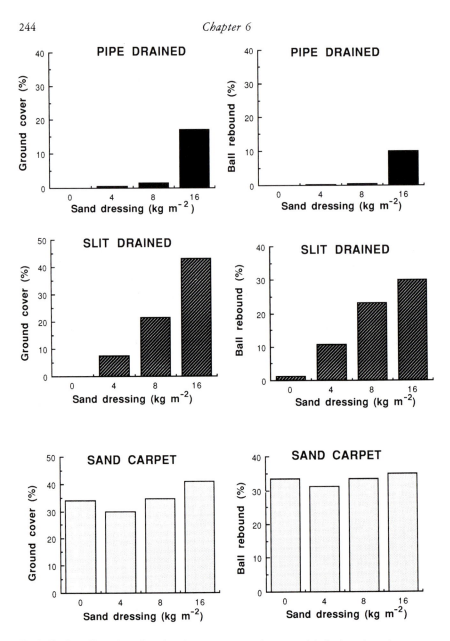

**Fig. 6.10.** The effect of sand topdressing rate on ground cover and ball rebound on pipe drained, slit drained and sand carpet constructions receiving simulated football-type wear (reproduced with permission of STRI, Bingley).

light applications during the course of the playing season. Increased rates of sand application affected the playing quality of all construction types except an all-sand total profile construction, particularly in terms of ground cover retention for pipe drained and slit drained pitches. For the pipe drained pitch, topdressing produced firmer playing conditions. For the slit drained pitch the higher rates of sand application prevented the slits from capping.

## Maintenance in the playing season

As the season progresses and the surface becomes increasingly compacted and sealed, spiking or slitting become important operations. Their key function is to pierce a sealed surface to a depth where the soil is less compact and has a greater macropore space. The holes they create allow gaseous diffusion in and out of the soil and increase water infiltration rates. Spiking creates no benefits unless the tines reach a zone with macropores. Since the benefits from spiking relate to the condition of the surface it is irrational to adopt a fixed frequency routine for the operations. Fixed freqency programmes can also lead to the operation being carried out in such wet conditions that the damage through smearing and compaction exceeds any possible benefit.

When waterlogging or ponding occurs in local and small areas, forking by hand is necessary. Machinery must be kept off the pitch.

Apart from the marking and mowing requirements described under Presentational maintenance, there are two other critical aspects of playing season maintenance, that of divot replacement and levelling the surface. Divot replacement should be automatic but it is particularly important on sand-based pitches because it helps prevent the onset of erosion.

Unevenness caused by foot imprints is a normal consequence of use. A frequent response is to use a roller to flatten the surface. Rolling winter games pitches is a contentious issue and blanket approval or condemnation is inappropriate. Rolling wet loamy or heavier textured pitches will flatten the surface but it also smears and seals it reducing aeration and water infiltration. Rolling sand or sand-dominant rootzones gives benefits of levelling and firming the surface with minor detrimental effects. As a general rule rolling is 'last resort' maintenance which should not be necessary when playing conditions have not got out of hand. A chain harrow, dragmat, brush or lute are much more appropriate implements to restore a true playing surface.

## REFERENCES

Adams, W.A. (1977) Fertilizer use on sportsturf. *Parks and Sports Grounds* January 1977, pp. 62–70.

Adams, W.A. and Bryan, P.J. (1974) *Poa pratensis* L. as a turfgrass in Britain. In: Roberts, E.C. (ed.) *Proceedings of the 2nd International Turfgrass Research Conference*. American Society of Agronomy, Madison, Wisconsin, pp. 41–47.

Adams, W.A. and Gibbs, R.J. (1989) The use of polypropylene fibres (VHAF) for the stabilisation of natural turf on sports fields. In: Takatoh, H. (ed.) *Proceedings of the 6th International Turfgrass Research Conference*. Tokyo, Japan, pp. 237–239.

Adams, W.A. and Jones, R.L. (1979) The effect of particle size composition and root binding on the resistance to shear of sports turf surfaces. *Rasen Turf Gazon* 10(2), 48–53.

Adams, W.A., Tanavud, C. and Springsguth, C.T. (1985) Factors influencing the stability of sportsturf rootzones. In: Lemaire, F. (ed.) *Proceedings of the 5th International Turfgrass Research Conference*. Institute National de la Recherche Agronomique, Paris, pp. 391–398.

Adams, W.A., Gibbs, R.J., Baker, S.W. and Lance, C.D. (1992) Making the most of natural turf pitches. A national survey of winter games pitches with high quality drainage designs. *Natural Turf Pitch Prototypes Advisory Panel Report No. 10*. The Sports Council, London, 20 pp.

Adams, W.A., Gibbs, R.J., Baker, S.W. and Lance, C.D. (1993) A national survey of winter games pitches in the UK with high quality drainage design. In: Carrow, R.N., Christians, N.E. and Shearman, R.C. (eds) *International Turfgrass Research Journal* 7, 405–412.

Asami, T., Togari, H. and Ohashi, J. (1988) Analysis of movement patterns of referees during soccer matches. In: Reilly, T., Lees, A., Davids, K. and Murphy, W.J. (eds) *Science and Football*. E. and F.N. Spon, London, pp. 341–345.

Baker, S.W. (1989) Soil physical conditions of the rootzone layer and the performance of winter games pitches. *Soil Use and Management* 5, 116–122.

Baker, S.W. (1991a) Compaction: a problem of wear 'n' tear. *Turf Management* May 1991, pp. 11–13.

Baker, S.W. (1991b) Temporal variation of selected mechanical properties of natural turf football pitches. *Journal of the Sports Turf Research Institute* 67, 83–92.

Baker, S.W. and Canaway, P.M. (1990) The effect of sand top dressing on the performance of winter games pitches of different construction types. I. Soil physical properties and ground cover. *Journal of the Sports Turf Research Institute* 66, 21–27.

Baker, S.W. and Canaway, P.M. (1991) The cost-effectiveness of different construction methods for Association Football pitches. II. Ground cover, playing quality and cost implications. *Journal of the Sports Turf Research Institute* 67, 53–65.

Baker, S.W. and Gibbs, R.J. (1989) Levels of use and the playing quality of winter games pitches of different construction types: case studies at Nottingham and Warrington. *Journal of the Sports Turf Research Institute* 65, 9–33.

Bell, M.J. and Holmes, G. (1988) The playing quality of Association Football pitches. *Journal of the Sports Turf Research Institute* 64, 19–47.

Bryan, P.J. and Adams, W.A. (1971) Observations on grass species persisting on English League soccer pitches in spring 1970. *Rasen Turf Gazon* 2(2), 46–51.

Canaway, P.M. (1983) The effect of rootzone construction on the wear tolerance and playability of eight turfgrass species subjected to football-type wear. *Journal of the Sports Turf Research Institute* 59, 107–123.

Canaway, P.M. (1985) The response of renovated turf of *Lolium perenne* (perennial ryegrass) to fertilizer nitrogen. I. Ground cover response as affected by football-type wear. *Journal of the Sports Turf Research Institute* 61, 92–99.

Canaway, P.M. and Hacker, J.W. (1988a) The response of *Lolium perenne* L. grown on a Prunty–Mulqueen sand carpet rootzone to fertilizer nitrogen. I. Ground cover response as affected by football-type wear. *Journal of the Sports Turf Research Institute* 64, 63–74.

Canaway, P.M. and Hacker, J.W. (1988b) The response of *Lolium perenne* L. grown on a Prunty–Mulqueen sand carpet rootzone to fertilizer nitrogen. II. Playing quality. *Journal of the Sports Turf Research Institute* 64, 75–86.

Canaway, P.M., Bell, M.J., Holmes, G. and Baker, S.W. (1990). Standards for the playing quality of natural turf for Association Football. In: Schmidt, R.C., Hoerner, E.F., Milner, E.M. and Morehouse, C.A. (eds) *Natural and Artificial Playing Fields: Characteristics and Safety Features*. American Society for Testing and Materials, Philadelphia, pp. 29–47.

DES (1982) *Playing Fields and Hard Surface Areas*. Department of Education and Science Building Bulletin 28, HMSO, London, 70 pp.

Douge, B. (1988) Football: the common threads between the games. In: Reilly, T., Lees, A., Davids, K. and Murphy, W.J. (eds) *Science and Football*. E. and F.N. Spon, London, pp. 3–19.

Gibbs, R.J. (1988) The influence of winter sports pitch drainage systems on the measurement of water infiltration rate. *Journal of the Sports Turf Research Institute* 64, 99–106.

Gibbs, R.J. and Adams, W.A. (1990) The damage to a natural turf soccer pitch caused by children of two different age groups. *Zeitschrift für Vegetationstechnik* 13(3), 99–103.

Gibbs, R.J., Adams, W.A. and Baker, S.W. (1989). Factors affecting the surface stability of a sand rootzone. In: Takatoh, H. (ed.) *Proceedings of the 6th International Turfgrass Research Conference*, Tokyo, Japan, pp. 189–191.

Gibbs, R.J., Adams, W.A. and Baker, S.W. (1991) Making the most of natural turf pitches. Final results of a case studies approach: IV. Soil physical properties. *Natural Turf Pitch Prototypes Advisory Panel Report No. 9*. The Sports Council, London, 48 pp.

Gibbs, R.J., Adams, W.A. and Baker, S.W. (1992) Case studies of the performance of different designs of winter games pitches. II. Cost-effectiveness. *Journal of the Sports Turf Research Institute* 68, 33–49.

Gibbs, R.J., Adams, W.A. and Baker, S.W. (1993a) Changes in soil physical properties of different construction methods for soccer pitches under intensive use. In: Carrow, R.N., Christian, N.E., and Shearman, R.C. (eds) *International Turfgrass Research Journal* 7, 413–421.

Gibbs, R.J., Adams, W.A. and Baker, S.W. (1993b) Playing quality, performance and cost-effectiveness of soccer pitches in the UK. In: Carrow, R.N., Christian, N.E. and Shearman, R.C. (eds) *International Turfgrass Research Journal* 7, 212–221.

Hayes, P. (1987) Looking at tomorrow's needs. *Turf Management* November 1987, pp. 14–19.

Karlsson, I.M. (1988) Soil construction, drainage and maintenance for Swedish grassed parks and sports fields. *Acta Agriculturae Scandinavica* Supplementum 26, 99 pp.

Lawson, D.M. (1989) The principles of fertilizer use for sports turf. *Soil Use and Management* 5, 122–127.

Shildrick, J.P. (1984) *Turfgrass Manual.* The Sports Turf Research Institute, Bingley, 60 pp.

van Wijk, A.L.M. (1980) *A Soil Technological Study on Effectuating and Maintaining Adequate Playing Conditions of Grass Sports Fields.* Agricultural Research Report 903, Centre for Agricultural Publishing and Documentation, Wageningen, 124 pp.

Ward, C.J. (1983) Sports turf drainage: a review. *Journal of the Sports Turf Research Institute* 59, 9–28.

Winterbottom, W. (1985) *Artificial Grass Surfaces for Association Football.* The Sports Council, London, 127 pp.

# 7 GOLF COURSES AND BOWLING GREENS

## INTRODUCTION

Golf dominates the turfgrass industry both in research and development and within the commercial sector. It is played all around the world and the spread in its popularity is not closely linked with racial, language or Empire ties as it is with most games. Golf provides a variety of contrasting turf surfaces ranging from those which confer reliable and consistent ball roll characteristics to those which approximate to natural habitats. Golf is the focus for most commercial turf-related products and, in the words of Watson *et al.* (1992), it has become the trend setter for the industry, providing in the USA, at least, the largest amount of private sector financial support for turfgrass research of any segment of the industry.

An attempt to condense the wealth of knowledge that exists on turfgrass-related aspects of golf into one chapter is ambitious. However, it is emphasized at the outset that no attempt will be made to discuss the history and philosophy of golf course design, a subject written about by numerous other authors. The aim of the chapter will be to examine the principles of soil and turf management related specifically to golf and bowls.

Although golf and bowls have totally different visual appearances as games, they share features which justify their inclusion together in one chapter. The most important feature is that golf and bowls are classified as fine turf sports where the ball roll characteristics of the surface are more important than for any other game played on natural turf. Whether on a golf green or a bowling green, precise judgement of line and pace can only be exercised on a surface which is smooth and of consistent behaviour. Whilst variations in topography are used on golf greens as an added complication for the player, the basic design and maintenance requirements for golf and bowling greens are very similar. In addition, golf and bowls have characteristics with appeal to a wide group of people; no great athleticism is required and both games can be played by participants of all ages at various levels of competitive and social activity. Golf courses in particular can be tailored to provide for a large range of different abilities, catering for those with virtually no experience to those who have developed a natural talent to the utmost.

## NATURE OF THE GAMES

## Historical background

### *Golf*

Golf is widely believed to have originated on the Scottish coastal strips of grazed grassland on windswept sand dunes (links land) during the 14th century. The soil was free draining and low in plant nutrients and the turf subjected to summer drought. The basic principles of the game probably evolved from shepherds challenging each other to hit small stones into rabbit holes on areas of sheep pastureland using upturned crooks. Later on players would have dug their own holes, choosing an appropriate route over the Scottish links. If this analysis is true, then the origins of golf are more directly derivative than earlier thoughts of a continental origin which associated golf with hitting a ball across a piece of land with a club or stick, an activity from which games like hockey are thought to have evolved.

Like football, the game came under attack from the British monarchy for being a diversion to archery although by the early 16th century the game was granted royal patronage after citizens had been freed of military duty. From 1502, all the Stuart monarchs played golf. In these early days, as with most other sports, there were few rules governing the game. The lie of the land primarily dictated the nature of the game with players aiming for relatively flat areas where grazed fine grasses prospered. Storms or over-grazing created blowouts in the sand dunes which were further formed and maintained by sheltering or merely resting sheep. These are now reproduced as bunkers or sand traps. Fairways originated from areas of playable turf criss-crossing the dunes that were grazed by sheep and rabbits and which connected level areas suitable for tees or greens. Playable areas of grazed turf were bordered by shrubs and tall grass which is now recognized as rough.

In the early stages of development of golf there was minimal management in setting out courses. However, as the game developed in the early 1700s, people began to influence the nature of courses and a regular route became established around of a sequence of holes, although their number varied depending on the space available. The game spread inland in Scotland to acidic moorland or heathland areas which were used as the inland equivalent of coastal links courses. By the 19th century golf courses had spread into traditional parkland areas. These areas differed from the links and heathland/moorland courses in that the soils were more fertile with different grass species from those found on the coastland or upland courses. Water features in the forms of streams, ponds and rivers together with patches of mature woodland replaced the traditional natural links features of sand dunes and bunkers.

Despite the changing landscape that was used for golf courses, it was the 18 hole old course at St Andrews which set a pattern which other golf courses followed. Many of its features were faithfully copied. For example, sand bunkers became an essential feature of golf course design even on parkland courses. The old course at St Andrews came into being after the community of Leith petitioned the Edinburgh City Fathers in 1744 to provide a trophy for open competition, but it was not until 1754 that the Royal and Ancient Golf Club of St Andrews (the governing body of golf in the UK) was founded for their inaugural competition.

As with football, the Industrial Revolution brought about many developments in the game with radical changes to the design and manufacture of balls and clubs, which were now mass produced and therefore more readily available. Competitive playing increased, professional golfers became established and since the middle of the 19th century the game has been one of continued expansion with the basic fundamentals of the game remaining unchanged.

The game spread internationally in the 19th century not only through the colonization of countries destined to become part of the British Empire, but also with the travel of wealthy British visitors to fashionable watering places where an interest in the game was quickly taken up by local players. Royal Calcutta (1829), for example, is the oldest golf club outside the UK. The way the game was introduced to the USA is less certain which is surprising given its outstanding popularity there. The catalyst for golf was undoubtedly the formation of the St Andrews Club in Yonkers, New York, in 1888, although it is known that a club was founded in Charleston, South Carolina, as far back as 1876.

In 1986 it was estimated that there were approximately 23,000 golf courses worldwide with over half being in the USA (Watson *et al.*, 1992). However, even since that time, golf course construction has increased dramatically in western Europe, Japan and the USA. One of the most significant events to have affected golf in the UK in recent years has been the publication of a report entitled *The Way Forward* by the Royal and Ancient Club of St Andrews in 1989. This report estimated that 700 new courses were needed in Britain by the year 2000 in order to cope with perceived demand. The report came conveniently at a time when landowners in the European Community were being encouraged to reduce agricultural production and to diversify into non-agricultural ventures. In 1990, around 250 new golf courses opened in Europe, with France and the UK accounting for over half this growth (France – 73 courses, UK – 59 courses). Between 1989 and 1992, 176 courses were built in the UK with the majority 18 hole or larger.

As land and construction costs have increased, golf course projects built as single ventures have diminished in favour of integral country club/hotel/ executive housing developments. A further consequence of the recent boom in golf course construction is a need for more clearly defined design standards together with maintenance requirements. Many courses have been built in recent years which are unlikely to survive in the long term because technically unsound schemes of construction have be used and/or maintenance costs have been underestimated relative to sustainable income. Nevertheless, without doubt, the demand for golf courses will increase, worldwide for the foreseeable future.

## Bowls

Bowls is a truly ancient game. Although legend has it that the game was invented by a king with a desire to roll the heads of his recently executed victims, it was the famous Egyptologist Sir Flinders Petrie who discovered the existence of bowls as a disciplined game when he unearthed bowling artefacts in the grave of an Egyptian child buried in 5200 BC. The ancient Polynesians, including some who sailed to New Zealand in the 14th century, are also credited with playing a version of bowls with pieces of whetstone shaped precisely into an ellipse.

Although many bowls-like games exist throughout the world, the British version of bowls is thought to have originated at the time of Roman colonization. Evans (1992) gives a detailed account of how the game developed in the UK along with related games such as skittles and 10-pin bowling. In brief the game became so popular in the 15th and 16th centuries that once again the monarchy became obsessed with its potential threat to the common public's need to be proficient in archery. In a state of rigid disapproval, Henry VIII issued a statute in 1511 downcasting the game, which was not officially revoked until 1845. The statute did not however apply to the wealthy who were considered respectable enough to play the game without the undesirable association with taverns and gambling. The fact is highlighted by the legendary game between Sir Francis Drake and Sir John Hawkins in 1588 when the imminent arrival of the Spanish Armada prompted the reply by Drake that 'there is plenty of time to win the game and to thrash the Spaniards too!'.

The game of bowls developed slowly and steadily in the 17th and 18th centuries, particularly among the Scots. The Scots were more fortunate than the English in the general acceptance of bowls, because their game was not restricted by legislation nor was it associated with beer and ale houses like the English version. According to Evans (1992), the Scots too can be credited with two major developments in the evolution of the game, firstly the formation of a code of laws in 1849 and secondly with early contributions to the construction of bowling greens, as will be seen later.

Today, bowls is a well-established game in the UK which has also spread worldwide mainly through countries that were once part of the British Empire including Canada, Australia, New Zealand and South Africa, as well as to the USA. More than a million people of all age groups are thought to participate regularly in the UK which qualifies the game as a major sport. Technically the game should be referred to as 'lawn bowls', or 'bowling on the green' to describe and characterize fully the British version of the game. Evans (1992) estimated that there are about 5200 flat bowling greens and 3500 crown bowling greens in the UK.

In England the game of bowls has developed according to three main codes being played on two types of bowling greens (flat greens and crown greens). The dominant form of the game is the World Bowls Board game governed by the English Bowling Association (formed in 1903) which is played on flat greens. This form is also the only code of lawn bowls played in Scotland and Ireland and it predominates in Commonwealth countries, mainly as a result of the spread of the game by immigrants from Scotland.

The second form of the game is the English Bowling Federation code which is also essentially a flat green game. Apart from one or two other differences with the rules and regulations on ditch construction, both English Bowling Federation and English Bowling Association codes share a relatively common ancestry.

In the northern areas of England and Wales (including the Isle of Man) a third form of lawn bowls is played on a turf surface where the centre of the green is higher than the boundaries; the crown green code. Although this variation of lawn bowls shares a common ancestry dating back to the target hitting days of 5200 BC, the reasons for its origin are believed to have evolved as a result of difficulties in producing level greens in some of the poorer areas in industrial northern Britain, where it was traditional for greens to be associated with pubs and inns. It is also interesting to note that betting is widespread in the crown green game, unlike the flat rink game. The rules of crown green bowling are rather different from flat rink bowling and a separate Bowling Association (the British Crown Green Bowling Association, founded by amalgamation of the National Crown Green Amateur Bowling Association and the British Crown Green Amateur Association in 1932) governs this code of play.

# Appearance of play

## *Golf*

Unlike most other games on turf which have precisely marked out areas
for play, golf courses have a relatively unrestricted area for play and play is
conducted on surfaces of divergent character. Golf courses can be built on areas
as diverse as deserts, swamps, moors and landfill sites a feature which gives
the game limitless variety. Golf distinguishes itself from other stick or club
and ball games, in that having propelled a golf ball with as much power as
a player feels justified to use, the golfer must then use talents of extreme
delicacy and accuracy to place the golf ball into a small hole (108 mm in
diameter) marked by a flag, using as few strokes as possible. The ability to
play the game successfully requires not only a range of clubs, woods and irons
(maximum 14 permitted in the bag) to achieve different spins and trajectories
of the ball, but also a considerable element of strategic planning and ability
to get out of difficult situations.

A traditional 18 hole golf course is generally accommodated within 60
hectares in total with probably no more than 1.5–1.8 hectares utilized for
greens and tees. A typical hole comprises an elevated teeing ground from which
the first stroke is made, a rather narrow fairway of mown grass bordered by
coarser vegetation (rough) extending to the green. The green itself may display
a variety of topographic designs, but it is essentially a prepared turf surface
of about 400–500 m$^2$ into which the hole is sunk. *En route* to a hole there
may be a variety of hazards, for example streams, sand-filled bunkers, trees,
shrubs and ponds, positioned or manipulated in such a way as to impede and
penalize an inaccurate shot.

The abrupt interface between quite closely mown grass on fairways and
rough unmown grass is usually softened by semi-rough which is a strip of
slightly longer grass which forms a buffer zone between the playing area
and the wilder parts of the golf course. Both rough and semi-rough can be
extremely important areas for the ecological conservation and enhancement
of golf courses (Nature Conservancy Council, 1989).

A good standard playing length of an 18 hole golf course is 5670 to 5852 m
(Hawtree, 1983) with championship courses being nearer 6400 m. This total
playing distance is divided into holes of different length estimated to require
a stroke number between 3 and 5 to complete (the par number). An average
course will comprise four or five par-3 holes (120–180 m), nine or ten par-4
holes (320–430 m) and three or four par-5 holes (440–500 m).

## *Bowls*

The flat green game approved by the World Bowls Board for international
competition is played on a green at least 36.58 m by 36.58 m in size and

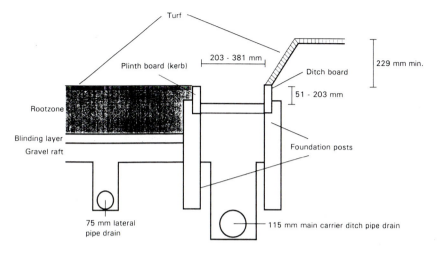

**Fig. 7.1.** Schematic representation of a modern flat bowling green design with a traditional grassed bank around the ditch. The profile is similar to the suspended watertable design in Fig. 3.9.

up to 40.23 m by 40.23 m. National games are allowed on greens as small as 30.17 m by 30.17 m. The green is surrounded by a ditch of between 51 mm and 203 mm deep and between 203 mm and 381 mm wide (Fig. 7.1). Greens are often enclosed by a grass bank sloping at an angle of 35° from the vertical, or have purpose-built vertical ditch units with a striking surface that protects bowls from damage. The bank should be not less than 229 mm above the level of the playing surface. The playing surface is the area comprising the green turf surface and the kerb, but it does not include the ditch.

The green is divided into parallel rinks of between 5.48 m and 5.79 m width (18–19 feet) which typically allows up to 12 teams to play simultaneously on a green at any one time on a standard sized green (six rinks each with two teams). Rinks may be reduced to 4.27 m width (14 feet) for domestic play which allows more rinks to be in use. A team may consist of one, two, three or four people. The basic tools of the game are the black or brown woods, whether they be made of composition, hardened rubber or lignum vitae (a hardened wood) and the smaller white jack. The general objective of the game is to have in position as many woods as possible nearer to the jack than the nearest wood of the opponent when all the woods have been delivered. Compared with other sports, the game is undemanding in physical fitness but requires high concentration and precise control of power and direction of wood delivery. A key feature of the wood is that it is flattened slightly on one side, and this imbalance, called the bias, allows the wood to roll in a curved path rather than in a straight line. This makes it possible to avoid blocking woods and adds interest to the game.

Like the flat green game above, crown green bowls is played on a square area of grass usually about the same size, but which can extend to 54.9 m by 54.9 m. The surface of the green slopes gently upwards from the sides to a central crown with the surface generally being more irregular than with a flat bowling green. The height of the central crown above the four corners varies with the size of the green, but is usually between 203 mm and 457 mm with 330 mm being the current recommended height for a 36.58 m by 36.58 m green. The ditch and green size requirements are far less strict than these for flat bowling greens and typical details of design are given by Evans (1992). The game is usually played in singles and play is not restricted to a rink.

As with the flat green game, the objective of crown green bowling is to rest woods nearer the jack than any of those of the opponents. The woods tend to be smaller than those used on flat greens and the rules governing the dimensions and weight of them are less rigid. Crown green woods are also biased, but an added difference to the crown green game is that the jack too is biased and it may be played in any direction. This random use of the green means that there is much more chance of interference from other games when two or more games are being played simultaneously.

# Type and location of wear

The type of wear experienced on golf courses and bowling greens is less dramatic visually than that found on winter games pitches. Of the four components of wear – tearing, smearing, crushing and compaction – it is the last two that cause the most damage on golf courses and bowling greens. Inevitably the consequences are worst where traffic is greatest, for example on golf greens at the pin position and on bowling greens around the edge of the green where players repeatedly begin play. It is not only wear by players which is important. Maintenance equipment for mowing, aeration, etc. has to be turned abruptly at the edge of bowling greens. The same situation often applies to golf greens because of adjacent hazards or topographic features. Skewing and multiple passes of equipment on golf greens can be avoided by good design. On existing courses it can be decreased by the use of pedestrian equipment.

Often the effect of wear is indirect. As compaction builds up, unfavourable soil conditions develop which then make it progressively more difficult to sustain good growing conditions. The end result is that compacted areas display open, slow-growing turf. The turf is more susceptible to stress through wear, drought and low nutrient levels. Also it is predisposed to invasion by moss and weeds and is susceptible to attack by pathogenic fungi.

Large areas of completely worn out (as opposed to diseased) turf are unusual on bowling greens unless they receive exceptionally bad management, but they can be quite common on golf courses for a variety of reasons. Firstly, golf is an all-year-round sport, unlike bowls, which is a summer game. Thus, in common with winter games pitches, soil on golf courses is frequently close to saturation, predisposing it to damage by smearing and compaction. Secondly, bowling greens tend to be maintained with small pedestrian machinery, whereas golf clubs which are under pressure to keep courses open all year often use the largest and fastest machinery that is practicable (e.g. ride-on mowers and tractor-drawn gang-mowers). Added to this is the pressure of golf trolleys and motorized golf carts. Both types of vehicle pass over well-travelled routes, many of which are also bottle-necks (e.g. entrances and exits to greens). The inevitable result of continually exploited traffic routes is worn grass and heavily compacted soils, particularly during late autumn and winter periods when grass growth is minimal. There are many examples of golf courses which are expected to sustain upwards of 70,000 rounds of golf per year when the indigenous soil (particularly on the greens) is only able to sustain 30,000–40,000 rounds. Renovation and maintenance of such courses are a major problem and reconstruction of greens, approaches and tees and the installation of fairway drainage to higher specifications may seem an endless activity.

Finally, there are two further aspects of wear in golf that are not experienced in bowls. The first is where, in the process of playing a shot, sections of turf are cut out by the club (divots) and thrown forward. This form of damage by tearing the turf is similar to that which occurs on winter games pitches and horse racing tracks. On golf courses, repair is time consuming because of the extensive areas involved. The second, which is less damaging overall, is when balls are pitched on the green with a high velocity of impact. In addition to creating a depression in the green (plugging) the bruising of turf often allows invasion by pathogenic fungi. Repairing ball marks on golf greens is the responsibility of players as is the replacement of divots. Golf is a game where repairing damage to turf is written into the rules of the game, and where there is an established etiquette to prevent unnecessary deterioration in turf.

## Grass species for golf courses and bowling greens

Golf courses incorporate contrasting turf surfaces ranging from tight, closely mown turf on greens through less closely mown tees and stretches of fairway to the coarse vegetation on areas of rough. The turfgrass species appropriate for these different components differ and so the areas must be considered separately.

Golf greens and bowling greens have similar turfgrass requirements. Although there are some differences in detail the two will be considered together.

*Golf and bowling greens*

Annual meadowgrass (*Poa annua*) is the dominant or co-dominant species on most golf greens and bowling greens in the UK. The same situation applies in other moist, cool temperate regions, for example western areas of North America including the States of Oregon and Washington in the USA and British Columbia in Canada. It is an unsown grass with characteristics which confer competitive advantages on worn and compacted areas of turf. In the wetter, western regions of the UK and in general where irrigation is used to avoid moisture stress to the turf, annual meadowgrass can compete effectively against any sown species in closely mown and especially intensively used turf.

Annual meadowgrass produces seed for most of the growing season even when the height of cut is very short. Also the seed has a wide range in dormancy period. Furthermore intact plants spread by rooting at nodes along prostrate stems in tight turf. Thus whether in severely worn turf or in intact turf, annual meadowgrass is a strong competitor (see Chapter 2).

A mistake which is often made is that all annual meadowgrass on golf and bowling greens is not only bad but equally bad. Greenkeepers are used to representatives from companies advertising the beneficial attributes of one cultivar of a species over another. This may be valid but the assumption that a successful unsown turfgrass species is universally and equally bad defies the laws of nature; especially since that species has been exposed to, and in its various 'strains' has survived, the diverse stresses imposed on golf and bowling greens. Annual meadowgrass is a very variable species and naturally selected strains provide high quality playing surfaces on many golf and bowling greens. Its relative susceptibility to *Fusarium* compared with bentgrasses is compensated for by its relative resistance to *Ophiobolus*. Nevertheless, shallow rooting, especially on compacted soils and sensitivity to drought and low nutrient status make it a turfgrass which demands cosseting.

For many years the standard seed mixture recommended for golf and bowling greens in the UK has been an approximately 80/20 mixture by weight of fine fescue and bentgrass. Because of the small seed size of bentgrasses (Chapter 2) this represents a seed number ratio of around 4:1 in favour of bentgrass. The origin of the mixture was an attempt to mimic the sheep-grazed fine turf on links and heathland and possibly the turf on the old ornamental lawns of stately homes which usually contained both fescue and bentgrasses. Cultivars of different fescue and bentgrass species have been and are used. Chewings fescue (*Festuca rubra* subsp. *commutata*) is the most widely used fescue but slender creeping red fescue (*F. rubra* subsp. *litoralis*) and hard fescue (*F. longifolia*) may be included in mixtures. Three bentgrass species are in common use: two browntop bents *Agrostis capillaris* and *A. castellana* and creeping bent (*A. stolonifera*). Most creeping bents in the UK are coarse and vigorous but elsewhere fine strains occur. 'Penncross', for example, is a very fine-leaved cultivar suited to warm temperate areas. Velvet bent (*A. canina*) is a species of potential value.

Thus far, no cultivars of fine fescues have been released which will both tolerate wear and compete successfully with bentgrasses and annual meadow-grass in the absence of stress through drought or low nutrient availability (see Chapter 2). In consequence the greater the usage in terms of intensity and length of playing season plus the type of maintenance necessary to achieve these levels of use, the less likely are fescues to survive.

Fine-leaved fescues contribute to the high quality of well-maintained and little used greens. They are virtually absent from most golf and bowling greens in the UK and their contribution on seeded greens is, by and large, as a 'nurse' grass for bentgrasses which are slower to germinate and establish. The lack of persistence of fine fescues has resulted in the use of bentgrass alone for seeded greens in temperate climates. The seed used ranges from a single cultivar to several cultivars of several bent species.

There are two aspects which separate bowling greens from golf greens. The first is the use of sea marsh turf for bowling greens. The second, also concerning the latter, is the use of the herbaceous species cotula (*Leptinella dioica* and *L. maniototo*) in New Zealand where the two species are native (Fig. 7.2).

Sea marsh turf used to be very popular throughout the UK and many bowlers attribute the quality of their green to that origin. The use of sea marsh turf was developed in Scotland and the border counties as part of a method of bowling green construction. Cumberland sea-washed turf still has a high reputation with some older players. The sea marshes around the UK, especially the northwest are grazed by sheep and/or wild ducks and geese, equivalent to close and regular mowing. This turf was recognized in the 18th century as a ready source of fine, weed-free turf suited to the requirement of bowling greens. From then until after the First World War it was the best and most consistent source of fine turf.

Sea marsh turf has two properties which have led to its virtually total demise as a source of bowling green turf. Firstly, the turfgrasses present, because of the site, had to be salt tolerant. They include two salt-tolerant meadowgrass-like species, *Puccinellia maritima* and *P. distans*, strong creeping red fescue (*F. rubra* subsp. *rubra*) and creeping bent (Hubbard, 1984). Experience has shown that some of the grasses are uncompetitive with annual meadowgrass. In consequence sea marsh turf greens usually begin to look patchy after two or three years. If maintenance is good, the turf improves as the uncompetitive grasses are replaced by annual meadowgrass and overseeded bentgrasses. The other disadvantage with sea marsh turf is that the growing medium is usually very fine sand or coarse silt depending on the site. The material may transmit water but remains saturated when drained (see Chapter 1) and thus can create persistent anaerobic conditions.

Cotula is a remarkably successful plant in New Zealand for creating good playing conditions on bowling greens. It is under trial in the UK and

**Fig. 7.2.** *Leptinella dioica*, one of two species of cotula widely used for bowling green surfaces in New Zealand (photograph courtesy of Brian Way).

it will be interesting to see what the future holds for this plant (Evans, 1992; Gibbs, 1992).

In tropical conditions Bermudagrass (*Cynodon dactylon*) is the most widely used species for golf and bowling greens. There are, nevertheless, other turfgrasses which are used and have potential, including species in the genera *Digitaria*, *Paspalum*, *Pennisetum* and *Zoysia* (Chapter 11).

## Golf tees

Tees are subjected to intensive treading causing compaction and scuffing or divoting on driving. The ill-effects of compaction can be controlled by appropriate rootzone and drainage design and the implementation of regular soil aeration treatments. Provided aeration is satisfactory, a wide range of turfgrasses will tolerate treading although speed of recovery from wear and tear varies. Scuffing and divoting cause the most damage and bentgrasses, because of the tightly knit turf they produce, tend to tear out in patches. In contrast, perennial ryegrass (*Lolium perenne*) which forms a weakly bound but strongly rooted turf is damaged less. It not only resists tearing out, but also recovers from damage more quickly than the finer turfgrasses. Modern cultivars are tolerant of the mowing height imposed on tees so perennial ryegrass is the most appropriate species. Cultivars should be selected which are wear tolerant, compact, tolerant of close mowing and cut cleanly. Tees are areas where the reinforcement of the rootzone with synthetic fibres would appear to be especially valuable.

Somewhat different circumstances apply in the tropics. Bermuda grass with its ramifying stolons and rhizomes provides a very stable and tightly bound rootzone tolerant of the wear and tear imposed on golf tees.

## Fairways

There is no standard prescription for turfgrass species on fairways. On heathland, links courses and areas of infertile pastureland a diverse range of turfgrass species and sedges occur naturally, determined mainly by soil pH and phosphate status (see Rodwell, 1992). The locally native species typically present on old pasture grazed by sheep will survive on fairways provided wear is not too severe and soil properties are not adversely affected by compaction and fertilizer input. When new turf has to be established in these types of situation bentgrasses should be chosen as the key species in seed mixtures.

In the situations outlined above, grass growth is slow and so is recovery from wear. As intensity of traffic on fairways increases, more aggressive grasses are needed to maintain good turf quality. In such circumstances perennial ryegrass is the obvious choice but timothy (*Phleum pratense*) and smooth stalked meadowgrass (*Poa pratensis*) are useful inclusions in seed mixtures for moist and dry situations respectively.

## Rough

The rough can give the course a distinctive character so that design is more relevant than performance. Nevertheless the degree of hazard presented must be considered.

Just as fairways can mimic grazed pastureland so the rough can be representative of wild moorland or heath or a hay meadow or whatever is locally appropriate. Within the concepts of the game of golf there is potential in the hazard component which few golf course architects exploit. The bunker reproduces a natural feature on links land but to the outsider this seems peculiar on an inland course. Golf courses in and adjacent to conurbations are green jewels not only in their vegetation but also in the insects, animals and birds they support. The rough is that part of a course which can reproduce with some artistic licence the local, natural, 'unspoilt' environment. No other game has this potential and, in this particular aspect, turfgrasses play a complementary rather than dominant role.

## PERFORMANCE CRITERIA

## Golf

The visual appearance of a golf course is, for many, the key index of playing conditions. Aesthetic appeal is however highly subjective and it is not uncommon for golf greens to be criticized without reference to the criteria which reward skill.

Surprisingly, golf is not as advanced with performance criteria as with other games like soccer, bowls and cricket. A possible explanation for this is that research and development involving golf course agronomy in the UK has been fragmented (see also Maintenance regimes . . . ). Indeed it has only been since the mid 1980s that research programmes have begun to assess the requirements of the surfaces for golf. Hayes (1992) noted that the lack of agronomic research contrasts greatly with the vast amount of research into equipment to play the game resulting from commercial competition (e.g. Cochran, 1990).

The diverse range of turf surfaces on which golf is played makes a description of performance criteria difficult. Hayes (1990) summarized the attributes of good playing surfaces for golf (Table 7.1). Some of the descriptive characteristics listed in Table 7.1 can be made quantitative, for example those relating to ground cover and drainage characteristics.

Most work on playing characteristics has concentrated on golf greens for it is these that by and large determine the playing quality of a course. Playing quality can be defined as the characteristics of the turf surface which make it suitable for the sport in question, as measured by relevant technical tests or as perceived by players. For golf greens, components of playing quality listed by Canaway (1990) include ball roll, ball rebound resilience (low energy, e.g. short chip), ball impact behaviour including the effects of spin (e.g. iron shot to the green) and green hardness.

The measurement of green speed is undoubtedly the most widely used

**Table 7.1.** Characteristics of good playing surfaces for golf.
(From Hayes, 1990.)

**Greens**
Fast, true, firm surfaces
Uniform grass cover (fine texture)
Dry surfaces
Suitability for year-round play

**Approaches and surrounds**
Dry surfaces
Freedom from wear patterns
Good cover of grass

**Tees**
Large, level, firm, dry surfaces
No wear pattern on and off tees
Uniform grass cover

**Fairways**
Free drainage
Suitability for year-round play
Good cover of grass
Lies uniform

assessment of performance criteria for golf greens. Green speed is governed by rolling resistance; the faster the surface, the lower the rolling resistance and vice versa (Canaway and Baker, 1992). Both golf and bowls use a similar concept of green speed but the methodology differs. In golf, green speed is measured as a *distance* rolled when a golf ball is released down a standard ramp, whereas for bowls, the *time* taken for a bowl to reach the jack is recorded.

The measurement of golf green speed is one of the few tests developed as a management aid. The original technique was developed by an American called Stimpson (the 1935 Massachusetts champion) and has been promoted as the United States Golf Association (USGA) stimpmeter for many years (Radko, 1977).

The stimpmeter is essentially an extruded aluminium bar, 914 mm (36 inches) long with a V-shaped groove extending along its entire length and which is supported to act as a ramp of fixed slope (Fig. 7.3). It is designed so that a golf ball meets the green surface at a standard velocity. The stimpmeter has been widely used in the USA for assistance with the preparation of greens by identifying variations in green speed whether these be within greens or between greens (Engel *et al.*, 1980).

The USGA published a green speed test comparison table in 1977 based on research using the stimpmeter which was the first published standard for

**Fig. 7.3.** A USGA stimpmeter being used to measure the 'speed' of a golf green.

**Table 7.2.** USGA green speed test values (metric equivalents).

| Distance rolled (m) | | |
|---|---|---|
| Regular membership play | Tournament play | Green speed |
| 2.59 | 3.20 | Fast |
| 2.29 | 2.90 | Medium-fast |
| 1.98 | 2.59 | Medium |
| 1.68 | 2.29 | Medium-slow |
| 1.37 | 1.98 | Slow |

the playing quality of golf greens (Table 7.2). Since that time the stimpmeter has also been used to quantify the effect of major variables on green speed and its consistency. Such variables include the effect of mowing procedures, rootzone construction, irrigation, and turf nutrition. For research work conducted in the UK, the stimpmeter has been modified to reduce errors in the measurement of ball roll, particularly where it is used on small plots of only 2 m × 2 m in size (Lodge, 1992b).

Although the management of the turf surface has a large effect on its green speed, the different grasses used for fine turf alter the playing properties of golf greens. Canaway and Baker (1992) found that slender creeping red fescue provided the fastest surface for both golf and bowls, and that annual meadowgrass was consistently the slowest. The browntop bents and chewings fescue species were intermediate and comparable in performance. Annual meadowgrass was equated with the medium–slow category in the USGA table and slender creeping red fescue was the only species to reach the medium–fast category. Of the other variables studied, surface moisture decreased green speed and a reduction in cutting height caused an increase. The effect of moisture was greater for the smaller golf ball than for the larger bowling wood.

Apart from golf greens, the only other areas of the golf course that have received some attention in the UK are bunkers. Baker *et al.* (1990) provided a series of qualitative guidelines:

1. The sand must not be so hard that the ball bounces out of the hazard, nor too soft so that the ball plugs to an excessive depth.
2. The sand must provide stable footing.
3. The sand must be stable on high angles of slope to reduce the maintenance requirements.
4. The sand must be free-draining and not be susceptible to wind movement.
5. Sand chipped on to the green must not interfere with the trueness of the putting surface nor contain a high lime content.
6. The colour of the sand must be aesthetically appealing (e.g. tan, light grey or white).

**Table 7.3.** Recommended and acceptable limits of sand size for golf bunkers on inland courses. (From Baker, 1990.)

|                   | % passing        |             |
| ----------------- | ---------------- | ----------- |
| Sieve size (mm)   | Recommended      | Acceptable  |
| < 0.063           | 0-1              | 0-2         |
| 0.125             | 0-2              | 0-5         |
| 0.250             | 0-40             | 0-50        |
| 0.500             | 35-90            | 25-100      |
| 1.000             | 90-100           | 85-100      |
| 2.000             | 100              | 96-100      |
| 4.000             | 100              | 98-100      |
| 8.000             | 100              | 99-100      |

Most of the above criteria are achieved by using a sand of an appropriate angularity and particle size distribution so that a golf ball impacts to give a 'fried egg' lie with half the ball's diameter buried beneath the surface (Beard, 1982). Baker *et al.* (1990) showed that excessive plugging tended to occur in coarse, uniform sands, in many cases making control of bunker shots virtually impossible because the plugging depth was greater than the diameter of a golf ball (42.5 mm). In contrast, shallow plugging depths of around 9 mm or less occurred with finer sands and those with a large Gradation Index. These sands left the ball lying on the surface such that a bunker did not act as a hazard.

Baker *et al.* (1990) agreed that the particle size range specified by the USGA Green Section's staff of 0.25 to 1.0 mm was suitable for bunker sands with the use of natural dune sand of 0.125 to 0.50 mm size being appropriate for the deeper and smaller bunkers associated with traditional links courses (Table 7.3). These workers further concluded that stability for footing and prevention of sand movement off the bunker face was best achieved by avoiding sands with greater than 60% of grains in the rounded and well-rounded shape categories. Angular sands allow bunker faces to be maintained with relative ease up to slopes of 35°, the maximum recommended by Beard (1982).

## Bowls

Unlike crown greens, the playing quality of flat bowling greens has been studied in detail in the UK (Holmes and Bell, 1986; Bell and Holmes, 1988) and in New Zealand on flat cotula greens (McAuliffe and Gibbs, 1993).

Four criteria are recognized to be important components of natural turf bowling surfaces for flat greens:

- A level surface
- A fast green speed
- An even draw on both backhand and forehand
- A good cover of fine turfgrass

Furthermore, a bowling green must be able to drain sufficiently fast to permit play after rain, but be relatively unaffected by drought.

In the past, the performance of a bowling green was described using qualitative criteria such as 'slow' or 'heavy' or 'fast', but now the quantitative criterion of green speed is used extensively. This is defined as the time in seconds for a biased bowl to travel from a bowler's hand and stop within 0.15 m of a jack located 27.4 m (30 yards) from the front edge of a bowlers' mat (English Bowling Association, 1986). The more seconds a bowl takes to travel, the faster the green. Thus a 9 sec green is slow, but a 14 sec green is relatively fast by British standards. In New Zealand even green speeds of 14 sec are considered unacceptably slow and speeds of more like 16–18 sec are preferred, on account of the different playing characteristics of cotula. It is not surprising therefore that bowlers from New Zealand described the English greens of the 1992 World Bowls Championships at Worthing as 'alien'.

The amount of 'draw', that is, the width of the curve taken by the wood as it travels towards the jack, is related to the speed of the green. The more the green allows the wood to trickle slowly before stopping at the end of its run, the more it will curve. The bias on the wood is a fixed and constant imbalance but the extent to which it causes the wood to curve will increase in relation to the time it takes the wood to cover a particular distance. Thus the slower the wood travels (but continues to travel) the more it will draw (Fig. 7.4). On very fast greens (18 sec or more), it is possible for the extent of the curve to cause problems during competitions because the woods are increasingly likely to cross rink boundaries causing possible interference or collision with bowls in adjacent rinks (Bell and Holmes, 1988). The unique feature of the draw, combined with green speed, are the two most important playing quality characteristics of flat green bowls.

Green speed increases with increasing hardness of the surface, closeness of cut and dryness of soil and turf. The average speed of flat bowling greens in Britain is about 11 sec with a typical variation of between 9 and 14 sec around the country (Holmes and Bell, 1986). In cooler and wetter conditions soil organic matter contents are usually greater and soil and turf are more frequently moist, thus green speed generally declines the further north and west one travels. For example, during the 1970 Commonwealth Games at Edinburgh, green speed rarely exceeded 9 sec whereas the average green speed in southern England in midsummer is likely to be nearer 12 sec. Green speed would seem to be of no practical value for assessing the performance of crown greens (Evans, 1992).

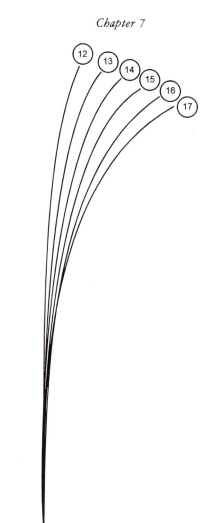

**Fig. 7.4.** Schematic illustration of different trajectories of a bowl showing how the extent of the draw changes with increasing green speed. Hypothetical greens speeds (in seconds) are shown in the bowl at the end of each arc.

With the widespread availability of laser technology for surveying equipment, it has been possible to make detailed studies of the levelness or evenness of flat bowling greens and to make recommendations based on survey results (Holmes and Bell, 1986; Bell and Holmes, 1988). Results from such surveys have shown that the levels on bowling greens in the UK vary greatly. None of the 72 greens surveyed by Bell and Holmes met the standards of Gooch and Escritt (1975) who recommended that the levels of newly constructed

**Table 7.4.** Variation in height ranges for flat bowling greens.

| Height range (mm) | UK[a] | NZ[b] |
|---|---|---|
| Average | 65 | 25-30 |
| Best green | 23 | 8 |
| Worst green | 226 | 63 |
| Planed green[c] | – | 10-15 |

[a] Bell and Holmes (1988).
[b] Walmsley (1990).
[c] Planing is a technique used in Australia and New Zealand where a rotary cutting machine shaves off the top 25-50 mm of a bowling green surface to re-establish surface levels and/or remove excessive thatch.

greens should be within ±6 mm of the average height of the green.

When compared with data from New Zealand, British greens have been shown to have a larger variation in surface evenness (Table 7.4). This difference is probably because poorer levels can be tolerated on slower greens.

Caution needs to be used when judging the levelness of a green solely on the basis of the difference between the highest and lowest spot height (the height range). It is possible to have a perfectly good playing surface with a 50 mm difference between the highest and lowest areas, provided there is a gradual fall from one end of the green to the other. Such a fall represents a gradient of less than 0.15% which is negligible. Variation around the average height (i.e. the standard deviation) and the minimum allowable difference between adjacent spot heights are more relevant criteria for the levelness of the surface. Following the work conducted by Bell and Holmes, which included an extensive survey of player perceptions of playing quality and evenness, it is possible to prescribe acceptable and preferable standards of surface evenness and green speed for flat greens (Table 7.5).

## DETERIORATION AND OPTIONS FOR UPGRADING OR RECONSTRUCTION

Deterioration of a golf course implies that the playing surfaces cease to provide one or more of the desirable characteristics described in Table 7.1. Attention is directed primarily to the parts of a course referred to in Table 7.1 which receive active play, nevertheless the prevention of serious deterioration of hazard features is an important consideration. Thus bunkers or sand traps, lakes and streams, woodland and shrubby plants require consistent maintenance. Not only this but there is a need for tree surgery, tree felling and replanting on a regular basis as part of long-term planning.

**Table 7.5.** Playing quality standards for flat bowling greens. (From Bell and Holmes, 1988.)

| Standard | Surface eveness for individual rinks[a] | | Green speed for individual rinks (sec)[b] |
|---|---|---|---|
| | Difference between adjacent spot heights (mm) | Standard deviation of spot heights (mm) | |
| Preferred maximum | 6 | 10 | |
| Acceptable maximum | 10 | 15 | |
| Preferred minimum | | | 12 |
| Acceptable minimum | | | 10 |

[a] Measurements should be carried out on a 2 m grid survey within the main playing area of the green (i.e. excluding the edges of the green), with the exception of the limits for the differences between adjacent spot heights which should apply to the outer 2 m edge of the green as well as the main playing area.
[b] All rinks playing during a competition should have green speeds within 0.5 sec of one another. Prior to testing of green speed, the rainfall shall not exceed 5 mm in the last hour, 10 mm in the last 5 hours, 15 mm in the last 10 hours, and 25 mm in the last 24 hours.


The golf course includes several types of distinctive area on which play occurs. The perception and nature of deterioration on tees, fairways and areas peripheral to greens are different from those on the greens themselves which are the main focus of attention.

# Golf greens

## *Deterioration*

The playing quality of a green is a subjective assessment. For example although the speed of a green can be measured with a stimpmeter in a quantitative manner a 'fastest is best' philosophy may be unacceptable. There is even greater potential for argument about the ideal 'holding' quality of a green. That is the way it affects the behaviour of a ball pitched with backspin. Preferences within what might be accepted generally as a good performance category for any attribute are subject to change in fashion (see Park, 1990). There has been a trend towards faster and less holding greens in recent years which are more demanding on player expertise. Their attainment from the greenkeeping point of view is mainly due to an upgrading in green maintenance strategy and the availability of better maintenance equipment.

There are three aspects of deterioration on golf greens:

1. Loss of surface trueness.
2. Loss of firmness of surface.
3. Unusable because of surface wetness.

The playing surface becomes untrue if the integrity of the turf is lost by disease attack or wear. Incompetent mowing and the presence of earthworm casts can be factors on poorly maintained greens. Trueness is affected by the presence of broadleaved weeds, of which white clover (*Trifolium repens*) and pearlwort (*Sagina procumbens*) are usually most important, and mosses. The effect of broadleaved weeds is worst in winter when they die back to leave gaps in the turf. It is often asserted that the patchy occurrence of different turf-grasses, for example annual meadowgrass and bentgrasses, can cause uneven-ness but this is probably attributable to an insufficient frequency of mowing.

A loss of surface firmness and the creation of a water-retentive surface are both attributable to thatch accumulation. The thatch layer remains saturated and spongy throughout the winter and following irrigation. It is the most widespread and serious problem on both golf and bowling greens. A soft, wet, thatchy surface favours annual meadowgrass and the dominance of the latter is often misinterpreted as a cause rather than a consequence of the former.

## Upgrading and reconstruction

Deterioration in green condition due to weed invasion or disease requires immediate action but this action can produce rapid benefits. A range of effective selective herbicides is available to control broadleaved weeds whose presence cannot be excused. Maximum control is effected in good growing conditions (temperature, moisture and nutrient status). We are entering a new era of rapid turfgrass disease identification by immunological techniques which should assist in combating the damage caused by diseases (see Chapter 5).

It is certainly easier to prevent a surface accumulation of thatch than it is to remove one. As a general rule it is possible to regain a rootzone surface with an acceptable organic matter content (around 10%) provided firstly that it is possible to scarify or verti-cut completely through the thatch layer and secondly that turfgrass rooting occurs below the thatch layer. In practice a 10 mm thatch layer can be dealt with but one deeper than 20 mm cannot, because neither of the criteria are met. Turfgrasses will withstand a tremendous amount of chopping and disruption. The removal of 10 mm of thatch cannot be achieved at one time. Removal requires a two-season programme of vertical mowing allied to topdressing to disintegrate a thatch layer and create a con-tinuity with depth of mineral rootzone. Hollow tine spiking (coring) will con-tribute to the amelioration but will not, alone, achieve removal of thatch.

On some courses the problem of thatch accumulation has been contributed to by the use of topdressing containing far too much organic matter.

Reliable specifications for the reconstruction of golf greens are being implemented more generally. There is no one set of ideal specifications but at least the basic principles of design are accepted. The most critical compo-nent is the rootzone. This is subjected to intensive traffic but must remain

**Table 7.6.** Recommended composition of sand/soil rootzones for golf greens.

| | |
|---|---|
| Percentage less than 125 $\mu$m | Less than 10% |
| Percentage in the range 125- 500 $\mu$m | Not less than 70% |
| Particle size with 50% smaller ($D_{50}$) | 330 ± 30 $\mu$m |
| Percentage organic matter | 1.5-3.5% |

**Table 7.7.** Physical properties of the final rootzone mix for a USGA golf green (USGA Green Section Staff, 1993).

| Sieve size (mm) | Recommendations (% w/w) |
|---|---|
| 2.0-3.4 ⎫<br>1.0-2.0 ⎭ | Not more than 10% of the total particle size in this range including a maximum of 3% fine gravel (2.0-3.4 mm) |
| 0.5-1.0 ⎫<br>0.25-0.50 ⎭ | Minimum of 60% must fall in the 0.25-1.0 mm size range |
| 0.15-0.25 | Not more than 20% in this range |
| 0.05-0.15<br>0.002-0.05<br>< 0.002 | Not more than 5% ⎫ Total shall not<br>Not more than 5% ⎬ exceed 10% in<br>Not more than 3% ⎭ this range |

| Physical property | Recommended range |
|---|---|
| Total porosity | 35-55% |
| Air-filled porosity (at −40 mbar) | 15-30% |
| Capillary porosity (at −40 mbar | 15-25% |
| Saturated hydraulic conductivity | 150-300 mm h$^{-1}$ (normal range)<br>300-600 mm h$^{-1}$ (high range) |
| % organic matter by weight | 1-5% |

Note that inorganic amendments (other than sand), polyacrylamides and reinforcement materials are not recommended at this stage.

permeable to water. The usual rootzone is a sand-dominant material produced by mixing sand with soil or sand with soil and a source of organic matter. A pure sand system has been used successfully, an example of which, the Purr-Wick system, is illustrated in Fig. 3.14.

In order to describe an appropriate constitution for a rootzone material the specifications should indicate the maximum amount of fine soil fractions acceptable, the general spread of particle sizes and the predominant size distribution. An acceptable range in organic matter contents should also be specified. Table 7.6 shows a recommended composition for UK conditions (see Chapter 1). The USGA specifications for golf green rootzones include both particle size and physical property parameters (Table 7.7). These specifications provide for a rather coarser and more free-draining framework than those in Table 7.6 and would be expected to increase irrigation requirement.

Provision must be made for the water draining through the rootzone to be conducted to a drainage outfall. The total profile construction is prescribed in detail in the USGA specification for putting greens (USGA Green Section Staff, 1993). The essential components are given below:

**1.** A subgrade contour which conforms to the general slope of the finished surface, excavated to avoid water-collecting depressions.

**2.** A subsurface pipe drainage system with a 100 mm minimum diameter main drain receiving 100 mm diameter lateral pipe drains spaced no more than 5 m apart. Perimeter lateral drainage is also recommended to facilitate drainage of water that may accumulate at the lowest end of the drained area. Fibre-wrapped honeycomb drains or other geotextile wrapped drains are not recommended (geotextiles are however permissible on unstable subgrades, but they should not cover the drainage trenches). A minimum 25 mm layer of gravel in the base of the trenches is recommended and a positive slope along the entire run of the drainage lines should be ensured.

**3.** A layer of gravel to a minimum thickness of 100 mm, conforming to the proposed final surface grade to a tolerance of ±25 mm. Soft limestones, sandstones or shales are not acceptable.

**4.** An intermediate blinding layer of 50–100 mm thickness depending on the particle size distribution of the rootzone relative to that of the gravel. With the USGA recommended gravel, the blinding layer is considered unnecessary. However a gravel with the majority of particles (at least 65%) in the 6–9 mm size range and with not more than 10% of its particles greater than 12 mm will require an intermediate blinding layer of material containing 90% of its particles in the 1–4 mm size range.

**5.** A thoroughly mixed and approved rootzone of a firmed 300 ± 12 mm depth. It is suggested that the rootzone be moist when spreading to assist firming and to discourage migration. Specified mixtures of sand, soil and organic sources have to be first approved by an official USGA recognized soil laboratory.

The typical cost of constructing a green to USGA specifications is around £20,000. Costs of the other components in a typical golf course construction in relation to these are illustrated in Table 7.8.

When the native soil is naturally freely drained and its structure has not been destroyed by earthmoving it is possible to avoid total profile reconstruction. The most appropriate scheme is to use the basic Prunty–Mulqueen design for a sand carpet but to place 150 mm of sand-dominant rootzone onto the slit drained base layer instead of sand alone (see Fig. 3.10). This type of design is also well suited to bowling green construction (see later).

**Table 7.8.** Estimated costs for golf course construction at 1992 prices. (From Hayes *et al.*, 1992.)

| Component | Cost (£) |
|---|---|
| Suspended watertable/USGA type green | 18,000-22,000 each |
| Tees | 3500-5000 each |
| Fairway pipe drainage | 8000 per hectare |
| Through the green areas (e.g. earthworks plus drainage) | 5000-20,000 per fairway |
| Tree planting | up to 40,000 |
| Automatic irrigation system (tees and greens) | 80,000-100,000 |
| New maintenance equipment | 120,000 |

## Approaches and surrounds to greens

Deterioration of the immediate periphery of greens is usually attributable to one or both of two causes. One is waterlogging through seepage, the other is through excessive and localized compaction. Greens are often cut back into a slope. The cut face is a source of seepage water from upslope and this must be prevented from flowing onto the green. A cut-off drain is vital outside the edge of the green and the surface of the green should be designed to slope back towards the cut face to prevent surface water running out onto the green. Excessive compaction is usually a result of poor design which causes a concentration of traffic by players and machinery accessing the green. When intensive use is anticipated then a construction design comparable with the green itself is advisable. One factor that should be borne in mind is that distinct collars on the perimeter of greens are especially susceptible to dry patch.

## Tees

The usual reason for excessive wetness and loss of turf cover on tees is a failure to appreciate the intensity of use they receive. In addition the tee surface is subjected to divoting. One advantage tees have over greens is that they can be mown at a height where cultivars of perennial ryegrass can thrive. Thus root development and turfgrass vigour for recovery are potentially better on tees than greens.

One way of reducing the intensity of wear on tees is to build alternative tees. Another is to increase the size of tees. On short holes divoting is worse because good players will pitch to the green off the tee surface to induce backspin rather than drive off a tee peg. For this reason tees for short holes should be wider and somewhat larger than for long holes. Hayes *et al.* (1992) recommended a total area of 300 m$^2$ for mid-to-short iron drive distances and 250 m$^2$ for long iron or wood drives. Hawtree (1983) considered that tees of around 450 m$^2$ were required for par-3 holes on intensively used courses.

Most golf courses have native soils which are inadequate for tees; however, tees do not normally warrant a total profile construction which may be adopted for greens. A 100–150 mm depth of sand/soil rootzone with the composition given in Table 7.6 would be appropriate over a slit drained base. Sand/gravel slits 50 mm wide and at 2 m centres would suffice. The rootzone could be produced on-site or off-site using native topsoil and an appropriate sand.

The divoting of tees is a serious problem and there is an increasing use of rootzone reinforcement materials. Rootzone formulations containing random mesh elements or fibres are now available and long-term trials are continuing (also see Chapter 5).

## Fairways

Fairways are the thoroughfares for play on golf courses and a typical 18 hole course has around 17 ha of fairway. The intensity of traffic on fairways is less than on greens and tees and the tendency to increase the width of fairways has compensated for an increase in the number of rounds played on many courses.

On courses used with moderate or low intensity, fairways usually contain grass and herb species typical of local low grade sheep pasture. On heathland courses, sedges are usually present and fine-leaved fescues are far more common than on the more intensively managed and used greens. The wide diversity of species found on nutrient-poor soils is perfectly acceptable but the slow-growing species present cannot tolerate a major increase in intensity of use. The greenkeeper's response to excessive wear is to sow more aggressive turfgrass species and to fertilize. This action can enable recovery but the greenkeeper must anticipate and be able to respond to new situations – for example a major increase in clover and maintenance more in keeping with a winter games pitch than an ornamental lawn.

Deterioration of fairways can occur through weed and moss invasion but most frequently it is due to an increase in the intensity of use. The difficulties encountered with an increase in usage is illustrated well by the experience of Pontardawe Golf Club. The club which has a heathland course is still quite small but membership increased by almost three-fold over the ten years up to 1992. Walkways became quagmires, the fairways were widened but then they came up to the soakaways in the rough used to accept drainage water from the original, narrower fairways.

Bottlenecks in traffic are potentially a major problem and can be resolved in one of two ways. One is to deliberately concentrate traffic on walkways etc. and create non-turf hard surfaces to cope with it. This is akin to tramlining in cereal production. The alternative is to spread traffic lines and avoid a concentration on set routes. The latter, which is typical of the approach of farmers moving vehicles over pastureland, is appropriate for low and moderate intensities of use. High intensities of use demand the former approach although both may be incorporated on the same course.

The drainage of fairways follows in general the philosophy of drainage for agricultural land. Important points are the installation of cut-off drains to prevent extraneous water entering the site and the pipe draining of obvious spring lines and wet spots. This apart, the aim should be to control watertable depth and thus increase the rate at which soils return to 'field capacity' after heavy rain. Drain spacing will range from 5 m to 20 m depending upon site and annual rainfall.

When drains are installed it is vital to cover the pipes to no deeper than 150 mm from the surface with 6–9 mm gravel as permeable fill. This will ensure that should surface wetness become a problem, sand/gravel slit drains could be installed to improve the situation.

# Bowling greens

## *Deterioration*

A good green speed and a smooth, flat surface are the key criteria which determine the quality of a bowling green.

The species of turfgrass on a green affect speed to some extent but the firmness and water-retentiveness of the rootzone surface are dominant factors. If the original rootzone design was appropriate then the most common cause of deterioration in common with golf greens is the development of a thatch layer.

The critical type of deterioration on a bowling green which does not apply to golf greens is the loss of surface flatness. On a short-range scale the smoothness of run of a wood depends upon having a turf without local undulations. Unevenness on the scale of a few centimetres causes 'bobble'. A smooth run can be achieved not only on a firm green but also on a green which is very soft and moist. Indeed a significant deterioration in the smoothness of run usually accompanies measures including hollow tining which are often used to ameliorate the soft and water-retentive surface caused by thatch. A more drastic approach to the removal of thatch is used in Australia and New Zealand. The thatch with surface rootzone is shaved off completely using a rotary cutting machine to a depth of 25–50 mm (Fig. 7.5). This machine can also be used to restore surface levels.

**Fig. 7.5.** 'Planing' a bowling green in New Zealand to re-establish surface levels and/or remove excessive thatch (photograph courtesy of Bill Walmsley).

Deterioration in levels over larger distances is caused by settlement or subsidence. Surface levels can now be expressed visually using computer graphics packages. The STRI uses such a system to present a three-dimensional view of the surface contours of the green from a 2 m grid survey. In New Zealand a similar surveying package is offered by the NZ Turf Culture Institute (Fig. 7.6).

It is practicable to correct discrepancies in levels up to about 20 mm by topdressing. Individual selective topdressings should not exceed 6–8 mm depth and these must be applied in the close season to permit stabilization. Hollow tine coring can be used in a complementary manner to lower high spots. The correction of major errors in levels requires the removal of turf. Laser levelling techniques are now available for use during construction and for the correction of deteriorated levels on greens. This type of technology has largely replaced variations on the levelling with strings or wires approach.

### Upgrading or reconstruction

Most of the problems which afflict bowling greens and the measures needed to ameliorate the situation are in common with those described earlier for golf

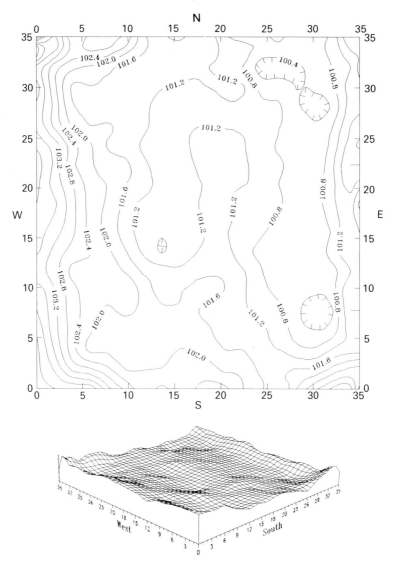

**Fig. 7.6.** Computer generated contour survey and three-dimensional image of a 'flat' bowling green. The vertical scale on the bottom image is exaggerated to show the extent of the levels problem. Individual spot heights are not shown on the contour survey. Contour heights are in centimetres.

greens. There are however some special features concerning bowling greens. The first is the legacy of the Scottish sea marsh turf construction system.

The Scottish construction system was advanced for its time because it not only involved growing sea marsh turf over a soil-less medium, but also utilized a comprehensive pipe drainage system. The profile was comprised of up to 300 mm of hard core (e.g. clinker, ash or broken stone) followed by a 25 mm to 75 mm layer of fine ash and a similar depth of sand before finally being covered with sea marsh turf of approximately 30–35 mm thickness (Evans, 1992).

The method was widely used for bowling green construction in the UK particularly during the 1920s and 1930s, although some modifications to the method involved the use of cheaper sources of turf. Sometimes seed was used instead of turf which necessitated the use of topsoil over the fine ash/clinker base. Many crown greens were also built using the Scottish construction system with the clinker and fine ash being used to create the desired contours.

The sea marsh/clinker system had many initial advantages as listed by Evans (1992), including:

1. Subsidence problems were minimized because of the firm clinker or stone base.
2. Good levels were relatively easy to achieve because of the ease of grading the final sand layer.
3. Cheap reliable sources of material were widely available at the time because they were industrial waste products.
4. The greens initially drained exceptionally well which was a particularly important consideration on the west coast of Scotland.
5. Worm cast problems were minimized as a result of the soil-less rootzone.
6. The use of sea marsh turf allowed uniform turf lifting and therefore an acceptable bowling surface could be established quickly and easily.

However, from a soil and maintenance perspective, the turf and construction system had three severe limitations. Firstly, greens were very prone to drought because of the poor water retention of the construction. By today's standards, this would not necessarily be a limitation with the advent of automatic sprinkler systems. Secondly, the greens frequently suffered from severe nutrient deficiency as a result of the soil-less nature of the rootzone. Excessive and unbalanced fertilization often resulted in increased disease incidence and a competitive disadvantage to sea marsh turfgrasses. Finally, the worst limitation resulted from the silt layer in which the sea marsh turf grew. Although this layer provided a satisfactory playing surface in the short term, the layer was prone to compaction which destroyed all macropores. As a result, physical growing conditions for the sea marsh turf became more difficult after two or three seasons' use. Moreover, the silty layer gradually became buried by subsequent topdressings applied in autumn as part of renovation maintenance. The overall result was an anaerobic buried layer acting as a barrier against root and water movement. Although use of the hollow tine fork invented by William Paul around 1920 alleviated the problem to some extent because sections of the silty layer could be physically removed, a buried sea marsh silt layer continues to be a problem on some greens.

The post-war decline in sea marsh turf and its construction system for bowling greens was replaced by a return to using traditional soil and improved cultivars of turfgrass seed as an embryo turf-seed-producing industry developed. Unfortunately, where soil was used as part of a total profile construction (often over a traditional Scottish foundation of clinker and ash), the rootzone depths were too shallow and many greens were covered with only a 50–75 mm deep layer of soil. Such constructions were droughty in summer and waterlogged in winter.

Deeper soil layers were recommended by the STRI in the 1950s with a minimum of 100 mm now being recommended for a turfed green and 150 mm for a seeded surface. Two further points require clarification regarding the evolution of modern bowling green construction. The first is that the currently recommended system in the UK is essentially a suspended watertable design (Fig. 3.9) where the hard-core clinker base has been replaced by graded stone, gravel or Lytag, and the ash layer has been replaced by a coarse blinding sand. The second point is that the rootzone is usually an off-site prepared sand/soil mix made to precise specifications typical of those used for golf greens (Tables 7.6 and 7.7) using a sand meeting the specifications given in Fig. 4.14. Sand/soil mixes are generally prepared to provide a minimum saturated hydraulic conductivity of 50 mm h$^{-1}$.

The organization of pipe drainage in general was discussed in Chapter 4. For bowling greens, a system of parallel lateral drains at 4.5–7.0 m spacings running diagonally beneath the gravel raft is the most widely used design (Fig. 4.8). Perimeter ditch drains are also advised for both flat and crown greens. Precast concrete curbing or glass reinforced cement curbing is now also widely available for ditch construction, replacing wooden edging and grass banks.

The particle size distribution appropriate for a bowling green rootzone is given in Table 7.9. Despite the demonstration that a coarse pore drainage layer causes a perched watertable, the depth of rootzones recommended over a gravel raft is still typically no greater than 150 mm (Evans, 1992). In contrast, the USGA Golf Green Specification recommends 300 mm for a similar design.

Reconstructed bowling greens are often turfed rather than seeded largely for historical rather than practical reasons. The only gain in time which can be achieved by turfing is if a green is turfed in autumn and is played on the following season. Apart from this, both systems involve the loss of one season's play and establishment from seed enables the creation of a uniform rootzone, with no discontinuities with depth.

Complete profile reconstruction can be avoided by the use of a sand-

**Table 7.9.** The mean particle size distribution of the rootzones of three bowling greens at a club in the UK hosting national competitions.

| Particle diameter | % by weight |
|---|---|
| 2-1 mm | 2 |
| 1-0.5 mm | 15 |
| 500-250 $\mu$m | 58 |
| 250-125 $\mu$m | 12 |
| 125-60 $\mu$m | 2 |
| 60-20 $\mu$m | 1 |
| 20-2 $\mu$m | 5 |
| < 2 $\mu$m | 5 |
| $D_{20} = 190\ \mu$m | % 125-500 $\mu$m = 70 |

dominant rootzone of 150 mm depth with sand/gravel slit drains in contact with the base of the rootzone at around 2 m centres. Such a design on a soil base avoids a capillary break layer and is economical to construct (see Golf green and tee construction earlier in this chapter). In essence it is a tailored compromise between the sand carpet and sand/soil rootzone systems described in Chapter 3. The size of a bowling green is such that no more than three pipe drains are required to deliver water from the slit system.

## MAINTENANCE REGIMES FOR GOLF COURSES AND BOWLING GREENS

### Environmental awareness

In the UK there is concern over the decrease in the area of traditional countryside. Golf courses are perceived as potential polluters and exploiters of natural resources. Regardless of the justification for these allegations, designers, managers and administrators of golf courses need to be aware of their responsibilities as custodians of important ecological habitats.

Environmental impact assessment is a virtual prerequisite for any golfing development and golf courses can range from being seriously damaging to the natural environment to providing major contributions to nature conservation. The potential benefits relate mainly to the areas of rough on golf courses. Currently these areas comprise around 40,000 ha in total in the UK.

The Nature Conservancy Council works closely with golf courses in safeguarding Sites of Special Scientific Interest (SSSI) that harbour rare plants or animals. Fifty top-ranking courses in the UK have SSSI status. In the USA, a similar situation exists whereby golf courses can register with the Audubon Cooperative Sanctuary Programme, an organization which promotes ecologically sound management and conservation of natural resources. There are several aspects of environmental control and natural resource management that will directly affect golf courses, both in the UK and overseas. The use of biocides and other chemicals will become more limited by legislation. Water use for irrigation will become more restricted and more costly. On the other hand, the conservation value of golf courses will become more appreciated as land is lost irretrievably to building, road construction and other uses.

Some countries in Europe have already put a blanket ban on certain pesticides making it much more difficult to control turf weeds, pests and diseases. In the UK, legislation is not so severe, although there has been a marked change in the legislation affecting the use of chemicals. This legislation affects all turf managers, not simply those working on golf courses. The main changes to pesticide legislation stem from two Acts passed in 1986 and 1990:

- **1986 Control of Pesticides Regulations (COPR), Food and Environmental Protection Act (FEPA).** This Act aims to control the products that can be used on turf areas, dealing also with safe storage, disposal and legally binding instructions on use. For example, golf clubs are now required to ensure that their greenkeepers have received adequate training and examination leading to a Certificate of Competence.
- **1990 Control of Substances Hazardous to Health Regulations (COSHH).** This Act is primarily aimed at protecting people at work by providing risk assessment and a reduction of exposure to hazardous substances. The provision of suitable training and information as well as records of exposures to hazardous substances are also requirements of the Act.

The recognition that greater control on the use of biocides and water is inevitable, has stimulated research to cope with the developing situation. The USGA has been foremost in funding research projects directed towards several relevant aspects and has sponsored a substantial review of current information (Balogh and Walker, 1992).

## Components of maintenance of golf courses and bowling greens

*Mowing*

The mowing practices considered in this section include grass cutting to a height related to the rootzone surface (horizontal mowing) and types of vertical mowing designed to lift or cut through the turf.

The principles governing horizontal mowing were presented in Chapter 2 and the general equipment required in Chapter 5. The closeness of mowing demanded by golf and bowls causes stress to turfgrasses. A high frequency of mowing is essential to minimize this stress. Mowing on alternate days is the guideline in the growing season but mowing frequency in general should be related to growth rate to ensure that no more than one third of the standing height is removed. A cylinder mower with ten or more blades per cylinder is necessary to give a totally unridged cut.

In the 1970s there was widespread adoption of mounted multiple mowers (typically triple) for mowing golf greens. The increase in speed of mowing was gained at the expense of increased turning and travel of a heavier vehicle adjacent to the perimeter of greens. Compaction and deterioration of turf at the edges of greens where a free run on and off was not possible have resulted in a return to the use of pedestrian mowers on prestigious courses. In essence the use of 'multiple' mowers creates special demands and constraints in green design. Pedestrian mowers are used on bowling greens.

Vertical mowers have brushes, tines, wires or cutting blades which pass through or cut through the turf and possibly the upper rootzone. The various types of machine serve one or more of three functions. The first function, which can be achieved by simple or complex rakes or brushes, is to lift prostrate tillers and stolons so that the height of cut approximates to the length of growth of shoots and leaves. Failure to achieve this control of growth results in the formation of crowns by individual plants in the sward and grass covering over, rather than tillering into, bare patches in the turf. The term 'turf groomer' describes this type of machine quite well because the main function is to comb rather than chop or cut. Turf groomers can be fixed permanently to cylinder mowers but, in UK conditions, it seems their action is required less than the frequency of horizontal mowing. Nevertheless, little damage to turf is caused by grooming equipment and use on a fortnightly basis or more frequently is recommended (Hayes *et al.*, 1992). Very frequent grooming results in significant amounts of fertilizer and topdressings being picked up by the mower.

The second function of vertical mowing is to chop through, and enable the removal of, excessive living and dead turf fibre on, and shallowly within, the rootzone. The purpose of this activity is to prevent or at least assist in the prevention of an accumulation of thatch. Excluding consideration of topdressing, the accumulation of thatch depends upon plant residue input and the rate of decomposition of plant residues (see Chapter 5). Minimizing input is therefore very important. The type of mechanical action required usually involves the use of rigid blades rotating on a horizontal reel and it is usually termed scarification. Damage to the turf depends on the depth of cut into the rootzone. Typically, autumn maintenance on a bowling green might involve four passes in different directions with the blades set about 10 mm into the rootzone. Use on golf and bowling greens in the growing season is typically less than once per month with blades barely penetrating the rootzone surface. An important use of this type of vertical mowing is immediately prior to topdressing, to assist its integration with the mineral rootzone (see later).

The first two functions of vertical mowing relate primarily, or in part, to thatch control. The third use of vertical mowing is as an aid in thatch removal. Removal is especially difficult because in some circumstances turfgrass roots are virtually restricted to the thatch layer. Provided it is possible to cut completely through the thatch layer, vertical mowing is a vital complement to both hollow tine coring and topdressing in a programme of thatch removal. Vertical mowing not only enables the mechanical removal of thatch but greatly improves the physical conditions, especially aeration, in the thatch, thereby increasing the rate of microbial decomposition. The mowing requirements of golf courses and bowling greens are summarized in Table 7.10.

*Fertilizer use*

The principles of fertilizer use were presented in Chapter 5. Recommendations for fertilizer use on golf greens in particular have been and still are a major source of controversy. Central to the conflict of opinions has been the alleged effect of phosphate fertilizer use on the success of annual meadowgrass, relative to bentgrasses and fine fescues. Whilst it cannot be denied that high available phosphate may favour annual meadowgrass, the situation is usually confounded by other factors. For example high phosphate usually reflects a high input of other fertilizer nutrients and an increasing percentage composition of annual meadowgrass has been shown to result from high fertilizer N input (Hayes *et al.*, 1992). The usual source of fertilizer phosphate is calcium phosphate which can prevent extreme soil acidity developing. Annual meadowgrass is more successful at less acidic pHs compared with bentgrasses and fescues (Lodge *et al.*, 1990).

Most golf and bowling greens have a much greater than necessary level of available phosphate. Part of the explanation is the better retention of phosphate in sand-dominant soils compared with N and K (see Chapter 1) but another cause is the fertilizer recommendations and compound fertilizer formulations which were in vogue until the mid 1970s. In 1975 it was virtually impossible to purchase, in the UK, one or a combination of compound turf-grass fertilizers which did not result in excess input of P relative to N and K. A transformation in fertilizer formulation began in the late 1970s when it was accepted that the main criterion was the ratio of fertilizer nutrients removed in clippings (Adams, 1977a, b). Thus fertilizers for turf with a nutrient ratio approximately to $4:1:3$ of $N:P_2O_5:K_2O$ became available and, in addition, N/K fertilizers began to be marketed which contained no P and which permitted complete control of fertilizer P input (Adams, 1983).

Fertilizer experiments on turf are often seriously flawed. Factorial experiments involving two or more levels of N, P and K alone and in combination became the paradigm of the agronomist. Rates of input, timing and frequency of input were often decided arbitrarily and no data were supplied on the soil status of the nutrients tested. Nitrogen inputs of in excess of $600 \, \text{kg ha}^{-1} \, \text{y}^{-1}$ were often used and significance was attached to the agronomic response when it was known that fine turf in the UK climate could utilize only about half of

**Table 7.10.** Typical mowing requirements for golf course and bowling green surfaces.

| | Maximum height of cut (mm) | | Type of mower | Preferred number of blades on cylinder | Removal of clippings | Frequency of cut during period of active growth | Approx. no. of cuts per year | Mowing pattern |
|---|---|---|---|---|---|---|---|---|
| | Active growth periods | Slow or dormant growth periods | | | | | | |
| Golf greens (horizontal mowing) | 5 | 7 | Pedestrian/triple | 9-11 | Essential | 3-7 times per week | 120 | Change direction at each mowing using at least four directions |
| Golf greens (vertical mowing[a]) | – | – | Pedestrian/triple | – | Essential | Every 7-14 days | 25 | Change direction at each mowing using at least four directions |
| Tees/golf green surrounds[b] | 12 | 18 | Pedestrian/triple | 6-8 | Preferable | 1-2 times per week | 60 | Different direction on each occasion or on every second cut |
| Fairways | 12 | 20 | Self-propelled or tractor-drawn hydraulic 5 or 7 gang unit | 6-8 | Unnecessary | 1-2 times per week | 60 | Mowing pattern should follow natural contours longitudinally |
| Semi-rough[c] | 37 | 37 | Self-propelled or tractor-drawn gang unit/rotary unit | 4-6 | Unnecessary | 0.5-1 times per week | 30 | Mowing pattern should follow natural contours longitudinally |
| Rough[d] | 75+ | 75+ | Self-propelled or tractor-drawn rotary unit/flail unit | – | Unnecessary | see next column | 1-12 | Mowing pattern blends with surrounding contours and landscape |
| Banks (tee, bunker) | 37 | 37 | Rotary/hover/cord trimmers | – | Unnecessary (except for large unsightly clumps) | 0.5-1 times per week | 30 | Mowing pattern blends with surrounding contours and landscape |
| Bowling greens (horizontal mowing) | 5 | 8-12[e] | Pedestrian | 9-11 | Essential | 3-7 times per week | 120 | Alternate diagonal mowing advisable during playing season; otherwise alternate diagonal + parallel mowing |
| Bowling greens (vertical mowing[f]) | – | – | Pedestrian | – | Essential | see next column | 2-8 | Angle of scarification should vary only slightly between multiple passes (e.g. in autumn) |

a Refers to Verti-cutting units.
b Golf green surrounds are strips of mown turf around each green (approx. 1.5 m wide) and they usually include bunker surrounds and mounds that cannot be mown with fairway mowers.
c Occasionally a further narrow band of semi-rough can be added as a further refinement, cut at a height intermediate between fairway and semi-rough.
d Variation of type of rough means that no hard and fast guidelines on mowing can be recommended.
e 8 mm would be the maximum whilst play is still taking place, with 12 mm being used out of season.
f Refers to scarification units.

**Table 7.11.** Guide to annual fertilizer requirements for golf courses and bowling greens.

| | N<br>(kg ha$^{-1}$ y$^{-1}$) | P$_2$O$_5$<br>(kg ha$^{-1}$y$^{-1}$) | K$_2$O<br>(kg ha$^{-1}$ y$^{-1}$) | Source |
|---|---|---|---|---|
| Golf and bowling greens (clippings removed, earthworms eliminated, sand-based construction, irrigated) | 100-160[a] | 40-80 | 100-140 | Adams, 1977a |
| Golf and bowling greens (soil rootzone) | 80-200[b] | 20[cd] | 60-150[ce] | Lawson, 1991 |
| Golf and bowling greens (sand rootzone) | 250[f] | 50[c] | 200[c] | Lawson, 1989 |
| Golf tees | 80-200[b] | 20[cd] | 60-150[ce] | Lawson, 1991 |
| Golf fairways | 80-120[g] | 0 | 0 | Lawson, 1991 |

[a] Minimum of three applications in March, June and August (e.g. 50, 50 and 20 kg N ha$^{-1}$ respectively), insurance trace element application every five years, controlled-released fertilizer and occasional Mg fertilizer recommended.

[b] Split into three or four applications starting mid April, with the last application being before the end of August (nitrogen content in final application should be of low N concentration).

[c] Requirement for potassium and phosphorus application depends on soil test results.

[d] Applied with first main application in spring.

[e] Given in two applications, one in spring and the other in mid to late summer.

[f] Applied over ten applications using conventional fertilizers (e.g. ammonium sulphate), reducing to 200 kg N ha$^{-1}$ y$^{-1}$ after four or five years (using six to eight applications).

[g] Routine fertilizer treatment generally unnecessary because clippings are returned. Localized applications however are required to sections of fairway receiving heavy traffic (e.g. approaches).

this N in useful metabolism. Nitrogen was often applied in typically three to eight equal fertilizer dressings over the growing season and although this may seem rational there is no objective evidence to justify it. Traditional fertilizer experiments are long outdated. The future lies in experimental designs which enable the fate of applied fertilizer to be monitored. Total accounting is an essential requirement. Fundamental to the response of turf to a particular fertilizer input is the availability status of that nutrient in soil. Available P and K and some minor elements can be determined with moderate accuracy and yet, amazingly, the value of soil analysis is still questioned. Fertilizer inputs of P, K and Mg must take account of soil analysis.

Guidelines on the level of annual input of fertilizer nutrients for golf courses and bowling greens is given in Table 7.11. Fertilizer recommendations are not and never will be precise prescriptions because of the various effects of soil and weather. A sensible approach is to ensure, overall, a 4:1:3 ratio input of $N:P_2O_5:K_2O$ and to modify this on visual appearance (for N) together with results from annual or biennial soil analysis. A typical annual application on golf and bowling greens with sand-dominant soils is, in kg $ha^{-1}$ of N, $P_2O_5$ and $K_2O$, 220, 50 and 150 respectively. The traditional autumn/winter plus spring/winter fertilizer formulations have little merit. The only benefit of a rather greater input of P in the autumn is if autumn maintenance allows P fertilizer to be luted into the soil (Adams, 1983). Otherwise it has a tendency to accumulate at or near the surface.

Fertilizer input should be biased towards the first half of the growing season when demand and response are greatest. No less than 60% of the annual input of N should be applied before mid June. This is especially important if IBDU or other slow release sources are used. No IBDU should be applied after mid June and this is especially important when irrigation is not used. Late-applied IBDU is likely to increase levels of available N in mid and late autumn which will either be utilized by turf to produce soft growth at low light intensities or be leached out as nitrate in drainage water.

Trace element nutrition was referred to in Chapter 5 but two additional aspects should be mentioned here. One is pH control and the other is application of Fe in excess of nutritional requirement. The soil pH on bowling greens and golf greens should be maintained between 4.5 and 5.5. At soil pHs below 4.5 the species of soil microflora able to survive decreases and the reduced biochemical diversity of the residual population results in slower breakdown of organic matter. Furthermore, even the acid-tolerant bentgrasses and fescues perform less well below pH 4.5. At pHs above 5.5, annual meadowgrass is favoured over sown species, take-all patch is more likely to occur and sulphide production or 'black layer' will occur more readily should anaerobic conditions persist (Adams and Smith, 1993). Before the 1950s the main reason for maintaining acidic conditions was to assist in the control of earthworms and broadleaved weeds. This situation will only recur if the biocides available for chemical control are banned. Since topdressings are applied on both golf greens and bowling greens annually or more frequently, these can and should be used to ensure that soil pH is maintained within the desirable range.

Iron sulphate, either alone or in combination with ammonium sulphate, has been used to improve grass colour and to control moss and broadleaved weeds for many years. Iron, especially as ferrous sulphate, is used for cosmetic purposes to give a deeper green colour to grass, to harden grass and as a soil acidifier. Its benefits and recommended applications procedure have been described by Kavanagh (1989). When iron sulphate is applied dry, subsequent foot traffic can cause turf blackening until it has washed off the grass. Liquid application at the rate of $5.5 \, g \, m^{-2}$ every six weeks or less frequently is recommended. It is not advisable to apply iron sulphate between June and mid August.

*Irrigation, dry patch and wetting agents*

Principles of irrigation for sportsturf have been discussed in Chapter 5. The main points are that irrigation should neither be used for cosmetic purposes alone nor to increase the holding of golf greens.

The major problem with the use of automatic irrigation systems for both golf courses and bowling greens is the inability of turf managers to calculate how much water has been applied. Applying '5 minutes' or '10 minutes' of water (i.e. the run time) is meaningless without knowing the flow rate of water from a sprinkler head. Moreover many systems are programmed for night-time operation to reduce interference with play and it is easy to assume that the 'correct' amount of water has been applied when the course manager or greenkeeper returns the next morning. Irrigation programmes cannot be effectively planned and executed without knowing the quantity of water that is applied. Furthermore input must take account of rainfall and evapotranspiration. Maximum anticipated evapotranspiration in the UK in summer is around 25 mm per week so irrigation systems should be capable of this level of input. It has been pointed out in Chapter 5 that because of the pore size distribution of sand-dominant soils, they experience a rapid fall to very low water potentials compared with most natural soils when the easily available water has been removed. Timing of irrigation is therefore very important.

The management of irrigation has a marked effect on the playing quality of golf and bowling greens. Moist greens are slower and there is an interaction between playing quality, irrigation management and the type of construction. Sand-dominant greens produce on average faster surfaces than soil greens and show less variation with moisture content. Soil greens are fast when dry but are for the most part softer and slower than sand-dominant rootzones.

Lodge (1992a) showed that the depth of pitchmarks made by a golf ball, which is related to holding, increased generally from a sand rootzone through sand-dominant soil to soil and that the increase in depth of pitchmark in response to irrigation showed the same pattern.

Dry patch is expressed as irregular areas of turf, visibly drought stressed with the characteristic that the drought stress is not alleviated by irrigation. The soil beneath the turf is dry and cannot be moistened readily (i.e. it is hydrophobic). Dry patch is usually localized in occurrence. Typically it occurs on 'high spots' even when the difference in levels may only be a very few millimetres – a situation typical of bowling greens. Soil desiccation is an essential prerequisite for dry patch development but the detailed cause of irreversible drying varies. Humified or partially humified plant residues become hydrophobic when dried. This came to light in a commercially serious way when soil-less composts began to be used for house plants. The problem was avoided by the use of a wetting agent in the formulation. Lipids and waxes responsible for creating the hydrophobic property seem often to be products of fungal metabolism (Parry, 1978; York and Baldwin, 1992). The water-repellent property may be confined to a very thin, thatch-affected layer at the surface or may be expressed more deeply through the coating of soil particles with waxy products (Plate 6). Sand-dominant soils have a low specific surface area and are therefore more susceptible to developing a water-repelling character than soils containing more clay.

Dry patch is very common in the UK especially in a droughty summer. It is easy to prevent and easy to cure using one of the several proprietary wetting agents which are commercially available. The main reason for it occurring as a 'problem' is the failure of greenkeepers and groundstaff to be alert to the symptoms in the turf and to confirm the diagnosis by taking a few small soil cores. The use of routine preventative treatments at one to two month intervals in summer is increasing but should be unnecessary for the competent greenkeeper.

*Aeration/decompaction*

The principles of decompaction/aeration were presented in Chapter 5 together with the types of equipment available and their mode of action. The problems of poor aeration and mechanical impedence to rooting due to soil compaction are in general similar on fine turf and coarse turf. The main difference lies in treatments which are acceptable to alleviate the situation since the requirement for surface trueness is stringent on golf and bowling greens. Because of their adverse effect on surface trueness, greenkeepers are reluctant to use some types of equipment necessary to relieve deep-seated compaction. Nevertheless it is possible to use a mini-mole plough or Verti-drain or equipment relieving compaction by injecting air or water on golf and bowling greens. Compacting forces on golf and bowling greens are expressed only weakly at depths greater than 100 mm so equipment capable of loosening below this depth is required infrequently. It should be remembered however that equipment used to relieve compaction at rather shallow depths (e.g. hollow tine coring) can cause compaction near its maximum working depth (Petrovic, 1979). Carrow and Petrovic (1992) have summarized much of the research related to cultivation techniques for turfgrass soils but there is little research published in the UK reporting the effects, beneficial or otherwise, of individual or combinations of decompaction methods. It is nevertheless possible to outline a logical strategy for aeration/decompaction for golf courses and bowling greens.

The most frequent need on golf and bowling greens is to improve (or restore) water infiltration and air entry into the rootzone. This does not require decompaction and can be achieved by physical penetration of the surface with solid tines which may be flat or round. Frequency of operation should be determined solely on need. In the growing season spiking should be required once per month or less often. A higher frequency is likely to be needed on golf greens from autumn to spring when rainfall is in excess of evapotranspiration. The depth of penetration required may be shallow (less than 40 mm) but this will depend upon the nature of the rootzone.

Hollow tine coring is the most widely used technique to improve aeration and relieve compaction in the top 100 mm of rootzones. On intensively used golf courses, treatment in autumn and spring may be necessary. On less intensively used golf greens and on bowling greens an annual treatment in autumn (with one or two passes) will suffice. Too frequent use results in a soft surface. An operation to relieve compaction below 100 mm should not be needed on annual basis. Indeed if it is there is an underlying problem which should be identified and rectified.

The intensity of foot traffic on tees is comparable with or greater than on greens so the compaction problems are similar. One difference is that since the turf usually has a larger root system the soil macropores are better protected. Hollow tine coring is probably the most valuable treatment.

Compaction problems are relatively minor on fairways except on restricted thoroughfares which may require special attention. Routine aeration maintenance using a slit tine spiker is normally confined to the period between late October and April. The purpose is to penetrate a surface which has become sealed. Decompaction at a depth exceeding 100 mm is rarely necessary but, when it is, a moderate amount of surface disturbance can be tolerated so that mini-mole ploughing is practicable.

*Topdressing*

Topdressing golf and bowling greens is essential as part of an integrated programme to prevent excessive accumulation of organic matter at the surface (see Chapter 5). Complementary to topdressing is the removal of excessive turfgrass debris by scarification and, in addition, the coring and spiking operations described earlier which help remove undesirable material from the rootzone and enable a better integration of topdressing with the mineral soil. These complementary procedures either alone or in combination should be carried out prior to topdressing to prepare the surface.

The topdressing requirement for golf greens and bowling greens is similar in quantity but the practice of applying frequent small dressings which is now commonplace on golf courses is rare on bowling greens. Typically bowling greens are topdressed either once in autumn as part of the end of season maintenance or twice in both autumn and spring. Golf greens may be topdressed at these times also, but usually nowadays, the total annual topdressing is split and applied in equal dressings on four to six occasions between spring and autumn.

Topdressings of sand (pure) became popular for golf greens in the USA in the late 1970s and straight sand is still often used in the USA and the UK.

There has however been a trend in more recent years favouring the use of sand-dominant soils. There are several reasons for this but all essentially relate to the increased surface area and surface active properties conferred by small quantities of clay and humus. Factors affected include turfgrass nutrition, the adsorption of biocides and the susceptibility to dry patch and possibly 'black layer'. The nature of the topdressings used increasingly on golf and bowling greens relates directly to the recommended rootzone composition in Table 7.6. Organic matter content should not exceed about 2% because control of organic matter is a key purpose in topdressing. Where sand alone is used the $D_{50}$ should be about the same as for a sand/soil mix but there should be at least 80% in the 125–500 $\mu$m range and less than 10% smaller than 125 $\mu$m.

Typically quantities of topdressing required annually in the UK for 'good practice' situations are around 2 kg m$^{-2}$ using a sand-dominant material with 2% organic matter. When scarification and grass debris removal are less rigorous, larger amounts are required. Split into four or six applications, topdressing can be carried out with minimal disruption of play.

Topdressings are usually applied with a mechanical spreader so the material used must be sufficiently dry to be free flowing.

Tees and fairways should not be left out in the topdressing programme. Both surfaces require selective topdressing of divot marks during the playing season (often pre-mixed with seed) plus an annual topdressing for tees once they are brought out of play for renovation. The same material as for greens should be used on tees, although for fairways it is not uncommon for a 50/50 sand/soil mix to be used. It is quite rare for fairways to receive a general topdressing.

## Presentational maintenance

Visual impact is the first and often lasting impression. The primary standard is the 'expected' one, conditioned by the experience of player and spectator. Over and above this is a presentation outside common experience. Excellence as it is perceived can be achieved in different ways. Hacker and Shiels (1992) focus strongly on the way sound techniques in maintenance and attention to detail can raise the general standard of greens, fairways and tees. They point out the significance of the areas of rough on a golf course reinforcing the comments made earlier in this chapter.

Outside the general concept of sustainable quality built on good grounds-manship there is also scope to create a masterpiece. There is no harm in taking every conceivable and affordable step to create an immaculate presentation provided it is recognized by all interested parties to be a short-term showpiece with little relevance to the playing conditions available to golfers throughout the year. The US Masters course at Augusta in Georgia is an example of a course where attention and resources can be focused on a very narrow time window for public viewing. Golf is not unique and Wimbledon falls into a similar category.

Bowling greens do not give the greenkeeper as much scope for inventiveness and self-expression as golf courses. High quality maintenance is the key but the surrounds, both immediate and wider, play an important role in presentation.

## REFERENCES

Adams, W.A. (1977a) Fertilizer use on sportsturf. *Parks and Sportsgrounds* 43(1), 62–70.

Adams, W.A. (1977b) The effect of nitrogen fertilization and cutting height on the shoot growth, nutrient removal and turfgrass composition of an initially ryegrass dominant sportsturf. In: Beard J.B. (ed.) *Proceedings of the 3rd International Turfgrass Research Conference.* American Society of Agronomy, Madison, Wisconsin, pp. 343–350.

Adams, W.A. (1983) Feeding turf on soil and sand. In: Shildrick J.P. (ed.) *Proceedings of the Second National Turfgrass Conference.* National Turfgrass Council, Bingley, pp. 87–93.

Adams, W.A. and Smith, J.N.G. (1993) Chemical properties of rootzones containing a black layer and some factors affecting sulphide production. In: Carrow R.N., Christians, N.E. and Shearman, R.C. (eds) *International Turfgrass Research Journal* 7, 540–545.

Baker, S.W. (1990) *Sands for Sportsturf Construction and Maintenance.* The Sports Turf Research Institute, Bingley, 58 pp.

Baker, S.W., Cole, A.R. and Thornton, S.L. (1990) The effect of sand type on ball impacts, angle of repose and stability of footing in golf bunkers. In: Cochran, A.J. (ed.) *Proceedings of the First World Scientific Congress of Golf.* E. and F.N. Spon, London, pp. 352–357.

Balogh, J.C. and Walker, W.J. (1992) *Golf Course Management and Construction: Environmental Issues.* United States Golf Association, Lewis Publishers, Chelsea, Michigan, 951 pp.

Beard, J.B. (1982) *Turf Management for Golf Courses.* The United States Golf Association, New Jersey, 642 pp.

Bell, M.J. and Holmes, G. (1988) Playing quality standards for level bowling greens. *Journal of the Sports Turf Research Institute* 64, 48–62.

Canaway, P.M. (1990) Golf green agronomy and playing quality – past and current trends. In: Cochran, A.J. (ed.) *Proceedings of the First World Scientific Congress of Golf.* E. and F.N. Spon, London, pp. 336–345.

Canaway, P.M. and Baker, S.W. (1992) Ball roll characteristics of five turfgrasses used for golf and bowling greens. *Journal of the Sports Turf Research Institute* 68, 88–94.

Carrow, R.N. and Petrovic, A.M. (1992) Effects of traffic on turfgrass. In: Waddington, D.V., Carrow, R.N. and Shearman, R.C. (eds) *Turfgrass.* Agronomy Series No. 32,

American Society of Agronomy, Crop Science Society of America and Soil Science Society of America, Madison, Wisconsin, pp. 285–330.

Cochran, A.J. (ed.) (1990) *Science and Golf, Proceedings of the First World Scientific Congress of Golf.* E. and F.N. Spon, London, 374 pp.

Engel, R.E., Radko, A.M. and Trout, J.R. (1980) Influence of mowing procedures on roll speed of putting greens. *USGA Green Section Record* 18(1), 7–9.

English Bowling Association (1986) *Official Year Handbook.* Marshment and White, Bradford-on-Avon, 312 pp.

Evans, R.D.C. (1992) *Bowling Greens: Their History, Construction and Maintenance*, 2nd edn. The Sports Turf Research Institute, Bingley, 186 pp.

Gibbs, R.J. (1992) Weedy king of the rink. *Turf Management* September 1992, pp. 17–21.

Gooch, R.B. and Escritt, J.R. (1975) *Sports Ground Construction – Specifications*, 2nd edn. National Playing Fields Association, London, 126 pp.

Hacker, J.W. and Shiels, G. (1992) *Golf Course Presentation.* Professional Sportsturf Design (NW) Ltd, Preston, 39 pp.

Hawtree, F.W. (1983) *The Golf Course: Planning, Design, Construction and Maintenance.* E. and F.N. Spon, London, 212 pp.

Hayes, P. (1990) Principles and practices for perfect playing surfaces. In: Shildrick, J.P. (ed.) *Minimum Standards for Golf Course Construction.* National Turfgrass Council Workshop Report No. 20, pp. 67–74.

Hayes, P. (1992) Golf turfgrass research in the UK. In: *12th National Turfgrass Conference*, Launceston, Tasmania, May 1992. Australian Golf Course Superintendents Association.

Hayes, P., Evans, R.D.C. and Isaac, S.P. (1992) *The Care of the Golf Course.* The Sports Turf Research Institute, Bingley, 262 pp.

Holmes, G. and Bell, M.J. (1986) The playing quality of level bowling greens: a survey. *Journal of the Sports Turf Research Institute* 62, 50–65.

Hubbard, C.E. (1984) *Grasses*, 3rd edn revised by J.C.E. Hubbard. Penguin Books Ltd, Middlesex, 476 pp.

Kavanagh, T. (1989) Iron in turfgrass management. In: Shildrick, J.P. (ed.) *Turf Nutrition 88. National Turfgrass Council Workshop Report No. 15*, pp. 60–66.

Lawson, D.M. (1989) The principles of fertilizer use for sports turf. *Soil Use and Management* 5, 122–127.

Lawson, D.M. (1991) *Fertilizers for Turf.* The Sports Turf Research Institute, Bingley, 45 pp.

Lodge, T.A. (1992a) A study of the effects of golf green construction and differential irrigation and fertilizer nutrition rates on golf ball behaviour. *Journal of the Sports Turf Research Institute* 68, 95–103.

Lodge, T.A. (1992b) An apparatus for measuring green 'speed'. *Journal of the Sports Turf Research Institute* 68, 128–130.

Lodge, T.A. Colclough, T.W. and Canaway, P.M. (1990) Fertilizer nutrition of sand golf greens. VI. Cover and botanical composition. *Journal of the Sports Turf Research Institute* 66, 89–99.

McAuliffe, K.W. and Gibbs, R.J. (1993) A national approach to the performance testing of cricket grounds and lawn bowling greens. In: Carrow, R.N., Christians, N.E. and Shearman, R.C. (eds) *International Turfgrass Research Journal* 7, 222–230.

Nature Conservancy Council (1989) *On Course Conservation: Managing Golf's Natural Heritage.* Nature Conservancy Council, 46 pp.

Park, E. (1990) *Real Golf: A Collection of Articles.* A. Quick & Co., Dovercourt, Essex, 170 pp.

Parry, R.M. (1978) The effect of pH on soil fungal and bacterial populations and the synthesis of hydrophobic substances. Unpublished Honours Dissertation, University College of Wales, Aberystwyth.

Petrovic, A.M. (1979) The effects of vertical operating hollow tine (VOHT) cultivation on turfgrass soil structure. Unpublished PhD Thesis, Michigan State University, East Lansing.

Radko, A.M. (1977) How fast are your greens? *USGA Green Section Record* 15(5), 10–11.

Rodwell, J.S. (ed.) (1992) *British Plant Communities. Volume 3, Grasslands and Montane Communities.* Cambridge University Press, Cambridge, 540 pp.

Royal and Ancient (1989) *The Way Forward.* Discussion document of British Golf Course Management, Royal and Ancient Golf Club of St Andrews Greenkeeping Panel, 33 pp.

USGA Green Section Staff (1993) USGA recommendations for a method of putting green construction. *USGA Green Section Record* Mar/Apr, 1–30.

Walmsley, W.H. (1990) Levelling methods – an overview. In: Evans, P.S. (ed.) *Proceedings of the Fourth New Zealand Sports Turf Convention.* Massey University, New Zealand, p. 42.

Watson, J.R., Kaerwer, H.E. and Martin, D.P. (1992) The Turfgrass Industry. In: Waddington, D.V., Carrow, R.N. and Shearman, R.C. (eds) *Turfgrass.* Agronomy Series No. 32, American Society of Agronomy, Crop Science Society of America and Soil Science Society of America, Madison, Wisconsin, pp. 29–88.

York, C.A. and Baldwin, N.A. (1992) Dry patch on golf greens: a review. *Journal of the Sports Turf Research Institute* 68, 7–19.

# CRICKET GROUNDS AND TENNIS COURTS

## INTRODUCTION

The games of cricket and tennis differ from other ball games played on natural turf in that the nature of the reaction on impact between ball and playing surface takes precedence over ball roll characteristics in determining playing quality. Acceptable ball bounce requires a hard, tightly bound soil. Clay is, in the main, responsible for conferring soil binding characteristics but strong binding is only achieved when the soil is sufficiently dry. Strong binding characteristics and free drainage are incompatible so that inevitably, in the UK, cricket and lawn tennis are summer games played when evapotranspiration exceeds precipitation. A general requirement worldwide is that normal soil moisture contents should be substantially in deficit of field capacity.

Cricket demands a high level of alertness and speed of response to specific situations but it does not, for most players, demand a high level of sustained physical exertion. A warm ambient temperature is therefore required and in temperate climates cricket is inevitably a summer game. Tennis involves sustained activity and can therefore be played under quite cool conditions.

Wear on a cricket pitch is both local and severe but a cricket square provides for several pitches which can make use of the same outfield area. Wear on a tennis court is also very localized with tearing and scuffing restricted to a very small proportion of the court. Damage to small parts of a grass court detract from and impair the playing quality of the entire court.

A cricket ground is large enough to have the appeal of both arena and parkland. A tennis court is much smaller and furthermore, for the practical requirement of ball retrieval, it has to be 'caged'. However, smallness has its advantages. A tennis court can be more easily enclosed within a building and synthetic playing surfaces, which have no biological limit on tolerance of wear, can adequately replace natural turf. It is not surprising therefore that only a very small proportion of the tennis is now played on natural turf. Because of this, expert knowledge and experience on the maintenance of grass tennis courts is shared by few people.

Although the reaction of ball with playing surface is an important feature of both cricket and tennis, the nature of the games is quite different and they are considered separately in this chapter. Cricket is dealt with first.

## CRICKET

## Historical background

The game of cricket originated in southern England and most probably on the downlands of Sussex and Hampshire. The establishment of any game goes through several distinct phases but the two most important are, firstly, acceptance into the local and regional culture and, secondly, the drawing up and adoption of the laws governing the game. Undoubtedly many ideas for games have fallen at the first hurdle. The sheep-grazed chalk downlands provided closely kept grass and the reasonably level terrain necessary for the early development of cricket in the 16th and 17th centuries. Nowhere else in the UK were there such large areas of land suitable for a game involving the rolling, hitting and fielding of a rather small ball.

Cricket was the first national game in the UK to have formal laws. These were drawn up in 1774 but it was the establishment of the Marylebone Cricket Club (MCC) which provided a national focus for the game and its administration. Their present ground, in St John's Wood, acquired by Thomas Lord, was occupied in 1814. The laws of the game have changed surprisingly little in the last two hundred years.

The change in bowling style from underarm to overarm which was sanctioned in 1864, redressed the balance between bowler and batsman which had come to favour the latter with the improvement in the trueness of playing surfaces. The higher delivery of the ball, with an overarm bowling action, gave much more scope for varying the reaction between ball and playing surface through variations in both speed of bowling and length of pitch of the ball. Hence the properties of the pitch, affecting its reaction with the ball, became much more important. Evans (1991) gives an informative and entertaining account of the development of cricket.

Words used to describe basic requirements for the game of cricket are sometimes used incorrectly or ambiguously. A cricket pitch is the prepared strip on which batsmen and bowlers play. It is not a wicket. There are two wickets one at each end of the pitch each comprised of three stumps and two bails.

The square is an area in the centre of the ground upon which several pitches can be prepared. The remainder of the ground is the outfield. This terminology will be used throughout the chapter.

## The two different playing surfaces of cricket

Although practice nets are often constructed on the outfield it should not be taken to imply that a good outfield would serve as a good square. The requirements of pitch and outfield are quite different and the means of achieving them are also different. The pitch, not surprisingly, receives closer attention by players, spectators and commentators on the game. Not only can its general character favour particular talents, but change in its character during the game can influence the result. In contrast an outfield is either good or bad and whichever it is will be appreciated similarly by both teams.

The playing character of a pitch, in decreasing order of importance, depends upon soil, preparation and turf. This is a general statement but it contrasts with an outfield where the turf is of primary importance. Pitch quality is determined by the bounce characteristics of the ball which depends upon the nature of the soil. Outfield quality depends upon the rolling behaviour of the ball which is determined, in the main, by the turf.

## The cricket pitch

A cricket pitch is a prepared 22 yard (20.12 m) strip on a square. The typical width of the strip is 10 feet (3 m) so that a square, 24 yard (22 m) × 24 yard (22 m) would have space for seven pitches. This size of square is adequate for most cricket clubs. Lord's square has space for 18 pitches. The square itself is often comprised of the same or similar soil. However, with an increase in the practice of reconstructing individual strips on a square, many county grounds have squares which contain pitches with quite different soil profiles.

### Development of pitch preparation

If you have an opportunity to visit the Long Room at Lord's, have a look at the early paintings of cricket matches. The pitch is indistinguishable from the outfield. From the earliest times when the game was played, there is evidence that efforts were made to produce a flat pitch with short grass. Cylinder mowers did not become generally available until the second half of the 19th century and up until the late 1800s close cutting involved a choice between sheep or scythe. Rolling, albeit with quite light rollers, was practised well before the widespread use of cylinder mowers.

Cricket can be an extremely dangerous game for the batsman, and to a lesser extent for the wicket-keeper, on an uneven pitch. To play a forward defensive shot to a ball from a fast bowler which skims past an ear is unnerving. Helmets and face masks were unheard of in the early decades of this century when the major leap forward in player competence occurred. The preparation of flat docile pitches was the main aim at that time and two materials came to prominance to achieve this in the UK, cow dung and Nottingham marl.

Cow dung mixed into a thin slurry with clay (heavy soil) or marl was 'matured' in a bin for a week or so before watering on the pitch. The dung gave plasticity which enabled the moist soil to be rolled out to form a flat surface. We now know that the organic matter mixed intimately with clay moderated its binding characteristics so that even well-prepared pitches never became very fast. Dung therefore took the potential fire out of strongly binding clay soils to provide easy paced, even bounce but lifeless pitches. The dung and marl formula was not restricted to the top levels but spread throughout the game into village cricket.

Marl is a biologically inert quarried material which contains calcium carbonate and sulphate. It is a product of tropical weathering and is coloured pink-red by small amounts of the iron oxide hematite. It is an enigmatic material because it is dense – that is heavy in relation to its volume – but it does not bind strongly when dry. Its clay content is around 30% or less, which is not particularly high. The denseness of marl, together with the fact that it is not especially either clayey or sticky, enables it to be rolled out to produce a flat but lifeless pitch surface.

Since marl does not bind strongly it is prone to dusting during play. This was a particular problem during the 1920s and 1930s when it was common practice to apply light dressings of marl in the spring when it could not possibly have had time to integrate with the surface before the playing season began. This dust provided 'bite' for the ball and spinners could turn the ball very sharply. Marl was used on its own or in conjunction with dung to produce flat easy paced pitches and was also used as a late pitch dressing to favour particular talents. Opposition to the use of marl strengthened progressively and in 1939 the MCC proposed that only a small proportion of marl in combination with clay (i.e. heavy soil) should be used for topdressing and that topdressing of any material after the beginning of March should be forbidden (Advisory County Cricket Committee, 1939).

Widespread use of marl as a topdressing continued up until the 1960s and past usage can be seen in the profiles of cricket squares where marl often occurs as virtually intact layers, buried by casts of earthworms and more recent topdressings.

The proposals of the MCC in 1939 concerning pitches marked a change in requirement from pitches which were simply safe to play on, to ones which enabled a full range of cricketing skills to be exploited. In particular there was a desire to increase the pace of county pitches. The early post-war years saw little improvement. Nottinghamshire had two attempts to improve the square at Trent Bridge in the early 1950s. The first involved removal of the top inch (25 mm) of soil and its replacement. This failed to produce any improvement so that subsequently a 12 inch (300 mm) depth of soil was removed and replaced by soil from Eton under the supervision of Bill Bowles who was groundsman at Eton College (Marshall, 1953).

The Advisory County Cricket Committee set up a special sub-committee in 1962 to consider ways and means of providing faster pitches. The report of the sub-committee (Advisory County Cricket Committee, 1963) included many recommendations but identified three key factors:

1. The right texture of soil.
2. Pre-season rolling.
3. The need to allow the pitch to dry out completely.

Some idea of what pre-season rolling should achieve was stated but there was no clear indication of what the 'right' texture of soil was except that 'a heavy binding soil (not clay)' was recommended. Also there was no suggestion of any measurement which a groundsman could make to help him assess whether he had achieved the desired objective.

With the sub-committee's report the scene was set for the development of a scientific base to the construction, maintenance and preparation of cricket pitches.

### Soil composition and its relationship to pitch performance

Prior to the 1960s there was very little analytical information on the character of cricket pitch soils. Piper (1932) reported a detailed study on the Australian test grounds which showed them to have very high clay contents in comparison with pitches in the UK. Moreover the predominant clay type was smectite (Chapter 1) which is again unusual in UK soils. A research programme at the University College of Wales (UCW) began in 1966 and, supported by the MCC, set out to identify the factors affecting the pace of cricket pitches. An immediate problem was that no one had identified a property of a pitch which could be measured reasonably easily and which was related to its pace. Three properties, soil bulk density, penetrometer resistance and vertical ball rebound height were measured immediately after the end of county cricket matches at 15 county grounds but only the last of these was closely related to pace as perceived by players and umpires.

County cricket pitches, whose ball rebound characteristics had been measured, were sampled down to 150 mm and laboratory assessments carried out to establish a soil property closely correlated with rebound height when

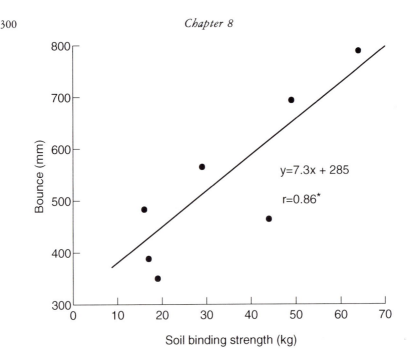

**Fig. 8.1.** The relationship between the mean vertical rebound height (bounce) of a cricket ball dropped from 4.9 m and the soil binding strength of the top 25 mm of seven English county cricket grounds examined in 1966.

a ball was dropped vertically from 16 feet (4.9 m). This investigation led to the initiation of the 'motty' test or Adams/Stewart Soil Binding test, ASSB (Stewart and Adams, 1969). Motty was coined from the term used in the south Derbyshire coal and clay mining area to describe the balls of clay used by children for marbles before the glass variety was available. The motty test requires little or no equipment, yet has provided groundsmen with a measurement of considerable practical value which integrates the various factors influencing soil binding strength.

Soil strength measured by the motty test was correlated with rebound height to the extent that around three quarters of the variation in bounce measured on different county pitches was explained by variations in soil binding strength measurements. The earliest, unpublished data for pitches on seven grounds including three test match grounds are summarized in Fig. 8.1. The binding strength measurements were for the top 25 mm of pitches. No improvement in the relationship was gained if the strength of soil at greater depths was included. It is undoubtedly because the pitches examined were prepared well and their potential pace was achieved that the close relationship was demonstrated.

Having established that soil binding strength was related to the measured

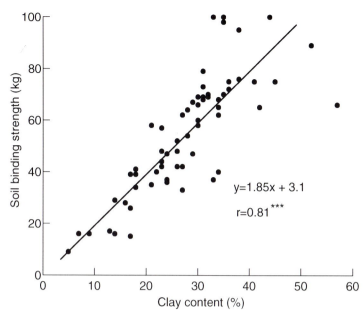

**Fig. 8.2.** The relationship between soil binding strength and clay content for 61 topdressings and soils from cricket pitches and tennis courts.

bounce on pitches, which was itself related to perceived assessments of pace, the next step was to identify the soil constituents responsible for conferring soil binding characteristics. Clay content was a key candidate. Figure 8.2 shows the relationship between clay content and soil binding strength for topdressings and soils from cricket pitches and tennis courts. Clay content is a good predictor of soil strength but at least two other factors need to be considered. One is organic matter content and the other is type of clay mineral.

Organic matter was found to be negatively correlated with soil binding strength. This means that topdressing loams which often have a low organic matter content, have a much greater measured binding strength than the pitches themselves with accumulated organic matter. Attempts were made in the UCW programme to examine the separate effects of fibrous as distinct from humified organic matter. It was clear that fibrous organic matter, that is live and dead roots and other plant fragments, decreased binding strength much more than amorphous organic matter although the data were not published (Stewart and Adams, 1968).

It is evident from Fig. 8.2 that either the soils examined had similar clay mineralogy or that mineral type is not particularly important in affecting binding strength. Little evidence is available on the effect of clay type on binding strength but the data in Table 8.1 throw some light on the matter.

**Table 8.1.** Properties and characteristics of some cricket pitch topdressing loams in 1986.

| Description | Qualitative clay mineralogy | | | | | % silt 20-2 μm | % clay less than 2 μm | % organic matter | pH | ASSB (kg) |
|---|---|---|---|---|---|---|---|---|---|---|
| | Kaolinite | Illite | Vermiculite | Smectite | Chlorite | | | | | |
| Rigby Taylor heavyweight loam | ++ | + | ++ | - | - | 50 | 32 | 6.8 | 6.0 | 85 |
| Binder's loam, Ongar, Kent | + | + | + | + | - | 19 | 25 | 4.6 | 7.1 | - |
| Surrey loams, GOSTO 175 | ++ | + | trace | ++ | - | 24 | 33 | 7.4 | 7.2 | 83 |
| Garford, Cambridgeshire | ++ | ++ | ++ | - | - | 15 | 24 | 3.3 | 7.2 | 90 |
| Mendip loam, Somerset | + | ++ | ++ | - | - | 27 | 30 | 9.0 | - | 58 |
| [a]Gresley clay, S. Derbyshire | +++ | + | - | - | - | 39 | 39 | 2.4 | - | 54 |

[a] Not generally available but included for comparison.

The Garford loam has a greater binding strength than might be expected from its clay content but it is low in organic matter. The mineralogy of the samples is reasonably similar except for the Gresley clay. This was used to upgrade the square at Grace Road, Leicester, and although not generally available, is included primarily because the clay is kaolinitic. From the analysis it seems reasonable to conclude that soils which contain kaolinite as the dominant clay mineral are likely to have a lower binding strength than other soils of similar clay content but which contain 2:1 minerals (see Chapter 1). In general, in the UK, clay content is a reasonable guide to binding strength provided organic matter is taken into account.

In addition to its influence on binding strength, clay content also affects the extent of soil shrinkage on drying. Shrinkage causes cracks to appear in pitches as they dry out and in general the greater the clay content the greater the shrinkage and the wider the cracks which develop. This is not the complete story since different clay minerals shrink to different extents and also the width of cracks depends upon the distance between cracks. The larger their separation the wider they will become on the same soil. Ideally the soil on a cricket pitch should be homogeneous and well integrated with depth but should have frequent lines of weakness in the horizontal plane which cause a large number of fine cracks to develop. A common problem of many of the pitches reconstructed on country cricket grounds since the mid 1980s has been the formation of unusually wide cracks separating large blocks of soil. An important factor in this has been undue haste to consolidate these pitches before adequate root ramification through the soil has occurred.

Provided crack edges are stable and do not crumble, cracking has no detrimental effect on pitch performance. Nevertheless, clearly visible cracks strike fear into many batsmen.

As stated earlier, clay mineral types differ in their potential for shrinkage on drying, with smectitic clays having the greatest potential. Despite differences between clay types, clay content is a good guide to likely shrinkage. Figure 8.3 shows the relationship between shrinkage and clay content for a range of soils. Two soils with high percentages of smectitic clays conform quite well to the general relationship. Current experience in the UK suggests that county cricket groundsmen are unhappy with soils which shrink to less than 75% of their driest mouldable volume.

## Factors affecting potential performance of cricket pitch soils

Cricket matches are of different length (time) and may involve one or two innings per side (team). There has been an increase in the number of matches with a restricted number of overs in recent years but much club and county cricket allows time for full innings to be played. Fifty, six-ball overs per side is the minimum in county cricket competitions but many village and small club

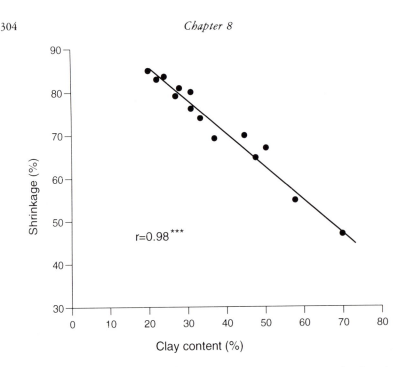

**Fig. 8.3.** The relationship between shrinkage (air-dry volume as a percentage of moist volume in driest mouldable state) and clay content. Twelve samples from five UK suppliers of cricket topdressings plus two smectitic soils with 58% and 70% clay from the Gezira, Sudan, and the Ethiopian highlands respectively.

competitions are run with a 20-over limit. Some minor leagues use eight-ball overs to decrease the number of changes of end. County cricket championship matches were three-day games for many years but became four-day in 1993. International test matches are five-day.

Pitch behaviour usually changes little in a limited-over game apart from a loss of greenness referred to later in the chapter, but substantial changes can occur in five-day matches. Change in the behaviour of pitches with time is a fundamental aspect of the game and 'natural' pitches provide for a range of cricketing talent which is not practicable with synthetic surfaces. Wear causes dusting through superficial damage to the surface and this increases friction and therefore turn. However, the surface must remain essentially intact otherwise bounce becomes dangerously erratic. Pace or bounce is either sustained or decreases with wear. In the past a major influence on pitch behaviour during a match was rain. This is now to a substantial extent avoided by the use of covers introduced primarily to decrease the interruption to play by adverse weather. The 'sticky dog' pitch is, in test cricket, a thing of the past.

Practical experience has shown that a clay content of around 30% and binding strength of approximately 75 kg are required to provide good pace and bounce and the wear tolerance demanded by a three to five-day game. Despite this simple statement, many pitches which have soils of adequate clay content are not fast and there are several possible explanations. Two reasons concern the soil itself, and relate to the level and distribution of organic matter and the occurrence of layering (discontinuity with depth). Other reasons relate to pitch preparation, particularly the degree of compaction or consolidation and water content.

Earthworms are controlled on most cricket squares and it was explained in Chapter 1 that this results in a tendency for organic matter to accumulate at the surface. If even a shallow thatch depth is allowed to accumulate, ball impact is cushioned and pace decreased. Moreover, as the thatch becomes humified it can create a surface which disintegrates on ball impact and bounce becomes unreliable and potentially dangerous. Even when organic matter is reasonably evenly distributed it can decrease binding strength and potential pace but a serious effect is not normally apparent until contents exceed about 10%. The control of surface organic matter requires an integrated maintenance programme involving scarification, plant residue removal and topdressing as described in Chapter 5.

There are two key causes of layering. One is when, through inadequate scarification and removal of plant debris, topdressing is applied which does not integrate with the mineral soil, but instead creates a compressed organic matter sandwich. The other cause is when a topdressing is applied which has very different swell/shrink characteristics from the existing soil. The usual reason for the latter is a major difference in clay content. Whichever of these is the cause of layering, its effect is to create discontinuities (with depth) which prevent soil binding with depth. Such discontinuities tend to be exaggerated by rolling and also become stabilized by roots which extend horizontally along cracks which open up when the soil dries (Arundell and Baker, 1984). Figure 8.4 shows an extreme case of layering where roots can be seen growing between layers. Although hollow tining and topdressing can be used to alleviate some of the effects of layering, once it has been allowed to develop it is impossible to rectify completely other than by reconstruction. Separation of layers on drying cushions bounce and is the main cause of low and uneven bounce on well-prepared pitches. It is also the main reason why pitches have to be reconstructed. The use of a topdressing which is incompatible with the native soil must be avoided because separation and disintegration of the topdressing can produce a very dangerous playing surface. Adams (1987) suggested a modified motty test to check the compatibility between topdressing and native soil. The crumbling of pitch surfaces in Australia due to the self-mulching properties of smectitic clay described by Harris (1961) does not occur in the UK.

Compaction through rolling is necessary in order to destroy the large pores created by aggregates of particles. The reason is that maximum mechanical strength is achieved as a soil dries when there is maximum surface area of contact between the particles of which the soil is comprized. Aggregates and

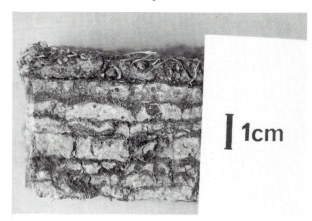

**Fig. 8.4.** The layering of soil topdressing on the Old Trafford square in 1977.

the spaces they create reduce the contact between particles. Aggregates are weakest when the pores within them are saturated and rolling achieves little if any compaction of a clay-rich dry soil because the aggregates are very stable. Good pitch preparation involves moulding the soil when it is in a plastic state – in effect kneading the soil to destroy aggregates. Thorough consolidation with depth takes time and pre-season rolling is intended to create a solid base to carry a prepared pitch. The creation of a shallow compacted zone of 10–15 mm which is a common feature of many school and club pitches produces a surface of moderate pace for limited-over games but the pace gets slower as it dries out. This is probably because the structured soil supporting the shallow compacted zone becomes more friable and loose as it dries.

An important concept is that rolling should create a soil with a bulk density close to its maximum. McAuliffe (pers. comm.) recommended that, ideally, the top 100 mm of the pitch should have a bulk density close to 80% of its maximum. Typical bulk densities of well-prepared cricket pitches are in the range 1.9–2.2 g cm$^{-3}$.

Compaction depends upon consolidation through rolling in conjunction with the natural shrinkage of soil on drying. A 2 tonne roller is only capable of pushing water out of soil pores down to a size which would remain saturated at around −100 kPa or 10$^4$ mm tension. This is very moist in relation to an air-dry state of less than −1500 kPa. Shrinkage as the soil dries further is a vital add-on to rolling. Stewart and Adams (1968) and later Cameron-Lee and McAuliffe (1989) observed that provided motties were allowed to become air-dried there was negligible effect of relative humidity (moisture) of the air on breaking strength. This suggests that once an adequate level of dryness is

**Table 8.2.** Bounce of a test match cricket ball dropped from 4.9 m on prepared pitches on the Melbourne Cricket Club ground, Australia, examined over a seasonal drying period in 1972.

| Date | Match | Mean bounce (mm) and range between three sections of the pitch |
|---|---|---|
| 12 November 1971 | Victoria vs Queensland | 572 (553–597) |
| 14 November 1971 | Victoria vs Queensland | 737 (no variation) |
| 12 December 1971 | Victoria vs West Australia | 838 (813–864) |
| 02 January 1972 | Australia vs Rest of the World | 828 (813–889) |
| 05 January 1972 | Australia vs Rest of the World | 914 (889–940) |

reached on a cricket pitch further desiccation is of no practical importance. Examination of the moisture state of well-prepared county pitches included in the UCW survey of 1966/67 suggested that water potentials of −1000 kPa to −1500 kPa were adequate to achieve potential pace. Uniform drying with depth is necessary to achieve potential pace and only small gradients in soil water content should occur over the top 100 mm (McIntyre, 1985). Good grass root exploitation of soil with depth and regular rolling during pitch preparation are required to achieve this. A clear example of the effect of progressive drying with depth on measured ball bounce was reported by the secretary of the Melbourne Cricket Club in Australia in a letter to V.I. Stewart in March 1972. Over the period covered in Table 8.2 soil moisture changed from a reported 'damp' to 'dry'.

*Turfgrasses and pitch quality*

In the preceding section no mention was made of the contribution made by the turf to pitch performance. This was deliberate because different turfgrasses have some effects in common but also there are marked differences between species in their effect on pitch character and performance and also on ease of maintenance.

The MCC special sub-Committee's report in 1963 referred to earlier, recommended a seed mixture of:

- 60% Chewings fescue (*Festuca rubra* subsp. *commutata*);
- 20% New Zealand browntop (*Agrostis capillaris*);
- 20% crested dogstail (*Cynosurus cristatus*).

How crested dogstail got into the recommended list of species is uncertain. A grassland ecologist could not have been consulted because this species never forms a turf although it is always present in 'low input' grassland on neutral soils. The Sports Turf Research Institute recommended against this species in the early 1970s and since then it has disappeared from recommendations and commercial seed mixtures. The reality of the turfgrass situation on county cricket squares in the early 1970s was reported by Adams (1975). In essence, annual meadowgrass (*Poa annua*) was the dominant species and this, together with bentgrasses (*Agrostis* spp.), accounted for two thirds of the average percentage ground cover. Data in 1980 based on 22 grounds which included all English test match grounds (TCCB course at Aberystwyth, 1980) showed that within ten years the percentage cover by perennial ryegrass (*Lolium perenne*) had doubled from 15% to just under 30%. The extent of this increase is astonishing considering that it was achieved by overseeding not reseeding.

The international focus on the selection and breeding of 'turf' as distinct from 'pasture' perennial ryegrasses in the early 1970s led to the introduction of cultivars which were well suited to the requirements of a cricket square. To appreciate this, one must recognize the key roles of turfgrasses on a cricket pitch. There are two, both of which have an indirect effect on pitch performance; one is to assist, through transpiration, even drying to depth (McIntyre, 1985), the other is through root penetration to assist the vertical integration of the soil. On both counts, in temperate climates, perennial ryegrass is better than fine-leaved bents and fescues. Not only this, but perennial ryegrass is an aggressive species so can compete effectively with the shallow-rooting annual meadowgrass. Thus an increase in the former is usually at the expense of the latter. Furthermore, and this is a vital asset in maintenance, perennial ryegrass is not a natural thatch producer and does not spread vegetatively when mown closely. Therefore perennial ryegrass can be scarified very harshly and suffer little damage. This attribute makes it easier to ensure that autumn topdressings integrate with the native soil.

In the tropical conditions of Australia, India, Pakistan, South Africa and the West Indies, Bermudagrass, often known by its local name (see Chapter 11), is used on cricket squares. This species which has both rhizomes and stolons 'knits' the soil both vertically and horizontally.

Despite the scientific and practical evidence that turf-type cultivars of perennial ryegrass were better suited to the requirements of a cricket square than cultivars of any other species, a Test and County Cricket Board report in 1983 still recommended the inclusion of chewings fescue and browntop. In practice, the recommendation for 60% perennial ryegrass in the mixture meant that only perennial ryegrass would be likely to establish (Chapter 2). Most county cricket groundsmen use only cultivars of perennial ryegrass for overseeding their squares. Seed of perennial ryegrass cultivars only, has been used on Lord's square for both overseeding and in establishing reconstructed pitches for the last decade. It is therefore surprising that seed suppliers do not offer seed mixtures for cricket squares which contain two to four cultivars of perennial ryegrass undiluted by cultivars of other less appropriate species.

# The cricket outfield

There are two key requirements of an outfield. Firstly, the ball must run smoothly along the surface and secondly it must be comfortable to run on, not being too hard and yet provide adequate traction for acceleration and stopping. Both of these requirements are achieved best in the UK, in summer, by a tight, fine-leaved turf and a shallow thatch layer. These circumstances apply on most county cricket grounds where the outfield is not used for winter games. This 'ideal' situation is incompatible with the requirements of soccer and rugby because the tight, rather shallow-rooting turf tears out with play in winter and is slow to recover.

When a ground can be 'dedicated' to cricket, the slow growth, tight turf and thatch-creating tendency of bentgrasses and fescues can be exploited. Soil pH can be allowed to fall into the mid 4s and the fertilizer input required is small. The depth of thatch and its ill-effects on rooting and water retention can be controlled by annual hollow tining and irregular topdressings of sand.

Cores from the outfields of all six English test grounds were examined at the 1980 TCCB groundsman's course at Aberystwyth. All had a distinct thatch layer which ranged between grounds from 5 mm to 25 mm. In addition, all had a dominance of fine turfgrasses with the Old Trafford and Headingley outfields containing virtually 100% bent/fescue. Bentgrasses were always dominant and their presence is easier to sustain.

Where winter games are played on the outfield the situation is quite different. Rugby is played on the St Helen's outfield at Swansea and here perennial ryegrass is dominant and there is no thatch. When cricket outfields double as winter games pitches, which is usually the case on Local Authority playing fields, the winter game dictates soil drainage and turfgrass requirements. Only perennial ryegrass can tolerate wear in winter and recover sufficiently quickly to produce a satisfactory outfield turf.

On grounds only used for cricket, outfield drainage is not usually a problem because, over the playing season, soil moisture is in deficit (Chapter 1). In general a pipe drainage system is required only to control the watertable. Nevertheless, because of the movement of rollers and other machinery on and off the square, decompaction of the outfield topsoil may be necessary from time to time. When the outfield is used as a winter games area construction and maintenance must comply with much more demanding requirements for soil drainage (Chapter 3).

# Playing quality standards

The pitch and outfield have different requirements so must be assessed separately.

*Pitch quality*

The key requirement is that a pitch should be 'safe'. This is extremely difficult to assess from measurements on the pitch because the sites causing grossly deviant bounce are usually small in size and number. In school and village cricket, where the games are of relatively short duration, dangerous pitches result from inadequate preparation. Visual inspection provides the best assessment of the 'flatness' of the surface provided for the game.

At higher levels of the game, 'dangerous' bounce (i.e. irregular bounce unrelated to the pace and trajectory of the ball as bowled) usually results from the disintegration of the pitch surface which is evident visually but difficult to quantify numerically.

Pitch quality in terms of both pace and evenness can normally be gauged from the mean and variation in rebound height from 4.9 m dropping height. A rebound height of 600 mm (12%) is a useful criterion because below this value pitches can be considered to be easy paced or slower (Stewart and Adams, 1969). A groundsman can get a semi-quantitative value of pace by throwing a ball down onto the playing surface. On a fast pitch the ball can be bounced at least head height whereas on pitches which are moderately paced or slower it is not possible to bounce a ball above about waist height irrespective of the how hard the ball is thrown. Dropping a ball from 4.9 m is tedious and an impact hammer may provide a more easily achieved assessment of pace and evenness of bounce since Lush (1985) showed that the deceleration of a 'Clegg impact soil tester' was correlated with rebound height. Recent data from New Zealand give support to this possibility (Table 8.3).

Inadequate pace can be a consequence of three main factors:

1.  unsuitable soil;
2.  inadequate preparation (rolling);
3.  insufficient drying out.

**Table 8.3.** Mean 0.5 kg Clegg impact test results (in gravities) for The Basin Reserve, Wellington (20 readings per cricket pitch) (from McAuliffe and Gibbs, 1993).

| Date and match | Mean | Standard deviation | Player assessment |
|---|---|---|---|
| December 1990, Wellington vs Otago | 428 | 137 | Variable bounce |
| February 1991, NZ vs England (one day match) | 553 | 93 | Good bounce height, acceptable uniformity |
| December 1991, Wellington vs Canterbury | 425 | 58 | Consistent bounce |
| February 1992, NZ vs England Test Match | 372 | 44 | Slow, low but consistent bounce |
| March 1992, NZ vs England World Cup Match | 413 | 60 | Slow but consistent bounce |

An ability to assess pace and consistency of pace is useful but unless the causes of any inadequacy can be identified, such tests are unhelpful to the groundsman. The inconsistency of bounce on returfed pitches has been recognized since the late 1960s (Stewart and Adams, 1969) but other causes of inconsistency are more difficult to diagnose. More work is needed to establish measurements which can provide reliable information on pitch performance and can be interpreted to account for or explain pitch performance.

The friction between ball and pitch determines the change in the component of velocity along the pitch (pace), lift and potential turn. A ball bounces higher than expected when the surface of the pitch deforms under impact to produce a ridge in front of the ball which acts like a ramp (Daish, 1972). This happens on a pitch which is soft through inadequate compaction or drying or both and the plastic deformation of the pitch is shown by the impact marks which remain. The ball slows considerably on bouncing because of the energy absorbed by the surface. On a very hard and fast pitch, deformation of the pitch is small and elastic and no ball impact marks remain. Friction is low and there is a relatively small loss of kinetic energy by the ball which accounts for the correlation between ball rebound height and perceived pace.

A dusting of topdressing on the pitch prior to a game increases friction and therefore assists turn without affecting measured vertical rebound height. This is simple 'doctoring' which slows the pitch from a batsman's viewpoint relative to what would be expected from ball rebound measurements. It also allows a spinner to turn the ball from the beginning of the match. Dusting of the surface occurs normally as play proceeds and accounts for the increasing amount of 'turn' exhibited in the second and subsequent days of a match. An increase in the sand fraction of soil increases friction between ball and surface but an increase in sand is usually at the expense of clay which determines binding strength and potential pace.

Soil constituents can be abrasive. For example Surrey Loams Ltd at one stage had difficulty in screening small flints from their otherwise excellent topdressing. Mick Hunt (Lord's Head Groundsman) reported that an umpire brought a ball to him half-way through its allotted overs which looked as if 'it had been worried by an Alsatian'.

The foregoing points relate to the effect of soil on friction, but the turf itself can be important in affecting the behaviour of the ball. Green leaf tissue can lubricate the contact between pitch and ball and give the illusion that the ball 'gains pace' off the pitch. This is not possible, but if the ball bounces rather low in relation to the angle it hits the pitch (a shooter) it may, after bouncing, move more quickly towards the batsman. The situation is illustrated simplistically in Fig. 8.5 for a ball which loses no momentum on bouncing.

The difference in friction between the seam of the ball and playing surface and non-seam and playing surface is greater when fresh green leaf tissue is present than when the residual grass has dried out. The difference in friction can be used to cause a deviation from the flight line of the ball (known as

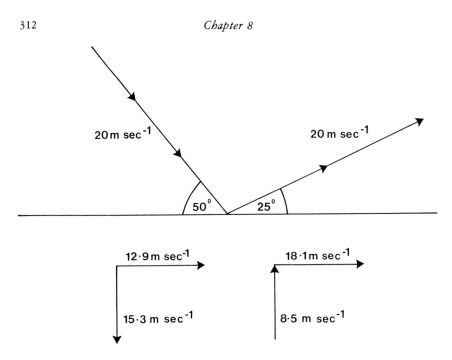

**Fig. 8.5.** An illustration of how a ball can appear to 'gain pace' on bouncing. A ball moving at a velocity of 20 m sec$^{-1}$ hits the ground at an angle of incidence of 50° and rebounds with the same velocity at an angle of 25°. Initially the ball is moving towards the batsman at a speed of 12.9 m sec$^{-1}$ but after bouncing it moves towards the batsman 40% quicker at 18.1 m sec$^{-1}$.

seaming). The potential for seaming is therefore most apparent early in a match especially if the pitch is 'green'.

Turf can have two additional effects on pitch performance caused by living and dead plant fibre at the soil surface. One is to cushion the vicious pace of a 'rock-hard' soil; the other is to increase the bounce height of moderate to slow paced deliveries.

The various factors affecting surface friction can be assessed visually by an experienced person and it would be tedious and complicated to devise and implement measurements which could be used in quality criteria. At present, in the UK, criteria for quality are restricted to assessments of bounce or related Clegg impact hammer measurements and consistency of bounce. These are essentially performance criteria and give no direct evidence on the cause of any inadequacy. Criteria which would provide evidence on poor preparation are possible. For example relative bulk density would indicate the level of compaction achieved in comparison with that which was possible and penetrometer resistance of the subsurface, which is affected by compaction and moisture content, could give evidence on whether pitch constitution or pitch preparation was at fault.

Other criteria which should at least be in design specifications are wicket to wicket and across pitch slope tolerances and the levelness of the pitch. Many old cricket squares have pitches like a Bactrian camel's back through the raising of wicket ends through poor returfing and/or topdressing practice.

### Outfield quality

Trueness of run of the ball is most important. This is affected by the nature of the turf/thatch as discussed earlier and short-range variations in topography. The pace of outfields is controlled by height of cut and moistness of grass. No criteria have been drawn up to categorize the quality of outfields but were this to be deemed worthwhile the two subjective categories of 'evenness' and 'speed' would seem to be adequate. McAuliffe and Gibbs (1993) have begun to address this issue.

## Causes of deterioration and options for correction

### Outfields

Deterioration of outfields can be due to:

1. Excessive weed invasion and poor mowing maintenance.
2. Deterioration in surface levels.
3. Uncontrolled thatch accumulation.

The first of these is readily reversed but the loss of levels (usually due to subsidence) may present problems. Minor and long-range undulations do not affect the game, but localized unevenness such as that caused by worm casts or settling over drain lines can affect the run of the ball and the latter can cause scalping on mowing. When local undulations have developed through subsidence which affect the game significantly, it is necessary to strip off the turf, relevel and replace the turf. Topdressing to redress levels is practicable, but one should not aim at more than 10 mm depth per application. Corrections requiring more than about 50 mm build-up should not be attempted by topdressing. Excessive thatch accumulation can usually be corrected by a combination of routine scarification and annual hollow tining in the autumn (to 80 mm depth) in conjunction with topdressing with sand or sand-dominant soil. Generally, deterioration in outfield quality can be reversed by rectifying poor maintenance practice.

### Squares

In contrast to deterioration of the outfield, deterioration of the square usually demands more comprehensive and expensive action. There are three basic causes of deterioration.

1.  Use of too weak (sandy) topdressing.
2.  Allowing thatch to accumulate.
3.  Creation of layering through poor topdressing practice or use of top-dressings of different character.

The use of a weakly binding topdressing will cause a prepared pitch to disintegrate, dust and break up on ball impact. The use of a topdressing which is heavier and shrinks more on drying than the native soil will cause crusting and shearing off of the topdressing. Provided these problems are identified quickly they can be rectified by vertical mowing and scarification to cut through and break up the shallow offending layer prior to topdressing with a compatible loam.

A shallow surface layer of thatch of up to about 3 mm depth can be broken up by deep scarification and to a large extent removed along with clippings and plant debris. Once the native mineral soil is exposed the surface can be topdressed and overseeded.

When layers which are liable to separate occur at depths greater than 10 mm it is impossible to reverse the situation completely. Hollow tining followed by topdressings luted into the holes can help tie the surface together but it is never completely successful. Pitch reconstruction is the only reliable solution.

## Construction and reconstruction of cricket squares and pitches

Intense compaction of clay-rich soil on cricket squares eliminates free-draining pore space. Drainage is therefore very slow. The use of a gravel or stone bed under a cricket square in conjunction with pipe drainage serves two main functions. One is to ensure that a watertable cannot rise up under the square, the other is to break pore water continuity so that the surface layer of soil can be dried out more easily. The control of a watertable can usually be achieved by a ring drain around the square and most cricket squares in the UK have no pipe drainage under them. Surface run-off is the main mechanism for getting rid of surface water in the playing season and therefore square design should provide for a slope of 1.5% to 2%. Also the ring drain should have permeable fill close enough to the surface to be accessed with a fork to help get rid of surface water.

Levels on the square must be maintained to facilitate run-off. A serious failing on many squares is the 'saddleback' problem referred to earlier which encourages water to lie in the centre of pitches.

*The base layer*

Most cricket squares are on native soil and do not have a drainage layer of stone or gravel in the subsoil. As indicated above, such a layer has benefits especially in wetter areas of the UK where it is difficult to dry out the native soil to an adequate depth because of the quantity of water it retains. It is not necessary to have a drainage layer under the entire square and it is often practicable to reconstruct individual pitches with their own drainage layer incorporating a pipe drain connecting with a perimeter drain.

A drainage layer under either individual pitches or the total square is not vital from the Midlands to southeastern England but it is advisable generally on heavy textured soils and especially in wetter areas where soil moisture deficits cannot be guaranteed throughout the playing season.

The base for a pitch or square must provide a firm foundation and ideally a free-draining one. This is best achieved by angular gravel – there is no need for material coarser than 6–10 mm in diameter.

*Pitch profile to the surface*

To achieve an adequate pitch performance three requirements must be met. These are:

1.   The surface soil must be of sufficient binding strength to provide adequate bounce and pace.
2.   It must be sufficiently deep to create a hard bound surface but must not be too deep to make drying out difficult.
3.   There must be no distinct interfaces between materials differing substantially in swell/shrink properties over the profile depth affecting bounce and pace.

The first requirement has been dealt with at least with regard to matches lasting several days. For club cricket where matches are of shorter duration, soils with clay contents down to around 25% are acceptable. Pitches with lower clay contents are easier to prepare and are less likely to develop shrinkage cracks. They are slower and more susceptible to dusting and breaking up. There is still some doubt about the second requirement but practical experience in the UK suggests that a depth of 70–100 mm is required although it is now evident that the nature of the soil below this depth can affect pitch performance. The third requirement usually receives the least attention but is important because layering must be avoided. Grading from a drainage layer of gravel or sand to a clay rich soil surface without incurring discontinuities requires a clear, practicable design and good on-site quality control. Figure 8.6 illustrates the reconstruction profile for three pitches on the Gloucestershire County Cricket square at Bristol. Each pitch was contained within a plastic envelope to control

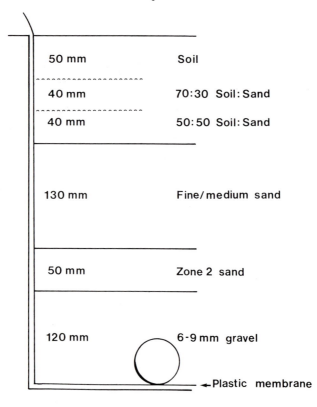

| | |
|---|---|
| 50 mm | Soil |
| 40 mm | 70:30 Soil:Sand |
| 40 mm | 50:50 Soil:Sand |
| 130 mm | Fine/medium sand |
| 50 mm | Zone 2 sand |
| 120 mm | 6-9 mm gravel |
| | Plastic membrane |

**Fig. 8.6.** The construction profile for three pitches reconstructed on the Gloucestershire County Cricket ground, Bristol, in 1986.

the watertable. A feature in the construction programme was the deliberate scoring of one layer before applying the next to decrease the risk of layer separation. The design differs from the experimental pitches at the Queen's Park Oval, Trinidad, described by Gumbs *et al.* (1982) primarily because the depth of uniform heavy soil used by the latter was 150 mm rather than 50 mm of clay-rich soil underlain by a soil dominant mix with sand. No importance was attached to avoiding discontinuities with depth in the Trinidad construction.

Since their construction in 1986 the pitches at Bristol have enabled good control over drying during preparation but prepared pitches have lacked pace and bounce. It appears therefore that the freely drained sand base has been close enough to the surface to cushion bounce and reduce pace. Whilst there is still some uncertainty about optimum depths, present knowledge suggests that the Bristol profile in Fig. 8.6 should be modified to increase the depth of strongly-binding soil from 50 mm to 80 mm and the depth of sand reduced from 130 to 100 mm. The depth of soil over sand (including transition layers) would be increased therefore to 160 mm.

In addition to total profile reconstruction, pitches will continue to be replaced, when the topsoil has become badly layered, by excavating the topsoil and replacing it with a homogeneous strongly binding soil. This will be successful provided the depth of soil removed/replaced is 120–160 mm and that the local climate will enable prepared pitches to dry out over most of the season. It is essential to score the base of the excavation to provide a 'key' for the introduced soil.

The soil used for refilling must be dry enough to avoid smearing and preferably quite dry to ensure that some aggregates will persist. Refilling should be carried out in stages of around 50 mm depth with treading, rolling and levelling at each stage. A dressing of $50\,\mathrm{g\,m^{-2}}$ of superphosphate should be applied 75 mm below the final surface. The amount of settling subsequent to refilling will be small if refilling has been carried out well. For pitches reconstructed within an existing square, an allowance of 10 mm depth of settling per 100 mm depth of fill should be used.

Turf should be established from seed and with an autumn reconstruction, seed must be sown before the end of September if the pitch is to be played over (not on) the following season.

### Grass establishment

The introduction of turf-type perennial ryegrass has meant that, provided pitches in the UK can be reconstructed and sown in mid September, grass will establish quickly enough to create a tight turf for the following season. Soil structure depends upon stabilization by roots so that in the first year when the grass is young, rolling should be kept to a minimum. Pitches should not be played upon in the first year because of the ill-effect this will have on long-term performance through inhibiting root development. The use of turf on cricket squares should be avoided to prevent creating horizontal discontinuity.

## Maintenance regimes

### Fertilizer use

Clippings will remove about $200\,\mathrm{kg\,ha^{-1}}$ ($20\,\mathrm{g\,m^{-2}}$) of N from a cricket square in a year and amounts of $P_2O_5$ and $K_2O$ corresponding to a ratio of $4:1:3$ respectively. Over the year, therefore, inputs of N, $P_2O_5$ and $K_2O$ in clippings removed systems should amount to around 20, 5 and $15\,\mathrm{g\,m^{-2}}$ respectively. Strategies for fertilizer use are explained in Chapter 5 but cricket squares have a few special features compared with the free-draining, sand-dominant soils used for many other turf playing surfaces. Firstly, the K economy of these clay-rich soils is good because of their large cation exchange capacity. Very little K is likely to be leached out of the soil. Not only this but the loam topdressings used on cricket squares often have a high status of available K. These potential benefits on K nutrition must be confirmed by soil analysis.

Topdressing with loam is carried out in the autumn and this is preceded by intensive scarification and spiking. The situation immediately prior to topdressing provides a good opportunity to get phosphate fertilizer below the soil surface. For this reason autumn-applied P allied to applications of N and K in the growing season is a logical strategy for fertilizer application to cricket squares.

Slow release sources of N fertilizer are expensive but an application in late spring will give a more even pattern of growth over the playing season compared with the surges produced by water soluble sources.

The type of maintenance on cricket outfields ranges from very intensive with clippings removed to gang-mower maintenance with clippings allowed to fly. The fertilizer input required for the former may be little different from a square but for the latter a small input will suffice provided the outfield is not used for winter games (see Chapter 5).

### Maintenance of soil physical conditions

The maintenance of acceptable soil aeration on a cricket outfield used in summer presents problems which are less acute than those on even a lightly used winter games pitch. A cricket square presents the problem of a deliberately compacted soil but where efficient rooting is necessary. Autumn is the only time when it is possible to loosen the soil on a square without the need for immediate re-compaction. This work may be subsequently complemented naturally by frost action but it is nevertheless vital. Despite advances in technology there is still no mechanized device which has superseded hand forking. The 'Verti-drain' is possibly the best implement available at the moment but this and others fall short of the efficiency of hand forking. Hollow tine spiking is a useful means of relieving surface compaction on a cricket square but compaction is likely to occur deeper than it is generally possible to reach with this implement.

Wicket ends are especially vulnerable to deep compaction and, if this is not corrected, shallow rooting and poor resistance to wear are an inevitable consequence.

### Pitch preparation

The foundation for good pitches is laid down by rolling in spring. Recommendations to start as 'early as possible' are still given but this is unnecessary and undesirable. A substantial amount of root development occurs in early March and this should not be inhibited by compaction in late February. Late March or even early April is quite soon enough.

The weight of roller should be built up gradually to around a 2 tonne single roller and rolling should be carried out in all directions across the square. Top-dressing must not be carried out in spring and although brushing and light scarification is necessary the use of slit tines must be avoided because the slits act as lines of weakness and cause a systematic pattern of cracks on prepared pitches.

Recommendations on pitch preparation were given by the Test and County Cricket Board in conjunction with the National Cricket Association (1989). These are stated below:

1.   Pitch preparation should ideally be commenced a week before the match. It is generally considered that water, either from rain or applied by the groundsman, is essential in the preparation of a pitch in order to given it a firm solid surface. If the square has been consolidated before the season, pitch preparation will be greatly assisted.

During a dry period it will be necessary to water the pitch. There is no fixed time for the latest watering that a pitch should receive, but this must be done well in advance of a match in order to ensure that the pitch is completely dry at the start.

2.   The rolling of the pitch should commence with the light roller when all surface water has disappeared. As the pitch dries, the weight of roller should be increased. The groundsman should use the heavy roller at every suitable opportunity prior to a match whilst any moisture content remains, but the heavy roller should not continue to be used when all moisture has gone from the pitch.

3.   At the start of a match the pitch should be completely dry.[1] This applies not only to the top surface but to a depth of 75–100 mm. Weather conditions may make this difficult but, if the weather is fine or with the correct use of covers, it should usually be possible to obtain complete dryness. Any moisture remaining in the pitch may produce a 'green' wicket.

4.   During preparation, the pitch should be lightly scarified by hand rake or strong broom, care being taken not to disturb the soil. There should be no mat of grass, and after scarification the surface soil should be visible between blades of grass.

5.   The mowing of a pitch prior to a match should be as low as possible provided that the surface is not scarred or disturbed in any way. Ideally, a mower with a thin bottom blade should be used and one that has at least a ten-bladed cylinder.

Table 8.4 illustrates the programme of pitch preparation for the Headingley test between England and Australia in 1989.

---

[1]This is subjective but probably means that any grass on the pitch would be suffering drought stress, that is, approaching −1500 kPa (Chapter 1).

**Table 8.4.** Pitch preparation for the Headingley test between England and Australia in 1989 as recorded by Keith Boyce, the Head Groundsman (pers. comm.).

| Date | Type of roller (tonne) | Rolling time (min) | Comments |
|---|---|---|---|
| 29 May | 1.5 | 10 | Wicket ends given extra roll |
| 30 May | 1.5 | 10 | Watered well after rolling |
| 31 May | 1.5 | 10 (a.m.) 20 (p.m.) | Wire raked, mown, full grass cover, looks OK |
| 1 June | 1.5 | 20 | Covers on all day, wicket firm |
| 2 June | 1.5 | 30 early (a.m.) | Brushed, lower cut, mobile covers on part day |
| 3 June | 2.5 | 30 early (a.m.) | Pitch becoming hard but top 25 mm still moist |
| 4 June | 2.5 | 30 early (a.m.) | Top drying out-on target |
| 5 June | 2.5 | 30 early (a.m.) | Pitch drying, fine cracks evident |
| 6 June | 1.5 | 15 | Closer cut (three boxes off), surface hard, many fine cracks |
| 7 June | 0.3 | 20 | Brush and cut, almost dry |
| 8 June | 0.3 | 20 | MATCH DAY brush and cut, pitch OK |

## Renovation after a match

The pitch should be watered and spiked with a sarel roller to pierce the surface. A light nitrogen fertilizer dressing may be applied. These procedures aid pitch recovery but even more important is attention to the wicket ends. These must be repaired to true levels and seeded immediately. The use of seed pre-germinated in a bucket for three days in a mix with topdressing loam will hasten establishment. Turfing is not recommended because of the tendency to shear off at the base.

## Re-use of pitches

The intensity of play on some squares requires that pitches are used for two or three matches. As a general principle it is better to play consecutive matches on a prepared pitch than to return after a 'rest'. The reason is that it is very difficult to rewet, uniformly, a pitch which has been prepared and dried out. Strong soils shrink and cracks open on drying creating small to large blocks of soil. On rewetting, water flows rapidly into and through cracks, wetting the surfaces of aggregates or blocks. The saturated surfaces trap air inside the blocks making rewetting very slow. A second preparation requires remoulding and if this is not possible because of inadequate rewetting, pitches are liable to break up and become dangerous.

## TENNIS

## Historical background

The name lawn tennis was coined to distinguish it from real tennis from which the game originated. Lawn tennis is a young game compared with cricket. The game is recorded as having been originated in 1859 by Major Harry Gem and Mr J.B. Perera who began playing an adaptation of real tennis on Perera's lawn in Edgbaston, Birmingham.

The earliest lawn tennis courts were rectangular or hour-glass shaped (1872–1877) but the shape and dimensions, standardized in 1880 by the All England Club and the MCC, have remained ever since. These are:

- Singles court 23.77 m × 8.23 m
- Doubles court 23.77 m × 10.97 m

The currently used net heights of 0.91 m in the centre and 1.07 m at the sides have remained since 1882. Ball weight and diameter have been changed slightly since then, but it is interesting that the nature of the racket used was not defined formally until 1978 (Tingay, 1983). With the change in title in 1977 of the International Lawn Tennis Federation to the International Tennis Federation it seems that the name lawn tennis will soon disappear.

## The playing surface

Although tennis began as a game on natural grass, Wimbledon is the only international tournament played on this surface and at all levels of the game little is played on grass. There are several reasons for the introduction of different types of non-turf surface but the main ones are:

1. Improved wear tolerance and uniformity of bounce.
2. Decreased interference by wet weather.
3. Reduced complexity and cost of maintenance.

Trickey (1991) has compared the character and performance of the main types of tennis playing surface (Table 8.5).

There is considerable competition amongst manufacturers of synthetic surfaces for sports and in tennis, as in other sports, the need to assess the relative merits of different playing surfaces has given a considerable impetus to the identification and measurement of aspects of playing quality.

**Table 8.5.** The playing characteristics of tennis court surfaces (from Trickey, 1991).

| Surfaces | Ball-surface | | | | | Player-surface | | |
|---|---|---|---|---|---|---|---|---|
| | Speed of court | Height of bounce | Trueness of bounce | Spin | | Sliding/firm footing | Traction (slip or non-slip) | Resilience (hardness) |
| | | | | Topspin | Slice | | | |
| Grass | Fast | Low | Variable | Little | Yes | Firm footing with partial slide | Slip | Soft |
| Shale | Medium | Medium | Variable | Yes | Yes | Sliding | Slip | Medium to soft |
| Porous macadam | Slow | High | Almost uniform | Yes | Little | Firm footing | Non-slip | Hard |
| Artificial grass | Fast | Medium to low | Variable | Little | Yes | Firm footing | Mainly non-slip | Medium to soft |
| Impervious acrylic | Medium | Medium | Uniform | Yes | Yes | Firm footing | Non-slip | Hard to medium |
| Continental clay | Slow | Medium | Almost uniform | Yes | Yes | Sliding | Non-slip | Medium to soft |

*Playing quality and its assessment*

In the games of cricket and soccer the performance of the best natural turf surfaces has been given the status of 'ideal' or at least 'desirable' in terms of quality, and non-turf surfaces have been rated in comparison with the best natural turf. With tennis this approach may be less justified because natural turf appears to have some less desirable characteristics by favouring particular talents, for example a powerful service. As yet there have been few studies on the testing and performance of turf and non-turf courts.

A tennis ball is softer than a cricket ball so that once a court becomes moderately hard it ceases to deform under the impact of a tennis ball but will with a cricket ball. Thus one would anticipate that bounce or resilience of a court surface would increase up to a hardness above which little if any increase would occur. This was found by Holmes and Bell (1987) who compared rebound resilience (%) with hardness using a Clegg impact hammer. An approximately constant rebound resilience of 58% occurred at hardnesses of around and above 200 gravities which is high for a natural turf court but not for a synthetic one. There was a linear relationship between rebound resilience and hardness at values less than about 200 gravities. Whilst courts differ in hardness and rebound resilience, the relevance of this to pace as perceived by the player is not as clear as it is with a cricket pitch. Indeed the hardness of a tennis court may be more relevant to player comfort in playing than to ball behaviour.

The friction between ball and surface is of major importance in determining both pace (i.e. the horizontal component of velocity) and the bounce height after impact of a tennis ball. Thorpe and Canaway (1986), in their examination of a variety of test methods to assess the performance of tennis courts, used two types of friction measuring method. One used a swinging pendulum and one a weighted sled with a tennis ball to surface contact. Some of their preliminary work pointed to the importance of friction in affecting perceived pace.

Performance testing of tennis courts is at an early stage and it will be some time before meaningful quality standards are developed and accepted.

*Soil and turf*

Tennis courts do not demand the high clay contents and binding strengths required on cricket pitches. A clay content of 18–20% is satisfactory (see Fig. 1.1) with an ASSB binding strength of around 40 kg compared with a typical clay content of 30% for county cricket pitches and an ASSB value of around 75 kg.

The grass has much more influence on the character of a tennis court than on a cricket pitch where it is shaved to the ground. A fine bent/fescue seed mixture has been used traditionally and the sward produced would appear to be ideal, being non-sappy and resilient. Unfortunately it is not practicable to sustain a good bent/fescue sward at even moderate levels of use because of its low tolerance to scuffing and tearing-type wear and slow recovery from wear. The non-aggressive nature of the grasses results in invasion by annual meadow-grass with consequent susceptibility to kicking out, to drought and to low nutrient input. Cultivars of perennial ryegrass are more vigorous and more tolerant of wear and these are now included in seed mixtures along with bent and fescue (All England Lawn Tennis and Croquet Club, unpublished papers for a two-day seminar held August 1991 at Wimbledon). The present 'half-way house' where bent and fescues are included in mixtures containing over 50% perennial ryegrass is difficult to justify and 100% perennial ryegrass would seem to be more logical (see Chapter 2).

## Construction and reconstruction

Since tennis demands soils which have a moderate clay content and binding strength, drainage is slow. The design of a construction profile should take into account the factors discussed earlier for cricket square construction. A drainage layer may or may not be necessary depending on site and geographical location. The construction in terms of guaranteeing an absence of horizonal discontinuities is less demanding than for a cricket square. Establishment from seed, not turf, is recommended.

## Nature of deterioration and typical maintenance programmes

The most serious deterioration is the accumulation of thatch. This must be avoided because it creates too soft a surface which affects bounce and the thatch is subject to tearing out with the turf during wear. When the latter occurs the grass may fail to recover. Another type of deterioration is the creation of bare areas through wear and loss of surface trueness. Soil compaction is a problem but it is usually less serious than on a cricket square.

The following annual programme for maintenance is recommended by David Rhodes, former Grounds Manager of the Northern Lawn Tennis Club, Manchester (pers. comm.).

### *Spring – late February to early May*

1.   Plan the opening of the courts, if you have a choice where to start, think about saving courts for key matches.

2.  Gradually reduce the height of cut in increments to 8 mm on opening and increase the frequency of cutting as the grass demands.

3.  Keep off the courts in frosty conditions.

4.  Make sure irrigation system is in working order.

5.  Lightly scarify with suitable machine in one or two directions (depending on thatch layer). At this time of year aim to promote upright growth and remove surface debris. Do not penetrate the soil as this will 'groove' the surface. If the thatch layer is not severe, a powered brush can be used.

6.  Roll to settle frost heave when the ground is dry on the surface and moist underneath. Sandy soils require less rolling and less benefit is gained by rolling than with clayey soils. Sandy soils must be rolled in quite wet weather to be useful. Do not use a roller over 2 tonnes on any tennis court.

7.  Over-sow any thin areas with a dimple seeder when the soil temperature is above 10°C, also use germination sheeting if necessary.

8.  Fertilize courts with appropriate formulation.

9.  Set-up the canvas, stop netting etc., around the courts. Mark-out each individual court only when the overall layout has been determined.

10.  Inspect the courts daily for weed, pest and disease problems and treat promptly.

## Summer – early May to late August

1.  Fertilize as determined by soil analysis and turf colour. Mow courts three or four times per week in alternate directions, preferably using a ten-bladed cylinder mower and turf groomer/brush. If courts are cut using a ride-on greens machine, grooming should be backed up fortnightly with verti-cutting, otherwise lightly scarify on a monthly basis with an appropriate machine.

2.  Mark-out three times a week to ensure clear bright lines. Thick liquid marking compounds are better than powders. Marking the base-line when it is worn is difficult. To overcome this, use either an aerosol spray or, if acceptable, lay a plastic, shale base-line tape.

3.  Irrigate when necessary.

4.  Periodic rolling may be useful to settle the court, though usually the weight of the players firms it adequately.

5.  Any weed, moss and worm killing that is necessary should be carried out.

6.  Hollows on base-lines should be repaired by lifting the hollow with a fork and topdressing underneath the hollow by pressing the topdressing down the fork holes – this stops the re-settling of hollows.

7.  Courts should be rested if they are not rotated as this will help towards a better quality playing surface. If they are rotated it is possible to fertilize lightly, over-sow and irrigate a court to re-open with full grass cover in 4–6 weeks.

*Autumn – late August to early October*

1.   Mow court ready for renovation and remove debris.
2.   Thoroughly wet the soil to a depth of 100 mm. Some procedure to improve aeration should now take place. This will be determined by the choice available. A 'Verti-drain' can be used to relieve the severe compaction. Any disturbance of the court surface should be repaired.
3.   Scarification should now be carried out. This must be rigorous and the majority of lawn tennis courts will benefit from being scarified using six passes at different angles. The blades should penetrate the soil 5–6 mm.
4.   Hollow-coring will aid aeration, thatch removal and soil exchange. This can be carried out annually or less frequently.
5.   The above operations should be performed over the whole court and then any severely worn areas can be prepared for seeding by hand. A pre-seed fertilizer should be applied.
6.   Sowing should be in two directions over all the court, and then hand sowing heavily worn areas. The overall application rate should be around 40 g m$^{-2}$. The following seed mixture is suitable:

- 70% perennial ryegrass;
- 10% chewings fescue;
- 10% slender creeping red fescue (*Festuca rubra* subsp. *litoralis*);
- 10% browntop bent

7.   Topdressing should be applied evenly over all the area, and in dry weather. Topdressing can be applied by hand or by powered machine. Always apply the recommended application rate (1–2 kg m$^{-2}$) immediately after seeding.
8.   Germination sheeting can be used if necessary over base-lines.

*Winter – early October to late February*

1.   Throughout the winter months work on the grass courts will be restricted by soil wetness and frost. Occasional spiking (not slits) may be beneficial if the surface remains saturated in wet weather.
2.   Mow when necessary. Perennial ryegrass will grow in mild spells and may require mowing at around 30 mm height.

## REFERENCES

Adams, W.A. (1975) Some developments in the selection and maintenance of turf-grasses. *Scientific Horticulture* 22, 22–27.
Adams, W.A. (1987) Modified motty test assesses soil suitability. *Parks, Golf Courses and Sportsgrounds* 52(8), 4–5.

Advisory County Cricket Committee of the MCC (1939) *Minutes of Meeting Held 28th March, 1939*, MCC, Lord's Cricket Ground, London.

Advisory County Cricket Committee of the MCC (1963) *Report of the Special Sub-Committee Appointed to Consider Ways and Means of Providing Faster Pitches (13th March 1963)*. MCC, Lord's Cricket Ground, London.

Arundell, P.A. and Baker, S.W. (1984) Photomicrographic examination of soil conditions of problem pitches at two county grounds in England. *Journal of the Sports Turf Research Institute* 60, 54–60.

Cameron-Lee, S.P. and McAuliffe, K.W. (1989) Evaluating the Adams/Stewart Soil Binding Test for cricket pitch soil selection. In: Takatoh, H. (ed.) *Proceedings of the 6th International Turfgrass Research Conference*. Tokyo, Japan, pp. 193–196.

Daish, C.B. (1972) *The Physics of Ball Games, Parts I and II*. The English University Press Ltd, London.

Evans, R.D.C. (1991) *Cricket Grounds: The Evolution, Maintenance and Construction of Natural Turf Cricket Tables and Outfields*. The Sports Turf Research Institute, Bingley, 221 pp.

Gumbs, F.A., Griffiths, S.M. and Shillingford, G. (1982) Design and physical features of the recently relaid cricket pitch at the Queen's Park Oval, Trinidad. *Journal of the Sports Turf Research Institute* 58, 16–20.

Harris, J.R. (1961) The crumbling of cricket pitches. *Australian Scientist* 1(3), 173–178.

Holmes, G. and Bell, M.J. (1987) Other sports. In: *Standards of Playing Quality for Natural Turf*. The Sports Turf Research Institute, Bingley, pp. 50–52.

Lush, W.M. (1985) Objective assessment of turf cricket pitches using an impact hammer. *Journal of the Sports Turf Research Institute* 61, 71–79.

Marshall, E.A. (1953) Operation wicket! In: *Details of all First-Class Matches Played by the County, Minor County Matters, Statistical Records and Other Information Likely to Interest Members*. Nottinghamshire County Cricket Club, Trent Bridge, Nottingham, pp. 15–17.

McAuliffe, K.W. and Gibbs, R.J. (1993) A national approach to the performance testing of cricket grounds and lawn bowling greens. In: Carrow, R.N., Christians, N.E. and Shearman, R.C. (eds) *International Turfgrass Research Journal* 7, 222–230.

McIntyre, D.S. (1985) Problems of the Melbourne test cricket pitch and their relevance to Australian turf pitches. *Journal of the Sports Turf Research Institute* 61, 80–91.

Piper, C.S. (1932) Some characteristics of soils used for turf wickets in Australia. *Transactions of the Royal Society of Australia* 56, 15–18.

Stewart, V.I. and Adams, W.A. (1968) County cricket wickets. *Journal of the Sports Turf Research Institute* 44, 49–60.

Stewart, V.I. and Adams, W.A. (1969) Soil factors affecting the control of pace on cricket wickets. In: The Sports Turf Research Institute (ed.) *Proceedings of the 1st International Turfgrass Research Conference*, Harrogate, pp. 533–546.

Test and County Cricket Board (1983) *Recommendations on the Preparation of Pitches Suitable for First-Class Cricket*. Report of the TCCB County Pitches Committee, Lord's Cricket Ground, London.

Test and County Cricket Board/National Cricket Association (1989) *Recommendations on the Preparation of Cricket Pitches*. TCCB/NCA, Lord's Cricket Ground, London.

Thorpe, J.D. and Canaway, P.M. (1986) The performance of tennis court surfaces. I. General principles and test methods. *Journal of the Sports Turf Research Institute* 62, 92–100.

Tingay, L. (1983) *The Guinness Book of Tennis Facts and Feats.* Guinness Superlatives Ltd, Enfield, Middlesex.

Trickey, C. (1991) *Tennis Courts.* Lawn Tennis Association Court Advisory Service, London, 224 pp.

# 9  HORSE RACING TRACKS

T.R.O. FIELD  *Ag Research Grasslands, Private Bag 11008, Palmerston North, New Zealand*

## INTRODUCTION

Horses have been raced for enjoyment and sport since they were first domesticated and ridden by humans. Since the 17th century, a concerted effort has been made to breed superior individual horses whose performance will enhance the standing of their owners and breeders. Within the sport of horse racing two industries can be distinguished: breeding, in which the breeders rely both on bloodlines with past performances and progeny testing on the track to enhance their position; and racing, with owners, trainers, jockeys and betters making up a highly visible part of a large industry concerned with racing horses purely for fun or profit.

Turf tracks for horse racing are therefore the concern of a wide variety of people. Owners of horses and the clientele who wager money on races are extremely interested in any influence of the turf on the competitive ability of individual horses. Trainers and jockeys have the additional concern over any interaction of the racing surface with the safety and health of horses. Racing administrators and course supervisors want the best racing surface to attract the better horses which in turn excite public interest and provide the best financial returns. This chapter addresses their concerns by providing an overview of factors that determine the condition of the turf racing surface, that is, the racetrack. As well as being of interest to all sections of the racing industry, turf racetracks fulfil a valuable role as a specialized study area for students of turf science.

In the modern world, natural turf is not the only option for horse racing surfaces. However, as this chapter will make clear, well-maintained turf has many advantages over using dirt, sand or artificial media. In recent times there

329

has been a clear move by administrators of the sport to improve turf tracks on all continents, from Europe to Africa to Asia and the Americas, so that they can attract the best horses in the world to race in prestigious events. These animals often have high potential value as breeding stock and owners and trainers select their racing venues carefully.

Techniques to improve turf racing surfaces have been based largely on experience gained from other coarse turf areas. Published research results indicate that the meagre research effort has concentrated on the veterinary rather than agronomic side of racing. In 1986 the New Zealand Racing Authority began funding a research programme to upgrade racetracks in that country. This chapter makes use of many of the observations and results recorded during the subsequent four years of research, and integrates those results with experiences and observations published for other parts of the world.

## NATURE OF THE SPORT

Throughout the world, horses are raced in many forms and on many different surfaces. The two major international codes are thoroughbred and standardbred racing. Standardbreds (trotters and pacers) are infrequently raced on turf, whereas in Europe, South Africa, Australasia and South America, thoroughbreds (gallopers) are raced almost entirely on turf surfaces (Table 9.1). In North America thoroughbreds are raced mainly on dirt surfaces but there are still many horse races on turf. This chapter will be restricted to discussing turf surfaces for racing thoroughbred horses.

Racetracks have a long history. In England, turf areas have been set aside for racing for many centuries. In the New World, English colonists constructed the first turf track at Newmarket, in the USA, in 1650. In 1819, The Jockey

**Table 9.1.** The international extent of racing on turf.

| No. of races | Country | No. of turf tracks | Year |
|---|---|---|---|
| 24600 | Australia | 410 | 1989 |
| 6370 | France | 200 | 1988 |
| 6280 | Britain | 59 | 1989 |
| 4190 | South Africa | 15 | 1989 |
| 3620 | USA, Canada | 38 | 1989 |
| 3345 | Italy | 13 | 1989 |
| 3320 | New Zealand | 57 | 1989 |
| 3218 | (West) Germany | 15 | 1989 |
| 1730 | Ireland | 28 | 1989 |
| 1710 | Japan | 10 | 1989 |

Club at Newmarket, in England, is recorded to have become involved in the active management of turf when they first committed resources to improving their gallops (Vamplew, 1976).

The shape and size of older racecourses often reflect their origin. For example in England the racecourse at Chester, an ancient Roman town, is reminiscent of a coliseum (Fig. 9.1). By contrast, Newmarket has long runs similar to those experienced when coursing hares or hunting foxes, sports popular with the royalty in the area (Magee, 1989). Modern racecourses, of which Ayr in Scotland is an example, are usually more or less oval in shape often with chutes to provide straight races or at least a reasonable length of straight run during which riders manoeuvre for position before a bend. Retention of variety in racetrack shape and contour is important for racing, as different racetracks favour different types of horses with, for example, long-striding horses often unable to cope with tight-turning tracks.

Most racecourses were originally laid out to fit the natural lie of the land. Although generally located on reasonably flat ground to allow spectators to witness most of the action, the amount and location of rises and falls over the course of a race can vary greatly, even within oval racecourses. Like variation in shape, variation in contour tests different attributes of both horse and rider. Only in recent times have bends been cambered to provide safer racing and allow horses to be better balanced when negotiating turns.

Natural turf for racetracks is grown on the full range of soil textures, from sands to heavy clays (Fig. 9.2). On the two English tracks, Newmarket falls among the soil textures found on New Zealand tracks, while Ascot is at the sandiest extreme. On the Tokyo racetrack, turf is grown in a sand textured topsoil of a man-made profile. The growing medium on many other turf tracks recently constructed in North America and Asia fall within the sand classification, with those at Hong Kong and Santa Anita, California, incorporating amending material in the sand to improve the water-holding and/or strength characteristics. Soil texture has an important influence on the performance of the turf surface, especially in winter.

## Seasonality of racing

The seasonality of racing is usually linked to climatic constraints within countries. In cold temperate regions there is little racing during the coldest and wettest months, while in the subtropics to tropics monsoon influences may make racing impossible in some seasons.

The structure of racemeetings varies widely among countries. In the USA a racemeeting may consist of many consecutive weeks of five or more racedays each week. If a turf track is available, two or three races per day may be raced on turf on the main two or three betting days each week. At Golden Gate

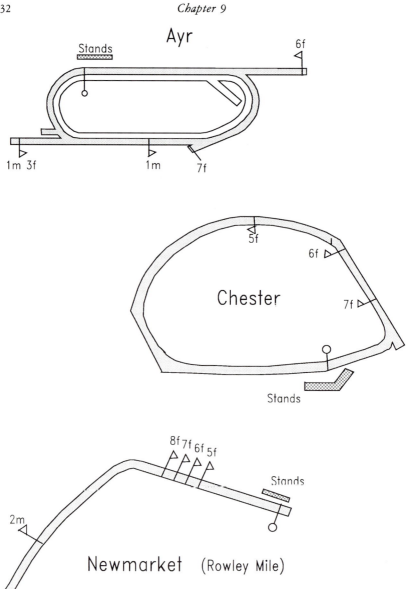

**Fig. 9.1.** Contrasting designs of three racecourses in Britain.

Fields in San Francisco, horses are raced on the seven eighths of a mile turf track over a broken meeting lasting from late March until June. Each year 115 to 120 races are scheduled on turf, with 10 to 14 races each week during the season rising to 31 in each of the last two weeks.

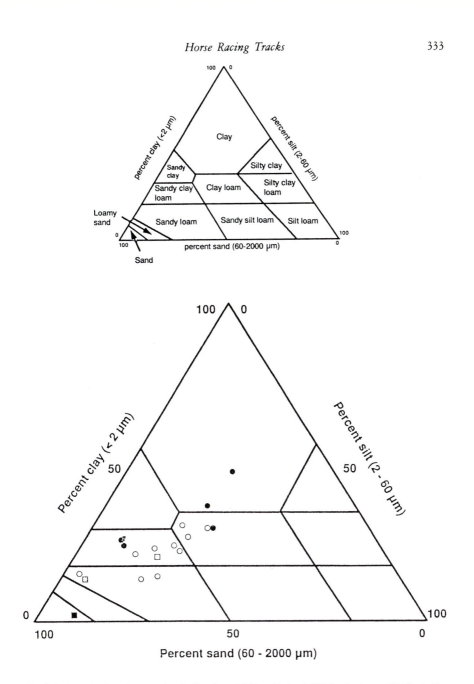

**Fig. 9.2.** Particle size classes of soils found on 15 New Zealand (NZ Grade A, ●; NZ Grade B, ○) □; two English (Ascot, ⊡; Newmarket, □) and one Japanese (Tokyo, ■) racetracks.

A less intensive structure but a similar overall arrangement is used at Longchamp in Paris. Races are all run on turf within two seasons of eight and 12 weeks. Racing is held on up to three days each week with a total of approximately 35 racedays, each of seven to eight races, every year. This high intensity of racing on turf is made possible by the availability of a wide racing surface with alternative tracks round the home bend.

In Britain races are held on tracks either in the flat (March to November) or National Hunt season (November to March). Racecourses with tracks for both flat and jumping races will race all year round. The number of fixtures (days) raced on a flat track ranges from six at Cartmell to 30 at Ayr.

In much of South Africa, Australia and New Zealand racemeetings are more or less evenly distributed through the year. The most intensively used tracks have between 20 and 30 racedays, with up to three days in a week in any one meeting, usually with intervals of two to four weeks between meetings. A movable running rail is used to spread damage sustained from racing across the track, and intervals between meetings allow time for recovery of the turf.

## Type and location of wear

Thoroughbred horses impart large forces to the turf and so have the potential to cause extreme damage and wear. During any one race 10 to 20 horses usually pass over a section of the track once if not twice. Over a programme of eight to ten races portions of the track can be subjected to extreme damage (Fig. 9.3). The nature of the damage can be best understood by looking at the interaction of the horse's hoof with the ground.

During horse locomotion each hoof is in contact with the ground during the support phase (Fig. 9.4), in a pattern and duration that are determined by the gait. For all gaits, the support phase consists of a more or less discernible sequence of the three stages of deceleration, load-bearing and propulsion.

During the first stage, horizontal deceleration occurs when the hoof first makes contact with the ground. Some deformation of the surface without fracture is required to prevent slipping. On very hard tracks, there is little deformation and horses may slip if light rain wets the surface or turf growth is lush. Vertical impact forces are at a maximum during the second load-bearing stage, causing compression or deformation of the soil in contact with the hoof.

In the third or propulsion stage, the tip of the hoof rotates into the soil as horizontal forces come into action propelling the horse forwards. This stage ends with a flick of the hoof when it is unweighted, an action which may create divots when the soil deforms or fails.

Wear from horses has therefore two major components, arising from shear and compression forces. On firm or hard tracks, turf plants can suffer bruising

**Fig. 9.3.** Surface damage on Trentham (NZ) racetrack after a midwinter raceday.

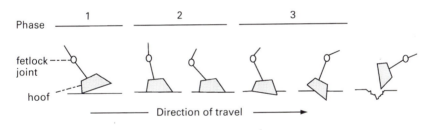

**Fig. 9.4.** Stages in the support phase of the gallop.

with little disturbance of the growing medium. As soil water content increases and tracks become softer, turf suffers shear damage during hoof deceleration and divots can be cupped out during the propulsion phase. Long-term damage to the soil is seen as a breakdown of soil structure resulting from smearing and compression of moist soil beneath the hoof. Typically, the soil from 70 to 150 mm below the surface forms a compacted layer on racetracks. On very wet tracks more immediate damage comes from horses' hooves that may sink up to 200 mm into the growing medium.

Wear is not evenly located round racetracks. Horses gallop within two to three metres of the running rail for all but the final stages of most races when they spread out across the track as jockeys seek the best opportunities for their mounts. As horses travel round bends, damage to the turf is greater because centrifugal forces increase the pressure applied by each hoof. Damage also tends to be more severe where horses are pulling up after completing the race. This is usually the same area that horses gallop over on the way from the saddling area to the start. Different racing distances can also result in various levels of overlap.

Extreme localized damage can occur when horses jump obstacles (hurdles or steeples) during the course of a race. This type of race is usually programmed for the winter season when the ground is soft. Where the obstacles are fixed, the action of the horses in propping for take off, jumping and landing can wear a trough in front of the jump and build up a mound at the landing point. On dual-purpose racetracks, where the obstacles are often movable, a high intensity of damage may be incurred around the jumps each raceday.

## Grass species

Grass species differ in their ability to tolerate wear. Species profiles change when one compares tracks or areas of tracks along the gradient from little or no use to intensive use. Under only occasional use heath-type species will persist on sections of racetracks, giving way to pasture-type species as galloping use increases. In New Zealand we have observed a dominance of browntop bent (*Agrostis capillaris*) and fine fescue (*Festuca rubra*) on sections of steeplechase tracks used only intermittently, with little input of fertilizer or irrigation. Tracks immediately adjacent to these which are galloped on more frequently are dominated by wear-tolerant pasture-type species such as perennial ryegrass (*Lolium perenne*). Bents and fescues were widely used in repair seed mixtures in the past, but because they do not tolerate high wear, they have been dropped in favour of the more persistent perennial ryegrasses. Following further compaction, especially where tracks are kept moist, annual meadow-grass (*Poa annua*) and/or rough stalked meadowgrass (*P. trivialis*) may become dominant. *Poa* species and annual meadowgrass in particular are shallow rooting and undesirable.

There have been great advances in breeding dwarf cultivars of perennial ryegrass for coarse turf areas, but these more expensive releases have been little used on racetracks. Until management and technology skills improve, much of the extra cost of the seed will be wasted as establishment of seedlings is poor where seed is being introduced to repair existing swards. Where new areas are being sown down, turf ryegrass cultivars promise improved turf quality. The very hardy dwarf tall fescue (*Festuca arundinacea*) cultivars also have many of the desired attributes of an ideal racetrack grass: they have a strong, hard foliage that is more wear-tolerant than that of many ryegrasses, and a coarse, strong root system that can maintain a high soil strength in wet conditions. A mixture of dwarf tall fescue and turf-type ryegrasses has provided a high class turf surface since 1989 for racing greyhounds on a sand-based track at Manukau in New Zealand.

In tropical and subtropical climates Bermudagrass (*Cynodon dactylon*) and Kikuyugrass (*Pennisetum clandestinum*) are used along with other indigenous species such as *Zoysia* in Japan. Most warm season species have the disadvantage of undergoing a dormant period during which they brown off (see Chapter 11). Although warm season species still provide a wear-resistant turf while dormant, they are usually oversown with a cool season species such as perennial ryegrass to give an attractive green colour.

The stoloniferous/rhizomatous growth habit of Bermuda grass confers additional advantages on it as a turf species. Firstly, on the evidence of its growth on the old soil-based track at Sha Tin, Hong Kong, the rhizomes appear to help Bermuda grass grow actively in extremely compacted soils, and, secondly, the dense underground network of rhizomes would be expected to add extra strength to wet soils.

## PERFORMANCE CRITERIA

Horses have evolved characteristics enabling fast running on turf surfaces of one form or another. Studies show that a cushioning, resilient surface is needed for continued leg health. Natural turf racetracks in prime condition can provide that elasticity when soils are at the dry end of the scale.

An ideal pattern of shock absorption, obtained from turf on an Australian racetrack, is compared in Fig. 9.5 with a less absorbent frozen turf in Canada. Both readings of impact force over time were measured with a resileometer. In the ideal situation deceleration following impact was spread over a relatively long period ensuring a low peak force (shock). After the rebound the remaining energy was absorbed smoothly. However, even on the best turf surfaces, a relatively low proportion of less than 3.5% of energy was returned to a mechanical hoof (Zebarth and Sheard, 1985). Tracks that are too dry or too hard, as shown for the frozen turf, transmit very large forces to the leg, expressed over short time periods. Under these conditions strain rates can be close to the critical threshold for bone fracture, and under normal racing conditions must result in microdamage and gradual fatigue weakening of the leg bones of the horse (Pratt, 1984). Artificial surfaces or dirt tracks can be groomed to provide cushioning similar to an ideal turf track for the first impact, but leave an imprint that will respond less sympathetically to successive impacts. Performance standards can and need to be developed so that irrigation can be used to keep hardness below maximum acceptable readings.

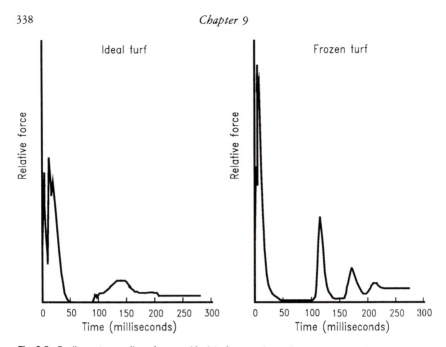

**Fig. 9.5.** Resileometer readings from an ideal turf on an Australian racetrack, and a frozen turf on a Canadian racetrack.

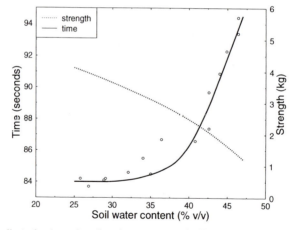

**Fig. 9.6.** The effect of volumetric soil moisture content (0-75 mm) on winning racetimes of maiden horses running over 1400 m, and on an ASSB test for soil strength for the Awapuni (NZ) soil.

Performance criteria for the wet end of the soil water scale are more difficult to determine. Racetimes increase rapidly once the soil has become wet, at which stage the strength of the mineral soil appears to decline a little more rapidly (Fig. 9.6). In New Zealand, horses will usually continue to be raced on very wet tracks unless splashing of water lying on the surface of the track makes racing unsafe by impairing the vision of jockeys. As tracks get wetter they become softer and more slippery to the horses' hooves, but it is difficult to define a limit at which conditions impose unacceptable stresses on the legs of galloping horses. Under wet soil conditions, low soil strength causing sinkage and sliding limits the effectiveness of propulsion and causes stress injuries on hind quarter muscles rather than on bones (Uren and Scott, 1982).

Prolonged racing on wet tracks leads to a breakdown of soil structure that has a serious long-term effect on turf performance. Therefore, racetracks can be considered to perform poorly if they remain excessively heavy through the winter period, and do not respond quickly to improved weather conditions. For a given soil type, this sort of performance indicates poor drainage properties usually caused by soil compaction.

## Track assessment

An accurate assessment of track conditions (going) would enable performance standards to be monitored both on and between racedays. Horse racetracks differ from nearly all other turf arenas because racetimes offer a direct, relevant but independent means of assessing the character of the surface. The winning times of comparable horses will reflect the characteristics of the going on different racedays with regard to their effect on overall speed. The assessment is not synonymous with quality which embraces a range of subjective criteria.

Presently, most tracks are assessed qualitatively and classified among several discrete categories, for example, firm, soft and heavy. Racetimes do not usually relate closely to these categories of going (Fig. 9.7, from Uren and Scott, 1982). Misjudgement of track condition only partially explains this overlap, seen even when single race class times distance combinations are examined (see times in Fig. 9.8). Errors in changing starting positions to allow for different running rail locations and in the line taken by jockeys on wet tracks are also possible sources of variation.

Uren and Scott (1982) examined the usefulness of developing a quantitative assessment of the conditions on turf racetracks. They concluded from their experiments on a Melbourne racetrack that assessment could be based on quantitative criteria but questioned the tangible benefit of such a measure. However, in their study they judged the likely benefits using indirect measures such as betters' success, rather than their value to owners, trainers and racetrack managers. If further work was to be carried out they recommended a range of instruments from penetrometers to shear vanes and impulse hammers be investigated.

Subsequently, the influence of soil and turf factors on impact and shear resistance of racetracks was investigated by Zebarth and Sheard (1985). They

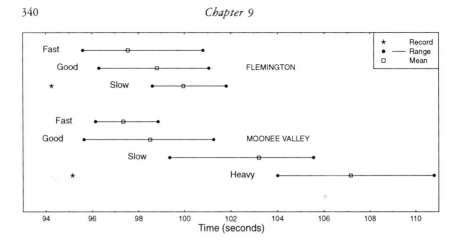

**Fig. 9.7.** Variability of winning times for 1000 m races at Flemington and Moonee Valley racetracks, Melbourne, Australia.

were able to show that racetimes at the Woodbine turf track in Toronto, Canada, were related to impact resistance, but again found considerable overlap with qualitative categories of going. Impact resistance was shown in their study to be mainly determined by soil water content.

A dynamic penetrometer was designed by Monsieur Branet of the Société d'Encouragement in Paris as an easy-to-use instrument for assessing racetracks. The hardness of the track is measured as the depth that a rod, one square centimetre in cross-section, penetrates the turf when struck by a weight of one kilogram falling a distance of one metre. The penetrometer is currently used on some racetracks in many countries including France, Japan, Australia and South Africa. In New Zealand we have shown that readings from the penetrometer (as Penetrometer Index – PI, corrected for initial surface roughness) can be an accurate predictor of racetimes for the racetracks monitored (Fig. 9.8). When soil water content is not markedly unstable, because of rain soon before or after the penetrometer has been used, the PI values appear to give a good prediction of likely racetimes. However, soils of different strength produce different relationships between PI and racetimes. At this stage our analyses indicate horses slow down appreciably more for an increase of one PI unit on strong soils than on weak soils. Further work in this area should enable us to calibrate penetrometers for different soil characteristics so that a single scale can be developed for widespread use. Such information would be especially helpful for interpreting the performances of racehorses across country boundaries, as the terminology for track conditions can be markedly different. At present PI information is useful for applying only to the track from which it was obtained.

Similar data are currently being collected with a shear vane tester. On one

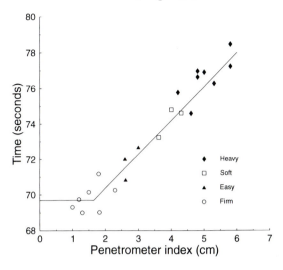

**Fig. 9.8.** The relationship between winning times, recorded by two-year-old horses running over 1200 m, and Penetrometer Index (PI).

soil type where the going can be very heavy (PI greater than 6), measured soil shear strength has proved useful in quantifying track performance after the soil has become so wet that suction places an upper limit on the penetrometer reading. On three separate days when the PI value was 6.7, the winners of races for maiden horses (non-winners) over 1200 m (6 furlongs) ran times of 77.8, 80.3 and 81.3 seconds. Shear vane readings of 4.3, 3.4 and 2.3 t m$^{-2}$ indicated that changes in track character continued to affect horse performance as soil strength declined beyond the limit detectable by the penetrometer.

Other more complicated instruments have been used to evaluate turf character. Electronic instruments such as the Clegg hammer or resileometer measure the deceleration of a weight dropped on to the track. A 'going meter' that works on similar principles has recently been tested by the Jockey Club in Britain as a means of deriving a standard objective classification of racetrack conditions in the UK. These instruments have not as yet been shown to be superior to the more easy-to-use penetrometer.

A calibrated instrument can aid in assessing track conditions both leading up to, and on raceday. Once performance standards are defined, these assessments can assist the track manager to modify soil conditions for racing. At Moonee Valley, Melbourne, Australia, irrigation is applied where necessary to produce a penetrometer reading of 2.2 on raceday (Fig. 9.9). In the lead up to the W. Stutt Stakes in 1990, rain fell during the week so that irrigation was not necessary.

In the long term, regular measurements of relevant track conditions should

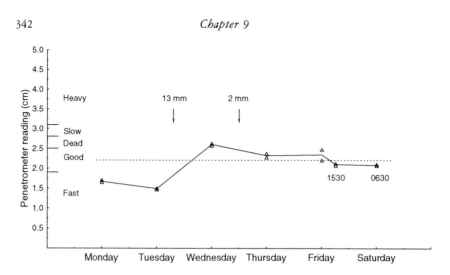

**Fig. 9.9.** Changes in 9.30 a.m. penetrometer readings and track classification leading up to the W. Stutt Stakes on Saturday 29 September 1990 at Moonee Valley; timing and amount of rainfall shown.

indicate whether management procedures are maintaining or changing the turf and soil characteristics which affect the health and performance of racehorses.

## SYMPTOMS, CAUSES AND CHARACTERISTICS OF DETERIORATION

## Symptoms

An ideal horse racing track will have an elastic and resilient surface, which will provide adequate cushioning and then return to its original state. The surface should also have soil shear strength properties that will provide adequate reactions in both the deceleration and propulsion phases of hoof action in the gallop. Racetracks in good condition have a turf cover that can fulfil these requirements in two ways: by building up sufficient density of stem bases and root systems mixed with soil near the surface to cushion the impact of the hooves when the soil is dry; and by having strong, deep root systems to add shear strength to the soil when it is wet. Dense networks of rhizomes beneath the surface may provide additional shear strength.

Soil compaction is the main impediment to achieving the ideal surface. When compacted, total porosity is reduced and the air-filled porosity can be very low. Compacted soils may have an increased bulk density which would be expected to lead to increased soil strength at an equivalent water content. In practice, increased bulk density leads mostly to higher water contents and,

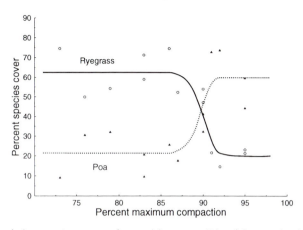

**Fig. 9.10.** Trends in percentage cover of perennial ryegrass (○) and *Poa* species (▲) with relative compaction measured between 75 mm and 150 mm depth on 12 New Zealand racetracks.

as a result, an increased tendency to shear expressed as a *lower* soil strength. Drainage is impaired in compacted soils so soils stay wetter longer and rooting is shallow. Racetracks that have suffered compaction will therefore have longer periods when track conditions are water-affected and may provide worse footing than better tracks at similar soil water contents. Neither of these effects are easy to detect in isolation, but the success or otherwise of growing desirable grass species has been found in New Zealand to be a symptom of subsurface conditions.

On 12 racetracks growing temperate species, the balance between deep and shallow-rooted grass species has been shown (Fig. 9.10) to be an indicator of deterioration of soil structure (Field and Murphy, 1989). Soil structure deterioration was estimated by calculating relative compaction defined as the ratio of soil bulk density found on the track to the maximum bulk density measured with that soil type at field moisture in the laboratory. Below a relative compaction of 90%, ryegrass was dominant. Above 90%, shallow-rooted *Poa* species dominated. Two of the worst *Poa*-dominant tracks had isolated areas where soil structure had deteriorated to such an extent that under wet conditions the turf virtually floated on a fluid soil and was unsafe for racing.

## Causes

The prime cause of soil compaction is traffic of both horses and vehicles under wet soil conditions. When soils are consolidated, voids and macropores between soil particles are broken down. This breakdown of soil structure occurs most readily when about 80% of the total pore space is water-filled. The maximum vertical force applied by a horse's hoof is of the order of 5000 newtons which increases by about 50% on the racetrack bends. When soils are wet, the hoof penetrates into the soil and turns over the surface. Where shear strength of the soil is not sufficient during the deceleration phase of the gallop, the subsurface layers will be smeared as well as compressed. Machinery traffic affects both surface and subsurface layers.

Any environmental or management influence that extends the period when soils are wet will increase the risk of further compaction. Poor drainage, resulting from a soil's natural characteristics or deficiencies in construction, will prolong wet conditions. Poor irrigation procedures may advance the onset of wet conditions in the autumn. A cycle of compaction leading to impaired drainage and therefore to more compaction often occurs.

## Characteristics

The location of compacted layers in the soil profile is shown from data gathered on New Zealand racetracks. In contrast to most sportsturf areas (Chapter 5) problems are usually located below 70 mm. Typical vertical profiles (Fig. 9.11) show that penetrometer resistance, relative to an uncompacted control beneath the running rail, increased most dramatically between 70 and 150 mm below the surface.

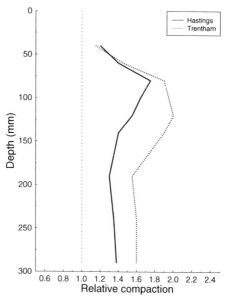

**Fig. 9.11.** Pattern of change in compaction with depth on two New Zealand racetracks (interpreted from 200 penetrometer readings on and adjacent to each racetrack).

The relative importance of compaction in this layer has been estimated from data from 14 racetracks. Bulk density between 75 and 150 mm depth showed an increase of 18% on average over bulk densities found in the surface 75 mm of mineral soil. Penetrometer resistance of the 75–150 mm layer was in turn nearly 20% greater than the less compacted layer between 225 and 300 mm depth. This localized increase in the packing of soil particles has significant implications for turf growth.

Total porosity of soil on the tracks tested ranged from 50% to 60% of soil volume at the depth of the compacted layer. This is still quite large compared with sand-dominant soils; nevertheless, the increase of 20% in soil bulk density corresponds to a loss of up to 10% of porosity. Much of this loss is likely to have occurred in macropore space, reducing air-filled porosity. Compaction within the 75 mm to 150 mm depth layer is therefore both a characteristic of deterioration of soil structure on racetracks and an indicator of potential deterioration of the turf racing surface as aeration of sub-surface soils affects plant growth. Air-filled porosity immediately above the compacted soil layer (40 mm and 70 mm), appears to be a very useful indicator of the compaction of a racetrack.

Percentage cover of desirable turfgrass species monitored on four New Zealand racetracks, measured before (winter 1988) or after (one year later in 1989) renovation, increased in response to a series of sub-aeration treatments relieving compaction (Table 9.2). Before renovation turf quality reflected soil aeration directly. The response to improvement in macroporosity was affected by other management factors, including difficulty in establishing improved species on the Ellerslie track. In contrast, the ryegrass content on the Awapuni racetrack improved markedly within six months of soil physical treatment. Nevertheless, longer-term observations indicate that a value of around 10% air-filled porosity in the soil at our test depth is critical for achieving adequate water movement through the soil and promoting a change to deeper-rooting species.

**Table 9.2.** Responses of four New Zealand racetracks to soil physical renovation.

| Racetrack | Soil type | Air-filled porosity (% v/v) | | Desirable species (% ground cover) | |
|---|---|---|---|---|---|
| | | 1988 | 1989 | 1988 | 1989 |
| Ellerslie | Clay | 6.1 | 12.5 | 17 | 45 |
| Te Rapa | Sandy clay loam | 8.8 | 13.6 | 66 | 79 |
| Awapuni | Sandy clay loam | 7.5 | 10.0 | 40 | 71 |
| Trentham | Sandy clay loam | 8.0 | 11.4 | 49 | 67 |

Soil bulk densities alone have not proven very useful in monitoring relief of compaction. Large seasonal changes in bulk density can introduce uncertainty into analyses other than where treated and untreated areas are compared directly.

## MAINTENANCE REGIMES

## Preparation for racing

Variation in placement of the running rail is an important management tool used to spread wear and shield areas for renovation. Racetrack widths vary considerably, with the options for rail position restricted by the requirement of the jockeys and stewards for minimum widths for racing. With full fields in New Zealand (18 horses) the accepted minimum width appears to be of the order of 16 metres.

Mowing and rolling are the main grooming procedures carried out in preparation for raceday. Racetracks normally cover an area of between 4 ha and 8 ha and are mown with either tractor-mounted gang or rotary mowers. The mowing height is high so front mounted mowers are preferred to avoid tyres flattening the grass in front of the mower. Variation in both direction and line of mowing helps maintain a more even sward. Mowing height is generally longer than on other coarse turf areas. On many racetracks a minimum height of 100–120 mm is considered necessary for raceday; longer grass is thought to provide extra cushioning in summer (even though experimentally this was not observed to be so by Zebarth and Sheard, 1985), and to help minimize damage in winter. Clippings are seldom removed on soil-based tracks, but their removal is recommended on sand-based tracks.

Tracks are often rolled before raceday when the soil is moist with the objective of increasing bulk density and firming up the surface. However, the consequences depend upon soil moisture content and particle size distribution. Heavy rolling of wet soils eliminates macropores. The result is a soil which may be marginally firmer but is likely to be slippery because soil particles are lubricated with water. If heavy rolling is to be carried out it is advisable to roll several days before raceday to allow time for some water loss by evapotranspiration. Although correctly timed heavy rolling may improve tracks in the short term, it will cause track deterioration if the soil does not dry sufficiently for remedial sub-aeration before it is raced on or rolled again. In the long term, heavy rolling is detrimental to soil conditions.

Regular spiking of a racetrack is recommended leading up to a raceday to ensure the surface is not sealed and remains permeable to water.

Irrigation can be used strategically to produce a desired state of track for racing. At Longchamp, irrigation management is used in a similar manner to that described earlier for Moonee Valley, in an attempt to produce tracks with a PI of about 2.5 for racing. This reading, on the sandy soil at Longchamp, corresponds to a good but not hard track that should suit all horses and minimize percussion injuries to their legs. Although this is a valid use of irrigation, if the track is not allowed to dry out between race meetings, the growth of shallow-rooted grasses will be encouraged. Over-use of irrigation can create a less wear-tolerant surface that becomes an extra problem for the manager. Continual watering and compaction do not appear to affect the competitiveness of Bermuda grass as much as perennial ryegrass.

Irrigation is only really useful during preparations for raceday if water can be applied to all the racetrack on any one day. At Haydock Park, in England, 10 to 15 mm of water can be applied to the whole track daily, so the track is watered only over the seven days preceding a raceday. This strategy ensures good to firm going that is even over the whole course. The course is allowed to dry out for about two weeks between racedays, encouraging ryegrass dominance. Operation of less efficient irrigation systems must be stopped earlier before raceday to avoid sudden changes in going around the track that may inconvenience some horses.

## Restoration after racing

When tracks are firm, damage from racing is limited to bruising of turf plants. This can make tracks unsightly if large amounts of wear-susceptible species such as *Poa* are present in the sward. Application of a small amount of fertilizer nitrogen and irrigation can easily overcome this degree of wear.

When tracks are soft or very soft, large divots can be turned over or thrown out by the galloping action. Damage is worse where shallow-rooted grasses predominate, as divots are physically thrown along the track. With strongly rooted turf, plants divots are often hinged by intact roots and remain in contact with the damaged area. On many racetracks these are replaced manually between races on raceday. After the meeting all divots are replaced, where possible, either manually or by brushing in the opposite direction to horse traffic. Where holes remain in the track they must be repaired. Ideally this should be carried out by loosening the soil around each hole to relieve compaction as well as levelling the surface. The high labour requirement of this strategy usually means the quicker method of filling holes with a soil/seed mixture or with sods from a turf nursery is used. The track is usually then lightly rolled with an implement like a Cambridge roller.

Movable running rails are a valuable aid for protecting damaged areas so that restoration and repair can be effected, and to spread wear. If the running rail is fixed, wear is restricted to the same areas at each meeting. Continual addition of soil to fill holes near the rail often builds up the track gradually, disrupting crossfalls and creating problems. The movable rail helps the manager overcome this source of problems.

## Seasonal maintenance

Recommendations for fertilizer use on racetracks are imprecise. A statistical study of racetrack data (Field *et al.*, 1988) indirectly established that the proportion of *Poa* species on tracks was positively associated with high levels of available soil phosphorus (P). Restricted P applications were recommended to attain a relatively low soil P availability, lower than recommended as optimal for agricultural soils where ryegrass and clovers are grown. Ryegrass cover was not associated with soil analysis for P or soil potassium (K), but did appear to be linked to fertilizer K applications. Highest quantities of ryegrass were found where fertilizer K was being used. This is in accordance with the modern perception in turf culture that K is important for stress resistance and persistence of desirable species.

Clover is absent from most racetracks, so nitrogen (N) fertilizer must be applied if vigorous growth is required. However, at high soil N availability, leaf growth is stimulated at the expense of root growth (Chapter 2). Fertilizer N allied to the return of clippings has led, on New Zealand tracks at least, to high total soil N levels (0.3 to 0.6%) with close C:N ratios (in the range 9.5 to 10). Only small inputs of fertilizer N are required on soil-based tracks and care must be taken to ensure desirable root growth is not prejudiced. Fertilizer N applications should be linked directly to specific expected outcomes. Applications of 25 to 40 kg N ha$^{-1}$ to aid recovery from damage, to boost young grass, and to thicken the sward in autumn and spring fall in this category. If fertilizer N is being routinely applied only to maintain colour, low rates should be used in each application (15–20 kg N ha$^{-1}$).

Sub-aerators are being used extensively to enhance aeration, relieve compaction and so to improve soil structure on racetracks. Implements used in New Zealand range from agricultural aerators and mole ploughs to specialized turf vibramoles and corers. Large solid tine corers (e.g. the 'Verti-drain') are popular in Britain, South Africa and Australia, where the core holes may be refilled with sand. Each implement has its role and its drawbacks, and a range will suit any combination of climate, soil type and racing intensity. Sub-aerators must be used when the soil is no wetter than moist, and appear to be more effective when used in spring than in the autumn.

## OPTIONS FOR RECONSTRUCTION OR UPGRADING

Existing racetracks are seldom completely reconstructed, but tracks are often partially reconstructed to meet either inadequate crossfall requirements on bends or to upgrade tracks suffering from limitations of the rootzone or total soil.

## Reconstruction to meet crossfall requirements

Optimum crossfalls for a given radius of curvature do not appear to fit neatly into engineering standards. By leaning into the bend, horses shift their centre of gravity to compensate for centrifugal forces to an extent that appears to depend on the type of terrain on which they are trained. New Zealand horses are trained mostly on flat ground and when crossfalls are of the order of 6% on a 1600–1800 m track they have difficulty holding a true line. Their apparent compensation causes them to fall down the camber causing tightening of horses running against the rail. Horses in the USA reportedly have no trouble running round bends with crossfalls of 8–10% (Siemans, pers. comm.). Most of these horses train on highly cambered dirt tracks.

Continual addition of soil during repairs after racing builds up negative cambers near the rail on many old tracks. Such crossfalls on bends can become very dangerous if the track is hard and grass is lush, or when light rain or dew falls on the surface to make it slippery. Under these marginal conditions, horses slip and racedays are cancelled. Between 1978 and 1980, a number of tracks in New Zealand have been partly or fully reconstructed to correct crossfalls. Two of the largest reconstructions converted negative crossfalls on all the bends to cambers of 3 to 4%. For horses trained on flat tracks, these crossfalls appear to allow them to move round bends fully balanced and therefore safely. New turf tracks constructed in the USA have crossfalls of up to 8%.

## Reconstruction to overcome soil medium limitations

Most often partial reconstruction is used to install extensive drainage systems where track wetness is limiting racetrack performance. Partial reconstruction can also widen tracks where narrow sections limit the range of movement of the running rail.

Very few metropolitan courses in countries with year-round racing can meet performance standards without extensive drainage systems. In Sydney, Australia, the Randwick track has been intensively slit drained with sand to move water into a highly permeable underlying deep sand subsoil. Although the track still becomes heavy, it is now recognized as providing one of the best wet-weather racing surfaces in Australia.

Slit drainage is very popular in the UK. The Ascot racetrack has been intensively slit drained by track groundstaff under the supervision of Gordon Hiscock. Haydock Park, which was well known as a track for 'mudlarks', was also extensively drained in 1989. A system of relatively narrow lateral drains at 5 m centres discharge into a collecting drain on the inside of the racetrack. On many parts of the course sand/gravel slits have been installed at 1 m centres to run parallel to the running rail and intersect the laterals.

Drains have the potential to cause problems on racetracks because of the large forces exerted through the horse's hoof. Problems most often arise when the soils above the drains behave differently from the surrounding medium in the period soon after installation. Sinkage may occur during rain if the backfill is inadequately compacted. On the other hand, the backfill may be firmer than the surrounding soil if it has been overcompacted. In a more unusual situation, an old drain collapsed during a race at Doncaster, England, after a long period of exceptionally dry weather in 1989. Improved drainage techniques using narrow slit drains (50 mm wide or less) or careful cutting, removing and replacing of sod, can reduce the problems of bad settling and poor grass re-establishment after installation.

Full reconstructions have been carried out mainly in Asia to upgrade the drainage and rootzone characteristics of the growing medium to cope with local extremes of rainfall. For example, the Japan Racing Association has taken an engineering approach and rebuilt grass tracks at all ten of its racecourses along similar lines. Turf is grown in sandy soil overlying layers of successively coarser sands then gravel aggregates.

In North America many turf tracks have been constructed in the last 20 years, mainly using sandy soils or modified sand growing media. Recent full reconstructions have replaced these media with more-or-less pure sands to improve control of soil water. The advantage of sand as a growing medium was considered by Zebarth and Sheard (1985) to outweigh the higher impact resistance and lower resistance to shear measured in a comparison with turf grown in soil. Weakness in shear strength has been overcome by amending the sand with interlocking mesh elements in the most recent sand reconstruction at Santa Anita.

Sand grades are selected for these tracks to give a specified pore size distribution once turf has been established. Sand does not maintain all the desired characteristics as the turf system ages. In a recent study (Table 9.3) we have found that over time the air–water ratios change as the apparent pore size distribution in the sand is changed (Murphy and Field, unpublished data). Increasing contents of organic matter parallel an apparent degradation of the macropores. Where sand is used as a growing medium in temperate areas we envisage increasing difficulty in maintaining a turf of healthy deep-rooted species as air-filled pores decline. Use of mesh elements, or fibres, to reinforce the medium may compensate for the effect of poor root growth on resistance to shear and therefore on racehorse performance. However, a turf made up of shallow-rooted species will not tolerate wear and provide the same quality racing surface as desirable turf species. Like soil-based tracks, sand media need regular surface and subsurface treatments to ensure an adequate air-filled porosity for the success and vigour of desirable turf species (see Chapter 5).

**Table 9.3.** Physical properties of sand-based racetrack growing media.

| Age and location | Organic matter (% w/w) | Bulk density ($t\,m^{-3}$) | Total porosity (% v/v) | Macro porosity (% v/v) |
|---|---|---|---|---|
| **1-5 years** | | | | |
| Santa Anita | 2.21 | 1.44 | 41.9 | 25.1 |
| Churchill Downs | 4.23 | 1.72 | 36.0 | 24.1 |
| Bay Meadows | 6.61 | 1.52 | 43.3 | 19.1 |
| **6-15 years** | | | | |
| Santa Anita chute | 9.53 | 1.17 | 54.9 | 10.7 |
| Golden Gate Fields | 12.7 | 1.01 | 57.7 | 20.0 |
| Turf Paradise | 13.0 | 1.30 | 51.1 | 11.6 |

Sand tracks appear to suffer from their own form of compaction not unlike the problems on winter games pitches with sand-dominant rootzones.

Sand-based tracks are at present the nearest thing to natural all-weather turf surfaces for racing. Because an average of 75 days of racing have been lost over each of the last 20 years in Britain, The Jockey Club has been at the forefront in seeking alternative non-grassed media for racing in cold wet weather. 'Equitrack', a polymer-coated sand, provides a free-draining surface that has some freezing resistance and added cohesiveness over pure sand. This surface has been in use since 1989 at Lingfield Park in England and Remington Park in Oklahoma, USA. 'Fibresand', a mixture of polymer fibres and sand, has been installed at Southwell, in Lincolnshire, England. This use of random fibres to reinforce media to support natural turf for horse racing, merits further study since their use is increasing in the UK especially in Association Football (Chapters 5 and 6). Finally, 'Pacemaker' is a bonded granular plastic mixture, mainly made up from the covering stripped from plastic-coated wires. This has been tested on training tracks. No doubt many other materials will be developed in the future.

All-weather media aim to maintain track conditions within a narrow range. If these tracks became widespread they would limit the scope for breeding racehorses and further narrow the genetic base of thoroughbred horses. The ability to handle various classes of track condition appears to be strongly inherited. Bold gallopers on firm to good going often will not stretch out on wet tracks. Conversely, some sires leave horses that are not competitive on fast tracks and only come into their own when going conditions are at their worst.

# Financial implications

Maintenance of the racing surface is one of the main costs on major racetracks. Poorly performing soil-based racetracks are not only dangerous but also incur high maintenance costs repairing damage to over-wet soils and wear-susceptible turf. Partial or total reconstruction to improve the water relations of the soil or sand medium reduces maintenance costs and has a number of other immediate benefits. For example, improved raceday turf leads to a better class of horse racing on tracks, safety is improved, and the risk of losing racedays because of wet weather is decreased.

The costs of partial or full reconstruction vary with its scope. The complete drainage system installed at Haydock Park cost £0.25 million (Harbridge, 1990). The full construction of Santa Anita cost US$3.2 million, of which US$2.5 million was for material and work associated with the track itself (Palmer, pers. comm). In turn, about one sixth of the this cost was for the mesh elements. These two examples set upper limits as they represent top-of-the-line options for partial or full reconstruction. By comparison, drainage of the important sections of Ellerslie racetrack in New Zealand, across the whole 42 m width at 10 m intervals, was estimated to cost the equivalent of about £50,000.

Any losses in racing caused by reconstruction must be included in the costs of reconstruction. Both major projects outlined above were completed in racing off-seasons. At Haydock Park, drainage was laid in three phases: Phases 1 and 3 during successive autumns, involved draining the flat course in two sections after the start of the jumping season; Phase 2 was drainage of the hurdle course in the intervening spring. All work was completed over 15 months without disrupting racing. At Santa Anita reconstruction was completed over 45 days, 32 for construction and 13 to lay the washed sod. The grass track was then fully functional for the next race meeting. In Australia, races have been held on washed turf at Moonee Valley only 30 days after laying.

In principle, maintenance costs may differ on an upgraded track because of different fertilizer and irrigation requirements, but should not be increased significantly. However, in many cases upgrading has become necessary because of inadequate past maintenance procedures. New maintenance regimes are needed to ensure that the improvement in track performance after reconstruction is not lost by neglect over the first ten years of racing. Therefore, reconstruction can often increase the cost of maintenance.

The total cost of reconstruction has to be weighed against the benefits from improved racing. In countries where a proportion of betting turnover is returned directly or indirectly to the club, there is an obvious financial incentive to improve the racing surface. Betting on individual races is very dependent on the number and quality of runners. Betting has also been found to be reduced where turf surfaces are known to be inferior, and the chances of upset results are greater. This has been seen in New Zealand, on two tracks of contrasting performance at the same racecourse. On Californian tracks the 'handle' (betting turnover) is usually quoted as being 10 to 30% greater where an equivalent race is run on turf rather than dirt. Additionally, more individual races can be held on an upgraded turf track, and fewer racedays lost through unsafe track conditions.

## CONCLUSIONS

Turf racing tracks are turf areas with specialized requirements to meet the patterns of use and damage incurred by horses. They also have peculiar management problems because renovation often has to be fitted in to apparently inadequate periods between meetings. Nevertheless, the problems, their symptoms and remedies are those of natural turf surfaces in general. Further observation and experimentation will be necessary to tune many common turf maintenance practices to the neglected area of racetracks.

## REFERENCES

Field, T.R.O. and Murphy, J.W. (1989) Ecological analysis of racetrack turf species. In: Takatoh, H. (ed.) *Proceedings of the 6th International Turfgrass Society Conference*, Tokyo, Japan, pp. 147–149.

Field, T.R.O., Murphy, J.W. and Hamlin, A.J. (1988) Racecourse uncertainties – the plants that cover racetracks. *Proceedings of the New Zealand Grasslands Association* 49, 107–110.

Harbridge, M.K. (1990) How to drain a heavy racecourse. *Turf Management* 10(4), 28–29.

Magee, S. (1989) *The Channel Four Book on Racing*. Sidgwick and Jackson Ltd, London, 208 pp.

Pratt, G.W. (1984) Racing surfaces – a survey of mechanical behaviour. *Proceedings of the American Association of Equine Practitioners* 30, 321–333.

Uren, N.C. and Scott, R.V. (1982) *Objective Assessment of Horseracing Tracks*. School of Agriculture, La Trobe University, Melbourne, 55 pp.

Vamplew, W. (1976) *The Turf*. Allen Lane, London, 288 pp.

Zebarth, B.J. and Sheard, R.W. (1985) Impact and shear resistance of turf grass racing surfaces for thoroughbreds. *American Journal of Veterinary Research* 46, 778–784.

# AMENITY GRASS FOR NON-SPORT USE

E.J.P. MARSHALL   *AFRC Institute of Arable Crops Research, Long Ashton Research Station, Bristol BS18 9AF, UK*

## DEFINITIONS AND FUNCTIONS OF NON-SPORT AMENITY TURF

Amenity grassland has been defined (NERC, 1977) as that having 'recreational, functional or aesthetic value, and of which agricultural productivity is not the primary aim'. In excess of 400,000 ha in the UK were identified at that time as non-sport grassland and classified into several functional categories. Categories covering non-sport use include man-made and semi-natural trampled open spaces, untrampled open spaces and some derelict land (Table 10.1). These areas, which are not used for formal sports, have increasing importance as demands for informal recreation increase. These demands include informal sports, picnicking, walking, the appreciation of nature and the conservation of wildlife. In the countryside, some of these demands may be met within designated areas such as country parks. Amenity grassland is often the most important feature in such areas, though aquatic and woodland habitats may also be present (Plate 7). Recently, increases in on-farm recreation have demonstrated potential uses for unproductive areas of farmland. With the advent of 'setaside' schemes, in which land is taken out of arable production and given over to other vegetation, opportunities exist for the creation of new amenity grasslands.

The functions of non-sport amenity grass are diverse, ranging from providing pleasant surroundings, picnic sites and play areas, to delineating road and waterways and maintaining particular botanical and faunal assemblages. Within any of the categories of grassland listed in Table 10.1, there may be several management objectives (Table 10.2). Domestic lawns, which may be classified as non-sport turf, need to be visually pleasing, yet capable of withstanding a degree of wear. Visual requirements in one situation may be for

**Table 10.1.** Functional and habitat categories of non-sport amenity grassland. (From NERC, 1977.)

| Trampled open spaces | | Untrampled open spaces |
|---|---|---|
| Man made | Semi-natural | |
| Domestic lawns | Golf rough | Cemeteries |
| Urban parks | Archaeological sites | Military airfields |
| Urban road verges | Rural road verges | Civil airports |
| Car parks | Waterway banks | Railway embankments |
| | Picnic areas | Motorway embankments |
| | Forestry recreation areas | Dam faces |
| | Nature trails | Derelict land |
| | Camping and caravan sites | |
| | Country parks | |
| | Country estates, private | |
| | National Trust land | |
| | Common land | |
| | Nature reserves, open | |

**Table 10.2.** Functions and requirements for non-sport amenity turf.

| Function | Requirement[a] |
|---|---|
| Informal recreation | Short grass swards; wear tolerance |
| Ground stabilization | Rapid establishment; soil consolidation |
| Visual interest | Either sward uniformity or visual contrasts; common flowers; contrasting management |
| Nature conservation | Botanical diversity; managed access |

[a] For all requirements, safety to the public and ease and economy of management are essential.

uniform swards, as for sports turf, but in other situations botanical interest provided by common flowers may be needed. A prime objective of managing public open space must be safety to users. Road verges, for example, need to be managed for safety to traffic, though nature conservation and visual amenity may also be important. In country parks, separate areas may need to be managed for visual impact, nature conservation and informal play, while golf courses include formal sports areas, as well.

There are two main situations to consider for the management of non-sport amenity grasslands, firstly the maintenance of existing grassland and secondly the creation or re-establishment of turf. As the uses of non-sport grass are diverse, so too are their botanical composition. Intensively managed and used turf, such as lawns, is often sown and thus contains artificial assemblages of species. Less intensively used grass may consist of indigenous species. Often, such swards are species-rich with considerable conservation value. An ecological understanding of grassland components and their functions, proposed by Grime (1980), is a prerequisite for effective management. To continue as grassland and not tumble-down to scrub or woodland, a sward must be positively managed. Maintenance may be frequent or intermittent, based on mechanical means or grazing and may, on occasion, include the use of agrochemicals. More selective manipulations may also be required. For example, particular weeds may need to be controlled or the vegetation composition manipulated. An overriding principle must be that the objectives of management are carefully considered before maintenance programmes are established.

## AMENITY GRASS COMMUNITIES

## Amenity grassland in the landscape

As an element in the landscape, as a recreational area and as a habitat, amenity grassland constitutes an important resource. For most people, the garden lawn is a source of much pleasure, as well as demanding considerable upkeep. For many urban dwellers, parks are an important, often the only, open-air resource for recreation and relaxation. People travelling in the countryside enjoy motorway, road and railway verges, and many visit picnic areas, country parks and private estates. These areas are usually dominated by grassland habitats.

Road verges, river and canal banks are linear habitats usually made up of grassland communities which enhance or ameliorate the visual impact of the major features in the landscape, such as roads, of which they form a part. They also form important wildlife habitats, supporting a range of plant, animal and bird species which otherwise would be absent (Burel and Baudry, 1990). Certain lengths of road verges in the UK are designated as Sites of Special Scientific Interest (SSSIs) on account of their floral composition. In such cases, special management agreements, designed to maintain the species of interest, are drawn up between local authorities and nature conservation groups. As well as acting as refuges in otherwise hostile environments, such linear features may act as corridors for animals and birds to move between other habitats (van Dorp and Opdam, 1987). This connecting function may be crucial for the maintenance of fluctuating populations.

Some extensive areas of amenity grassland contain examples of different types of agricultural grasslands. Before agricultural intensification, grassland management maintained diverse swards in meadows, pastures and grazing marshes. Changes in agriculture, reviewed by Marshall and Hopkins (1990), have resulted in the loss of most diverse grasslands. For example, few traditional hay meadows survive. Those that do, are often managed for conservation purposes, for example Iffley meadow, Oxford (Plate 8), which supports the snakeshead fritillary (*Fritillaria meleagris*). As their prime role may no longer be agricultural production, such sites might be defined as amenity areas, though public access may limited. In extensive areas of poorly drained grassland, such as in parts of the Netherlands and the Somerset Levels in south west England, management agreements may be made between landowners and statutory bodies (Park, 1988). Such changes in agricultural management give conservation objectives similar priority to those of production and can be defined as serving public amenity in the widest sense.

## Botanical components of non-sport amenity grassland

The grassland ecosystem contains a range of plant species with different growth forms and varied survival strategies. While monocotyledons constitute the major part of above and below-ground biomass (see Chapter 2), dicotyledonous species may form a visually important component of amenity swards. Under some circumstances, for example in ornamental lawns, such species are undesirable. In other situations, such as road verges, common flowers provide visual interest. In addition, grassland communities or individual species are the subject of conservation, either for their inherent value or for their associated wildlife.

Plant growth forms have been reviewed by Crawley (1986). Grassland plants are mostly polycarpic (flowering many times) perennials, which live for several years, usually flowering annually. Monocarpic (once-flowering) species may live several years, but once they flower, they senesce and die. Most Umbelliferae are monocarpic, often biennial, species. Spear thistle (*Cirsium vulgare*) is another example of a monocarpic species. Such species usually occur in amenity swards only where there is an element of disturbance. Where there is considerable disturbance resulting in areas of bare ground, annual plant species can occur. These opportunistic species rely for survival either on dispersal of seed from successful parents elsewhere, or on long-lived dormant seed in the soil. Wind dispersal is a common long-distance dispersal mechanism exhibited by such weedy species.

Classical studies by Chippendale and Milton (1934) have shown that the size and composition of soil seed banks under grass are not well reflected in the above-ground flora. Many components of grass swards have only transient seed banks (Grime *et al.*, 1988). Recent work on the re-creation of chalk grasslands after the cessation of agriculture, has demonstrated the slow rates of re-invasion from adjacent diverse areas (Graham and Hutchings, 1988). The propagules of weed species in the seed bank can have a major influence on initial colonization of new grass. In grasslands, most species rely on vegetative propagation and growth, with regeneration from seed playing only a minor part. The important implication of these findings is that once species are lost from the above-ground flora in grassland, they will be difficult to re-establish without introducing plants or propagules.

## Plant communities

Semi-natural and artificial grasslands are made up of a range of plant species. While the populations of individual species can show dynamic changes within the community, varying in time and space with micro-topography and environments, phytosociological studies have shown that species composition is remarkably stable. Such techniques allow the identification of constant species, indicators and rarities. This approach to classifying plant communities has been widely applied in Europe (Westhoff and van der Maarel, 1973) and can be used for impact assessment, for conservation management and in land use planning. In Britain, the National Vegetation Classification (NVC) project has identified more than 300 plant communities (Rodwell, 1992). A range of these are grassland communities, some of which are amenity grasslands created for recreational purposes. There are nine upland grassland and 13 mesotrophic grassland communities. In addition, there are several calcicolous grasslands and maritime grass communities which can be distinguished (Table 10.3). All these communities can occur in areas where the prime function is not agricultural production. Within each community, the major factors which influence the flora may vary. An understanding of the effects of these factors and their interactions is required, if the identified objectives of management are to be achieved.

### Major Factors Affecting Botanical Composition

### Climate, environment and soils

Climatic conditions influence the species and communities of plants capable of growing in particular locations. The broad groups of world vegetation formations and their associated climates and growth forms have been noted by Green (1981). The formations include the extensive grasslands of tundra, steppes (including prairie and veldt) and deserts. In Britain, grasslands include meadows, pastures and wetter fens and marshes. The individual environmental factors which affect plants have been reviewed by Crawley (1986). These include fire, drought, shade, waterlogging, disturbance of various forms, nutrient supply, soil pH and the action of humans and other animals.

**Table 10.3.** Some of the grassland plant communities recognized under the UK National Vegetation Classification (NVC) (Rodwell, 1992). The letters U1 etc. are the NVC code.

### Calcifugous and upland grasslands

| | |
|---|---|
| U1 | *Festuca ovina-Agrostis capillaris-Rumex acetosella* grassland |
| U2 | *Dechampsia flexuosa* grassland |
| U3 | *Agrostis curtisii* grassland |
| U4 | *Festuca ovina-Agrostis capillaris-Galium saxatile* grassland |
| U5 | *Nardus stricta-Galium saxatile* grassland |
| U6 | *Juncus squarrosus-Festuca ovina* grassland |
| U7 | *Nardus stricta-Carex bigelowii* grass-heath |
| U13 | *Deschampsia cespitosa-Galium saxatile* grassland |
| U24 | *Arrhenatherum elatius-Geranium robertianum* community |

### Mesotrophic grasslands

| | |
|---|---|
| MG1 | *Arrhenatherum elatius* coarse grassland |
| MG2 | *Filipendula ulmaria-Arrhenatherum elatius* tall-herb grassland |
| MG3 | *Anthoxanthum odoratum-Geranium sylvaticum* meadow |
| MG4 | *Alopecurus pratensis-Sanguisorba officinalis* flood meadow |
| MG5 | *Centaurea nigra-Cynosurus cristatus* pasture |
| MG6 | *Lolium perenne-Cynosurus cristatus* pasture |
| MG7 | *Lolium perenne* leys and related grasslands, including lawns and recreational grasslands |
| MG8 | *Cynosurus cristatus-Caltha palustris* flood-pasture |
| MG9 | *Holcus lanatus-Deschampsia cespitosa* coarse grassland |
| MG10 | *Holcus lanatus-Juncus effusus* rush pasture |
| MG11 | *Festuca rubra-Agrostis stolonifera-Potentilla anserina* inundation grassland |
| MG12 | *Festuca arundinacea* coarse grassland |
| MG13 | *Agrostis stolonifera-Alopecurus geniculatus* inundation grassland |

### Calcicolous grasslands

| | |
|---|---|
| CG1 | *Festuca ovina-Carlina vulgaris* grassland |
| CG2 | *Festuca ovina-Avenula pratensis* grassland |
| CG3 | *Bromus erectus* grassland |
| CG4 | *Brachypodium pinnatum* grassland |
| CG5 | *Bromus erectus-Brachypodium pinnatum* grassland |
| CG6 | *Avenula pubescens* grassland |
| CG7 | *Festuca ovina-Hieracium pilosella-Thymus praecox/pulegioides* grassland |
| CG8 | *Sesleria albicans-Scabiosa columbaria* grassland |
| CG9 | *Sesleria albicans-Galium sterneri* grassland |
| CG10 | *Festuca ovina-Agrostis capillaris-Thymus praecox* grassland |
| CG11 | *Festuca ovina-Agrostis capillaris-Alchemilla alpina* heath |
| CG12 | *Festuca ovina-Alchemilla alpina-Silene acaulis* dwarf-herb community |
| CG13 | *Dryas octopetala-Carex flacca* heath |
| CG14 | *Dryas octopetala-Silene acaulis* ledge community |

### Maritime cliff communities

| | |
|---|---|
| MC8 | *Festuca rubra-Armeria maritima* maritime grassland |
| MC9 | *Festuca rubra-Holcus lanatus* maritime grassland |
| MC10 | *Festuca rubra-Plantago* spp. maritime grassland |
| MC11 | *Festuca rubra-Daucus carota* subsp. *gummifer* maritime grassland |
| MC12 | *Festuca rubra-Hyacinthoides non-scripta* maritime grassland |

Vegetation structure and composition, which are determined by environment, are further modified by management.

The diverse soils on which grasslands occur influence the botanical composition, the patterns of growth and therefore the use, appearance and management of such areas. Amenity grass can occur on all but the most inhospitable soils, ranging from acid mountain soils to those on alkaline chalks and clays. The most botanically diverse semi-natural grasslands in temperate climates are those on alkaline soils, typically over chalk or limestone. Up to 40 species of higher plant can be recorded per square metre, in the short grazed swards of the English chalk downlands (Smith, 1980). Slightly acid soils (pH 4.5–6.0), for example the Culm measures of north Devon, England, can also support diverse grasslands. Although grasslands on neutral soils may be managed to create botanical interest, they are often species-poor, dominated by broadleaved, tall grasses.

Soil nutrients have a marked influence on plant growth. Different species have differing requirements and different abilities to exploit soil resources. In amenity grassland, robust growth is required for wear tolerance and thus sufficient nutrients should be available for growth in situations where wear damage is likely to occur. Where wear is of minor consideration, high nutrient status favours vigorous species adapted to rapid nutrient uptake. These will out-compete slower-growing species. Thus, high nutrient status favours dominance of the flora by a small number of plant species and a decline in species richness. False oatgrass (*Arrhenatherum elatius*) and stinging nettle (*Urtica dioica*) are examples of species capable of exploiting nutrient-rich soils and dominating the flora. Willis (1963) demonstrated that fertilizing dune soils resulted in declines in species diversity. The classic demonstration of the effects of fertilizer on grassland is the Park Grass Experiment set up at Rothamsted Experimental Station in 1856 (Brenchley and Warrington, 1958). The plots receive a range of nutrient and lime treatments and are still cut for hay 138 years later. There are dramatic differences in botanical composition and sward appearance. Species numbers have declined markedly in comparison to the original sward on plots receiving ammonium sulphate, and where an acid flora has developed. The effects on community structure may reflect different direct effects, including nutrient uptake and soil pH, as well as indirect effects on interspecific competition (Williams, 1978). Aerial deposition of nitrogen and other pollutants is now a major factor to consider in amenity and conservation areas. In the Netherlands, mean deposition of nitrogen was $20 \, \text{kg ha}^{-1} \text{y}^{-1}$ (van der Meer and van uum-van Lohuyzen, 1986). More recently, mean figures have reached $40 \, \text{kg N ha}^{-1} \text{y}^{-1}$ with point sources exceeding $140 \, \text{kg ha}^{-1} \text{y}^{-1}$. The effect of this deposition on soil nitrogen content can be large, with corresponding changes in vegetation. For example, grasses are invading heathlands (Aerts and Berendse, 1988). There may also be direct effects on plant communities by soil acidification.

ECOLOGICAL PROCESSES

## Succession

Grassland is usually described as a plagioclimax vegetation – that is, a structure which differs from the natural climax vegetation, being maintained by management or environmental influences. If those influences are removed, the vegetation structure and composition will change. Temperate grasslands are maintained as such by management, usually grazing. If grazing pressure is removed, grassland changes in a directional sense towards a climax vegetation, which, for most of Britain, is deciduous woodland. This succession is only partly predictable and may be fast or slow, depending on many factors. Under conditions where herbaceous vegetation cover is maintained, succession may be delayed (Ward, 1979). Certain grasslands, particularly the steppes, do not show successional changes towards forest or woodland, as climatic conditions constrain tree growth and survival. However, for most amenity areas, some form of management is required to prevent succession and to maintain open grassland.

Secondary succession is the term applied to vegetation change in older environments, as compared with that on virgin habitats such as volcanic rock. Typically, secondary succession occurs following a change in land use. The most widely studied successions are those following the abandonment of agricultural land, particularly in America (e.g. Bard, 1952) and eastern Europe (Osbornova *et al.*, 1989). Initial colonization is by annual plant species, followed by monocarpic and polycarpic perennials. Shrubs and tree species arrive later, depending on dispersal distances and the location of propagule sources. The process of secondary succession provides an opportunity for creating new habitats on farmland for informal recreation and wildlife (see later). During the course of succession, changes occur in the environment. Light characteristics at the soil surface and soil conditions may change. The most significant changes may be associated with increases in soil nutrients, caused by deposition, nitrogen fixation and weathering (Green, 1972). Acidification may also be significant in the longer term.

## Dominance and diversity

Where environmental conditions do not limit plant growth, plant communities are often dominated by a small number of competitive species. For example, on high nutrient soils with little disturbance, species such as nettle (*Urtica dioica*) establish and out-compete other species. This phenomenon can be exploited. For example, suitable resilient grasses such as perennial ryegrass (*Lolium perenne*), can be established where heavy wear occurs. However, dominance is negatively correlated with species diversity (Green, 1981). Therefore, the conditions which favour dominance should be avoided if botanical interest or conservation is the primary objective of management.

Low nutrient status may create a stressed environment for plants (Grime, 1979). Rapidly growing species are not able to survive, while shorter, slow-growing species may persist. Such stressed environments tend to be very heterogeneous, for example with variable soil depths. This may create opportunities for more species to occur. Where there are many limiting factors to plant growth, there are more opportunities for adaptations to particular conditions. Many species survive through their ability to exploit particular circumstances (Grubb, 1977). Gaps in the sward created by mole hills, hooves, dung or urine, may favour the regeneration of particular species.

## MANAGING NON-SPORT GRASSLANDS

## Objectives

As grassland is a plagioclimax, an intermediate but maintained stage before climax vegetation, its very existence is dependent on management. The varying forms of management, the timing of operations and their intensity, interact with the flora. Thus the maintenance of particular botanical assemblages and visual effects or sward heights, are dependent to a large degree on the form of management employed. As well as these visible effects, more subtle long-term environmental changes, particularly increases in soil nutrients and acidification, need to be considered in designing maintenance programmes.

The main forms of management are cutting and grazing (Wells, 1980), each with a variety of techniques, intensities and timings of implementation. Burning may also be employed under some circumstances (Green, 1980). Botanical composition can be manipulated with herbicides and with plant growth regulators; some of which are retardants of grass growth. Whilst the general principles of grassland management are well known, scientific information on the precise effects of particular management treatments on differing plant communities are generally unavailable. In many areas of conservation value, the common sense approach to management must be that the traditional methods of maintenance, which resulted in the particular communities now regarded as valuable, should continue. However, traditional management is often impractical. Moreover, changes in management may be necessary to combat successional changes in the environment. In assessing the needs for managing amenity grass and developing maintenance programmes, it is essential that the objectives of management are examined. As many sites are multi-functional, compromises may be required between different maintenance techniques.

## Mowing

As outlined in Chapter 5, the several types of mower which are available have different mowing actions and these determine their usage. Cylinder mowers are best suited to fine turf for sports use and top quality ornamental lawns. Rotary cutters are suitable for coarser grassland and are widely used for lawns and verges. Flail cutters are best used on tall grass and overgrown areas. For inaccessible areas and around posts and benches, a pedestrian-operated strimmer with a rotating blade or plastic cords may be used. The size of area to be managed and its location in relation to where machinery is stored, as well as the objectives of management, are the major factors in selecting the machinery to use. Extensive grass areas which need to be kept short can best be managed with tractor-mounted machines, such as gang-mowers. Where taller grass can be accommodated, then agricultural management for hay or silage may be practised, for example adjacent to airport runways.

It is vital that the objectives of management are defined. Grass that is to be kept short all year, often for safety reasons, will necessarily have minimal botanical interest. Where botanical interest is of prime concern, the timing and frequency of cutting should allow plants to flower and preferably set seed. In most non-sport grasslands the cuttings are left on, except where a hay or silage crop can be removed. The problem of disposal of cut material needs to be addressed. It is often impractical to dispose of the cuttings away from the site. However, if a reduction or some control of soil fertility is required, as a means of encouraging plant diversity, then clippings should be removed. Motorway verges in the Netherlands are mown two or three times a year and the cuttings removed. This encourages botanical diversity and reduces dry matter production (Melman *et al.*, 1988), so further reducing the need for cutting. Cuttings are composted or, if they are unsuitable for horticultural use, burnt. Where the effects of aerial deposition of nitrogen are significant, grass cuttings need to be removed. In some untrampled grasslands, fire may be an occasional necessity to remove plant litter and reduce organic nitrogen in the stand. This will check dominant species and diversify swards (Green, 1980).

## Grazing

While it might be inappropriate to use grazing in most amenity areas open to the public, it is the best way of managing enclosed areas of untrampled grasslands. The short herb-rich grasslands on chalk and limestone are a product of sheep and rabbit grazing (Smith, 1980). The old methods of sheep grazing on the downs of southern Britain, in which the sheep were penned at night in the valley bottoms, resulted in nutrient export from the hills. On the downs, this maintained a low-growing, diverse sward adapted to grazing and poor nutrient conditions.

The requirements for grazing stock are that they are enclosed, have sufficient food and water, shelter from sun and adverse weather, are regularly monitored and receive necessary veterinary treatments. The overriding considerations should be safety to the general public and safety to the animals. Road verges, airports and similar open areas where transport movements occur are not appropriate for grazing animals. However, country parks, deer parks and nature reserves can be successfully managed by grazing animals (Wells, 1980). In such areas, animals provide another feature of interest to the public.

Many animals can be used for grazing, ranging from horses, cattle and sheep to more exotic animals including deer and camelids. Within domestic stock, rare breeds can provide further interest. In all cases, suitable stock-handling expertise is required. The expense of this may be offset against income generated by stock-rearing and animal products. Managing the grassland then largely becomes an agricultural process, which is beyond the scope of this chapter. The various animals have different grazing actions, some cutting and others tearing the grass. Unlike machines, though, animals select material from the sward. Horses are particularly selective grazers, producing patchy vegetation in paddocks. Sheep also graze selectively but cattle are less selective. Horses and sheep graze grass closer to the ground than cows and cut plant material with their teeth, in contrast to cattle which tear it. Plants which are able to maintain their growing points below the grazing height survive best. Grasses, which have their meristem at ground level, are particularly well adapted to grazing.

Thus, different animals affect the sward differently. One of the important effects of stock grazing is that it provides areas of disturbance in the sward. These may be vital for the recruitment of some plant species. The provision of gaps may allow seeds to germinate and establish (Grubb, 1977). In addition, dung and urine patches will affect composition. However, if the vegetation cover is destroyed by over-grazing and poaching, then undesirable plant species, such as thistles, can become a problem. Interestingly, grazing can be used to control some problem species in areas of conservation interest. Sheep have been used to control ragwort (*Senecio jacobea*) in chalk downland nature reserves, cattle to control bracken (*Pteridium aquilinum*) and goats to prevent birch scrub invading dune systems (Bullock and Kinnear, 1988).

# Plant growth, weed, pest and disease control

Synthetic chemicals have been used as management tools in agriculture, horticulture and forestry for many years. In amenity grassland, herbicides are useful for controlling undesirable species, such as docks, thistles and ragwort. They also may be used to modify botanical composition in more subtle ways (Marshall, 1983) and to retard the growth of plants. Pest and disease problems can also occur in non-sport turf, for example leatherjackets and fusarium patch. Control methods have been reviewed by Baldwin and Drinkall (1992). The use of pesticides in the UK is controlled under the Food and Environment Protection Act 1985, Control of Pesticides Regulations 1986. This covers the storage, sale, transportation, application and disposal of materials. Under the Regulations and the associated Codes of Practice, users and supervisors need to be competent and must follow the directions given on product labels. A certification scheme is operated by the National Proficiency Tests Council (NPTC) for agriculture and horticulture. Similar regulations are operated in most other countries in order to ensure the safe use of chemicals.

Herbicides with clearance for use on amenity grassland can be grouped according to their mode of uptake in plants (through leaves or roots) and their selectivity between species. They also vary in their persistence in the environment and their speed of effect. Herbicides can be used before establishing new grass to control existing vegetation. They may also be used in mature turf. A full listing of products for amenity use in the UK is given in the *Amenity Handbook 1990/91* (BAA, 1990), which also gives useful information on handling, storage and application. Under agricultural conditions, products formulated for agricultural use may be applied. These are listed in *Pesticides 1993* (MAFF/HSE, 1993) and the *UK Pesticide Guide* (Ivens, 1994). Products available for domestic use on lawns are also listed in the *Directory of Garden Chemicals* (BAA, 1989) and in *Pesticides 1993*.

Examples of non-selective herbicides, which can be used to kill all existing vegetation, include paraquat and glyphosate. Glyphosate is translocated from the leaves and, thus, will kill deep-rooting plants. Paraquat, which has a contact effect, will kill shallow-rooting species, but will allow plants with rhizomes and tap roots to survive.

So-called selective amenity herbicides are used to control dicotyledonous species in grass. There are many herbicides available, each with a slightly different spectrum of susceptible species (Fryer and Makepeace, 1978). Many products, therefore, contain a mixture of active ingredients to control as many dicotyledonous species as possible. At present, herbicide use is limited to total vegetation control or to the elimination of dicotyledonous species. The selective removal of individual species is harder to achieve, as nearly all herbicides control at least most species in a taxonomic family. For example, the herbicide clopyralid is used for control of thistles, but also controls other members of the Compositae. The selectivity of most herbicides is relatively wide. Some selectivity can be achieved by modifying the application method to target just those plants that need to be controlled. For example, weed wipers were developed to exploit the height differences between weeds and crops (Lutman, 1980). In the amenity situation, selective application can be appropriate for controlling thistles, docks and ragwort. Overall elimination of dicotyledonous species is usually unwarranted, leading to a loss of botanical interest.

Attempts to use herbicides to reduce maintenance and increase botanical interest were reported by Marshall (1983). Variable results were obtained by using low doses of non-selective herbicides in autumn and spring to check coarse grasses. These novel uses of herbicides are not approved under the Pesticide Regulations. However, the selective removal of certain grasses from amenity swards remains desirable. For example, weed grasses which dominate newly sown areas need to be controlled. In mature swards, elimination of tall grasses may lead to more diverse swards and easier management.

Plant growth regulators are used to a limited extent in the management of amenity grassland. They can reduce mowing and maintenance costs. However, they are not suitable for fine turf, where rapid growth for the repair of wear damage and good appearance are important. Several products and active ingredients are marketed, including maleic hydrazide, mefluidide and paclobutrazol (MAFF/HSE, 1993). These have different modes of action and different characteristics. Maleic hydrazide and mefluidide are absorbed through leaves, act rapidly on grass growth and prevent flowering in grasses. Paclobutrazol is taken up by grass roots, is longer lasting than the former compounds, but does not suppress flowering. Mefluidide, in contrast to maleic hydrazide and paclobutrazol, can increase the botanical diversity of treated swards (Marshall, 1988) (Fig. 10.1). Where botanical interest is low, herbicides can be mixed with the retardants to control dicotyledonous species. Retardants are most useful for the maintenance of untrampled amenity grassland, where other forms of management are impractical. For example, there may be opportunities for their use on sown grass strips on farmland. However, some form of mowing is still required during the season, to produce an even sward and to prevent colonization by woody plants.

## CREATING NON-SPORT AMENITY GRASS

### Seed, seedbeds and ground preparation

New amenity grasslands can be created in the many situations outlined above. As with management, it is essential that the objectives of the use of the site are evaluated before construction. Thus, a domestic lawn will need to be treated differently from a wildflower meadow. The main methods used for creating

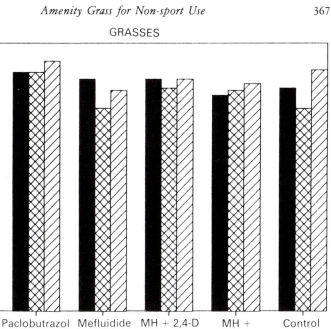

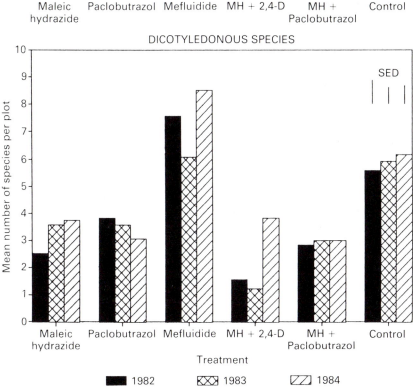

**Fig. 10.1.** Numbers of grass and dicotyledonous species in plots treated annually with growth retardants after three, four and five years.

grasslands are described in Chapter 5. A method unique to amenity grassland is the use of an imported topsoil alone with native seed load (Wells, 1984).

Seed mixtures are now commercially available for most amenity uses. Grass seed has long been available for agricultural, sports and domestic uses. Breeding programmes have provided many varieties for specific uses, including slow-growing amenity cultivars (Johnston and Faulkner, 1985). Manipulation of seed rates can influence the composition of swards, through competitive effects with germinating seeds from the seed bank (Parr, 1984). Wells *et al.* (1981) initiated the use of indigenous wildflower seed for the creation of herb-rich swards. This approach has been developed further, in the UK and elsewhere, giving dramatic results where establishment is successful. Commercial seed supplies are now available for a variety of soils and situations. In choosing a seed mixture, it is important that soil type, drainage, shading, final use and available management are considered, as well as the ecology of individual species. The germination requirements of species vary; some require ripening, others require scarification of the seed coat and others need a cold treatment (Grime *et al.*, 1981). Therefore, the timing of germination of species may vary, with some germinating in autumn and others in spring. Native seed should be used, to reduce any effects of introducing exotic genetic material into indigenous populations. If unsuitable species are selected, they will not survive. If unsuitable maintenance is carried out, the components of the sward may change. Commercial seed sources market conveniently packaged mixtures of wildflowers and grasses which can be selected for four main soils in the UK: clays, chalk and limestone, alluvial soil and dry acid soils. Herbs in the mixtures can also be selected for tall or short swards.

Soil fertility is of major importance in creating new swards. Nutrient-rich topsoil, which is undesirable where wildflower mixtures are to be established, may need to be ameliorated or removed. For creating wear-tolerant grass areas, high nutrient status may be beneficial. Ground for amenity grass establishment should be well prepared. Initially, existing vegetation should be removed, preferably using approved herbicides to prevent regrowth of plants from underground (BAA, 1990). Cultivation will be required to create a seedbed. Under some circumstances, drainage operations may also be needed. The weed seed bank in the soil should be considered. In disturbed soils, the annual weed species may germinate with the sown seed, competing for light, water and nutrients and reducing establishment. This can be partly resolved by creating a seedbed, allowing weeds to germinate and subsequently killing them with further cultivation or a contact herbicide, before seed drilling.

During establishment, seedlings are vulnerable to drought and waterlogging, so it is important to roll once seeds have been introduced, as this ensures good seed/soil contact. The use of a nurse crop, for example annual ryegrass, can also help to reduce weed establishment (Wells *et al.*, 1981). There are opportunities for using selective herbicides to remove unwanted plants, though many treatments may also affect desirable species. Graminicides and

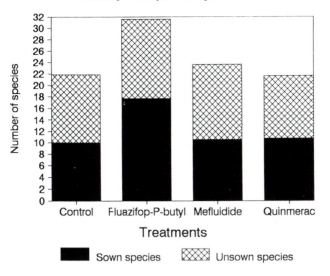

**Fig. 10.2.** Effects of herbicides applied once on the numbers of sown and unsown species in sown strips at the edge of an arable field.

newer compounds for broadleaved weed control are presently showing promise for use on strips of wildflower and grass sown at field edges on farms. Marshall and Nowakowski (1991) demonstrated increased establishment of sown species on sown field edges treated with fluazifop-P-butyl, a graminicide (Fig. 10.2).

## Managing new grass

The objective of managing newly established grass is the successful establishment of the components of the seed mixture sown. Grasses will tend to establish and grow quicker than dicotyledonous species, some of which (e.g. *Primula* species) may only germinate after over-wintering. Therefore it is recommended that for wildflower areas during the first growing season, the sward is mown several times to suppress unsown weed species and to reduce shading. Cutting every two months may be sufficient, especially if the cuttings are removed. If a nurse crop is used, which will persist if it sets seed, it must be cut to prevent it from dominating the sward. In the second season, the usual maintenance regime of one or two cuts or grazing, can be established. Cutting in the first year is particularly important where fertility is high but diversity is required (Wells *et al.*, 1981).

Wear on the newly established sward should be avoided, particularly if intensive public use is anticipated. It is also important that the first cut after sowing is delayed, usually for a minimum of eight to ten weeks, to allow the grass root system to establish.

In some circumstances, for example on farmland, it may not be possible to cut as frequently as outlined above. Good establishment of the sown species is then crucial, as is the timing of any cuts. Removal of cuttings is important as this minimizes the return of annual weed seed to the soil. The selective control of weed species before and after establishment, with suitable herbicides may be required. Traditional management with sheep may also be required to encourage plant diversity (Gibson *et al.*, 1987).

## Diversifying existing grassland

Simple modifications to existing maintenance regimes can often increase botanical interest in swards. For example, less frequent mowing may allow some dicotyledonous species to flower. However, to increase grassland species diversity, a resumption of traditional meadow management is recommended, with emphasis on the removal of grass as hay. The arrival of new species in areas will depend on propagule dispersal (or the seed bank), germination and seedling success. Creation of suitable gaps in the sward, perhaps by harrowing or grazing after mowing, may be essential. However, experience shows that often the most desirable species are poor dispersers and absent from the seed bank (Graham and Hutchings, 1988).

It is possible to introduce species into existing grassland, either by planting individual plants or by introducing seeds. A major factor in the establishment of new species, apart from the inherent ecological suitability of the new environment, is competition from the established sward. Introduced mature plants are less likely to be affected by such competition than seedlings. Nevertheless, some success has been achieved by using slot-seeding techniques. The machinery, which was designed for renovating agricultural grass, cuts ribbons of turf from the existing sward, sows seed and usually sprays a herbicide along the edges of the slot to suppress grass growth and reduce competition as the seedlings emerge. Once established, the swards need to be cut to allow new species to thrive. Both techniques can be used to increase botanical interest in many untrampled situations, like road and motorway verges.

## NEW OPPORTUNITIES FOR AMENITY GRASSLAND

## Urban and housing developments

Recreation is recognized as a vital part of modern living. Lack of opportunity for informal recreation is thought to aggravate social problems and antisocial behaviour. Amenity grassland, as part of open 'green space' in urban areas, has a major recreational role. Provision of greens, parks and wildlife areas should be made in all new housing developments. Increasingly, planning authorities require these facilities in new building schemes. There can be surprisingly large populations of wild fauna and flora in urban areas. Sukopp and Werner (1987) have reviewed the opportunities for wildlife in towns and cities, giving some ideas for ideal city design. As well as parks, gardens and playing fields, there are often large areas of derelict land in cities. These areas have particular importance for wildlife, as demonstrated and promoted by the Urban Wildlife Groups in the UK, maintaining populations of plants, animals and birds. Some of these areas can be further developed as grassland for recreation or as habitats for wildlife.

In the UK, the construction of new towns has been accompanied by the establishment within local authorities of Landscape Departments responsible for design and management. This has stimulated new approaches to landscape design. As well as imaginative tree and shrub plantings, new approaches to grasslands have also been developed. In particular, there is significant interest in creating botanically diverse swards for visual impact in new amenity areas and for rehabilitating derelict sites (Taylor, 1984). A major problem in the maintenance of irregularly mown grassland in urban areas is that any space with tall herbage is liable to be treated as wasteland and used as a tip by the local residents.

## Setaside and other opportunities on farmland

Under present agricultural conditions, the decrease in arable farmland provides new opportunities for the development of amenity grasslands. Within the European Community (EC), policies are in place to reduce arable production and other commodities in surplus. These policies include setaside and extensification. Under setaside, farmers are encouraged with financial incentives to take land out of arable production. Different member states of the EC have introduced similar programmes for setaside. Within the UK, setaside payments may be made if land is fallowed (with no crop sown) (MAFF, 1991), either on a rotational basis each year or as permanent fallow for five years in the first instance. Fallow land may be left to pass through secondary succession or may be sown. It should be mown at least once a year. Permanent fallow, now supported, offers the opportunity to create new grasslands, which should not be used for agricultural production, such as dairy or beef production. Setaside areas may be entire fields or strips at least 20 m wide at the edges of fields. As well as playing a role in conservation, these new swards (Plate 9) can provide access for farm machinery and areas for on-farm recreation, including walking, camping, riding and shooting. Thus these new grass areas fall under the widest definition of amenity grassland.

Extensification of agricultural management seeks to reduce inputs of energy, fertilizers and agrochemicals into farm land. The aim is to reduce overall production by limiting intensive inputs. This policy is already in action in Environmentally Sensitive Areas (ESAs), where farmers are encouraged to maintain traditional management methods and even re-create old agricultural habitats. The management agreements are not mandatory within the designated areas until signed by individual landowners, who then receive payments according to the level of extensification adopted. Similar agreements are made on designated nature reserves and Sites of Special Scientific Interest.

The development of diverse wildflower and grass areas on farmland is being investigated by several researchers. Smith and MacDonald (1989) have examined successional changes on strips at field edges of different width and established with different seed mixtures. The use of wildflower seed mixtures in this situation offers considerable scope for creating habitats that are of value to wildlife, the public and possibly for adjacent crops (Marshall and Nowakowski, 1991). Rare annual cornfield weeds, such as corncockle (*Agrostemma githago*) and cornflower (*Centaurea cyanus*), may be included in seed mixtures to provide added colour during the first season. While habitats at the edges of arable fields may harbour weeds, pests and diseases of arable crops, they can encourage beneficial insect populations, including crop pollinators and pest predators (Wratten, 1988). They may also act as weed barriers, preventing the spread of plant species into adjacent crops (Marshall, 1990).

## FUTURE RESEARCH NEEDS FOR NON-SPORT AMENITY TURF

There are many areas of our understanding of the functioning, establishment and management of non-sport amenity turf which are imperfect. Areas which are of particular importance at present and which require further research include:

- The efficient utilization of applied fertilizers, where this is required, to avoid losses to the environment and eutrophication of ground and surface waters.
- Methods to remove nutrients from semi-natural grasslands or to limit nutrient accumulation, in order to maintain species-rich swards.
- Improved methods of establishing or re-establishing turf, including information on appropriate seed mixtures, rates, and after-management for particular soils and situations.
- Development of appropriate management to replace traditional methods for species-rich swards.
- Selective cultural or other methods of controlling invasive or injurious plant species.
- Identification of the key processes that maintain species compositions in

different swards, particularly competition/management interactions and dispersal.

• Identification of any major threats to turf which may arise from climate changes.

Overall, there is a need for an information resource for managers, which will allow the identification of turf types, given their current use, location, composition and management, and their potential uses and development with different management strategies and thus the efficient targeting of management prescriptions. While this book will achieve this for sportsturf, the need remains for non-sport turf where the diversity of turf types currently leads to a matching diversity of information sources for managers.

## REFERENCES

Aerts, R. and Berendse, F. (1988) The effects of increased nutrient availability on vegetation dynamics in wet heathlands. *Vegetation* 76, 63–69.

BAA (1989) *BAA Directory of Garden Chemicals 1989/90*. British Agrochemicals Association Ltd, Peterborough, 39 pp.

BAA (1990) *Amenity Handbook 1990/91*. British Agrochemicals Association Ltd, Peterborough, 58 pp.

Baldwin, N.A. and Drinkall, M.J. (1992) Integrated pest and disease management for amenity turfgrass. In: *Aspects of Applied Biology 29, Vegetation Management in Forestry, Amenity and Conservation Areas*. Association of Applied Biologists, Wellesbourne, UK, pp. 265–272.

Bard, G.E. (1952) Secondary succession on the Piedmont of New Jersey. *Ecological Monographs* 22, 195–215.

Brenchley, W.E. and Warrington, K. (1958) *The Park Grass Plots at Rothamsted 1856–1949*. Reprinted 1969. Rothamsted Experimental Station, Harpenden, Hertfordshire.

Bullock, D.J. and Kinnear, P.K. (1988) The use of goats to control birch in dune systems: an experimental study. In: *Aspects of Applied Biology 16, The Practice of Weed Control and Vegetation Management in Forestry, Amenity and Conservation Areas*. Association of Applied Biologists, Wellesbourne, UK pp. 163–168.

Burel, F. and Baudry, J. (1990) Hedgerow network patterns and processes in France. In: Zonneveld, I.S. and Forman, R.T.T. (eds) *Changing Landscapes: An Ecological Perspective*. Springer-Verlag, New York, pp. 99–120.

Chippendale, H.G. and Milton, W.E.J. (1934) On the viable seeds present in the soil beneath pastures. *Journal of Ecology* 22, 508–531.

Crawley, M.J. (1986) Life history and environment. In: Crawley, M.J. (ed.) *Plant Ecology*. Blackwell Scientific, Oxford, pp. 253–290.

Fryer, J.D. and Makepeace, R.J. (eds) (1978) *The Weed Control Handbook. Volume II. Recommendations*, 8th edn. Blackwell Scientific, Oxford, 532 pp.

Gibson, C.W.D., Watt, T.A. and Brown, V.K. (1987) The use of sheep grazing to recreate species-rich grassland from abandoned arable land. *Biological Conservation* 42, 165–183.

Graham, D.J. and Hutchings, M.J. (1988) A field investigation of germination from the seed bank of a chalk grassland ley on former arable land. *Journal of Applied Ecology* 25, 253–263.

Green, B.H. (1972) The relevance of seral eutrophication and plant competition to the management of successional communities. *Biological Conservation* 4, 378–384.

Green, B.H. (1980) Management of extensive amenity grasslands by mowing. In: Rorison, I.H. and Hunt, R. (eds) *Amenity Grassland: An Ecological Perspective*. John Wiley, Chichester, pp. 155–161.

Green, B. (1981) *Countryside Conservation*. George Allen & Unwin, London, 249 pp.

Grime, J.P. (1979) *Plant Strategies and Vegetation Processes*. John Wiley, Chichester.

Grime, J.P. (1980) An ecological approach to management. In: Rorison, I.H. and Hunt, R. (eds) *Amenity Grassland: An Ecological Perspective*. John Wiley, Chichester, pp. 13–55.

Grime, J.P., Mason, G., Curtis, A.V., Rodman, J., Band, S.R., Mowforth, M.A.G., Neal, A.M. and Shaw, S. (1981) A comparative study of germination characteristics in a local flora. *Journal of Ecology* 69, 1017–1059.

Grime, J.P., Hodgson, J.G. and Hunt, R. (1988) *Comparative Plant Ecology. A Functional Approach to Common British Species*. Unwin Hyman, London, 742 pp.

Grubb, P.J. (1977) The maintenance of species-richness in plant communities: the importance of the regeneration niche. *Biological Reviews* 52, 107–145.

Ivens, G.W. (ed.) (1994) *The UK Pesticide Guide 1994*. CAB International, Wallingford, Oxon, and the British Crop Protection Council, Farnham, Surrey, 561 pp.

Johnston, D.T. and Faulkner, J.S. (1985) The effects of growth retardants on swards of normal and dwarf cultivars of red fescue. *Journal of the Sports Turf Research Institute* 61, 59–64.

Lutman, P.J.W. (1980) A review of techniques that utilise height differences between crops and weeds to achieve selectivity. *Spraying Systems for the 1980's. British Crop Protection Council Monograph No. 24*, 291–297.

MAFF (1991) *Set-Aside: A Practical Guide*. Ministry of Agriculture, Fisheries and Food, London. Leaflet No. SA6, 24 pp.

MAFF/HSE (1993) *Pesticides 1993. Pesticides Approved Under the Control of Pesticides Regulations 1986*. Ministry of Agriculture, Fisheries and Food/Health and Safety Executive. Reference Book 500. London, HMSO, 443 pp.

Marshall, E.J.P. (1983) A feasibility study of the use of chemicals for rural amenity areas. *Agricultural Research Council Weed Research Organization, Oxford. Technical Report No. 71*, 80 pp.

Marshall, E.J.P. (1988) Some effects of annual applications of three growth-retarding compounds on the composition and growth of a pasture sward. *Journal of Applied Ecology* 25, 619–630.

Marshall, E.J.P. (1990) Interference between sown grasses and the growth of rhizomes of *Elymus repens* (couch grass). *Agriculture, Ecosystems and Environment* 33, 11–22.

Marshall, E.J.P. and Hopkins, A. (1990) Plant species composition and dispersal in agricultural land. In: Bunce, R.G.H. and Howard, D.C. (eds) *Species Dispersal in Agricultural Habitats*. Belhaven Press, London, pp. 98–116.

Marshall, E.J.P. and Nowakowski, M. (1991) The use of herbicides in the creation of a herb-rich field margin. *Proceedings 1991 Brighton Crop Protection Conference – Weeds*, pp. 655–660.

Melman, P.J.M., Verkaar, H.J. and Heemsbergen, H. (1988) Species diversity of road verge vegetation and mowing regime in the Netherlands. In: During, H.J., Werger, M.J.A. and Willems, J.H. (eds) *Diversity and Pattern in Plant Communities*. SPB Academic Publishing, The Hague, pp. 165–170.

NERC (1977) *Amenity Grasslands – The Needs for Research*. Natural Environmental Research Council, Swindon, 64 pp.

Osbornova, J. *et al.* (1989) *Succession in Abandoned Fields: Studies in Central Bohemia, Czechoslovakia*. Kluwer Academic, Dordrecht.

Park, J.R. (ed.) (1988) *Environmental Management in Agriculture. European Perspectives*. Belhaven Press, London.

Parr, T.W. (1984) The effects of seed rate on weed populations during the establishment of amenity turf. In: *Aspects of Applied Biology 5, Weed Control and Vegetation Management in Forestry and Amenity areas*. Association of Applied Biologists, Wellesbourne, UK, pp. 117–125.

Rodwell, J.S. (ed.) (1992) *British Plant Communities. Volume 3. Grasslands and Montane Communities*. Cambridge University Press, Cambridge.

Smith, C.J. (1980) *Ecology of the English Chalk*. Academic Press, London.

Smith, H. and MacDonald, D.W. (1989) Secondary succession on extended arable field margins: its manipulation for wildlife benefit and weed control. In: *Proceedings 1989 Brighton Crop Protection Conference – Weeds*, pp. 1063–1068.

Sukopp, H. and Werner, P. (1987) *Development of Flora and Fauna in Urban Areas*. Council for Europe, Strasburg. *Nature and Environment Series 36*, 67 pp.

Taylor, G. (ed.) (1984) *Creating Attractive Grasslands. Diverse Swards – The Dos and Don'ts*. National Turfgrass Council Workshop Report No. 5, Bingley, 92 pp.

van der Meer, H.G. and van uum-van Lohuyzen, M.G. (1986) The relationship between inputs and outputs of nitrogen in intensive grassland systems. In: van der Meer, H.G., Ryden, J.C. and Ennik, G.C. (eds) *Nitrogen Fluxes in Intensive Grassland Systems*. Martinus Nijhoff Publishers, Dordrecht, pp. 1–18.

van Dorp, D. and Opdam, P. (1987) Effects of patch size, isolation and regional abundance on forest bird communities. *Landscape Ecology* 1, 59–73.

Ward, L.K. (1979) Scrub dynamics and management. In: Wright, S.E. and Buckley, G.P. (eds) *Ecology and Design in Amenity Land Management*. Wye College, Ashford, Kent, pp. 109–127.

Wells, T.C.E. (1980) Management options for lowland grassland. In: Rorison, I.H. and Hunt, R. (eds) *Amenity Grassland: An Ecological Perspective*. John Wiley, Chichester, pp. 175–195.

Wells, T.C.E. (1984) An ecologist's view: floristic possibilities – a scene-setting review. In: Taylor, G. (ed.) *Creating Attractive Grasslands. Diverse Swards – The Dos and Don'ts*. National Turfgrass Council Workshop Report No. 5, Bingley, pp. 8–19.

Wells, T., Bell, S. and Frost, A. (1981) *Creating Attractive Grasslands Using Native Plant Species*. Nature Conservancy Council, Shrewsbury, 35 pp.

Westhoff, V. and Maarel, E. van der (1973) The Braun–Blanquet Approach. In: Whittaker, R. (ed.) *Handbook of Vegetation Science. V. Classification and Ordination*. Junk, The Hague, pp. 617–726.

Williams, E.D. (1978) *Botanical Composition of the Park Grass Plots at Rothamsted 1856–1976*. Rothamsted Experimental Station, Harpenden, Hertfordshire, 61 pp.

Willis, A.J. (1963) Braunton Burrows: the effects on the vegetation of the addition of mineral nutrients to the dune soils. *Journal of Ecology* 51, 353–374.

Wratten, S.D. (1988) The role of field boundaries as reservoirs of beneficial insects. In: Park, J.R. (ed.) *Environmental Management in Agriculture. European Perspectives*. Belhaven Press, London, pp. 144–150.

# WARM SEASON TURFGRASSES

J.R. WATSON  *Vice-President and Agronomist, The Toro Company, Minneapolis, Minnesota, USA*

## INTRODUCTION

The previous chapters of this book have focussed on the design, establishment and maintenance of turf facilities in temperate climates with few references to tropical areas. Most of the information on soil design and turf culture which has been covered is directly applicable to the tropics. Soil properties in relation to their constitution and drainage design criteria, together with the basic principles of turfgrass establishment and fertilizer use are applicable universally. Some aspects which might be thought to differ do not do so appreciably. Typical of this category are aspects of maintenance concerned with soil organic matter management and thatch control. Provided soil water status is adequate in the tropics, turfgrass growth is both greater and much less seasonal than in temperate regions. More plant residues are produced but since soil temperatures are greater, microbial decomposition of these is more rapid. In consequence soil organic matter contents are not necessarily greater in the moist tropics and neither is the trend to thatch accumulation.

Despite the similarities in turf culture between temperate and tropical areas, there are some key differences. There are for example a number of turfgrass pests and weeds in the tropics which are restricted to those areas. It is not within the scope of this book to cover these aspects. Furthermore, although many diseases of turfgrasses are common to both temperate and tropical areas their relative importance in turf culture varies.

# Adaptations of Turfgrasses in Tropical Areas

A feature of a tropical climate is that although there are variations in temperature through the year, seasonality of rainfall becomes a much more important aspect of climate. Because of high temperatures, evapotranspiration is high and aridity or drought is a more widespread problem. When irrigation is not practicable drought tolerance in turfgrasses has a much higher priority than in temperate regions. When irrigation is practised in arid and semiarid areas the potential for soil salinization increases. Salt accumulation in soils results from inadequate flushing out of soluble salts either because it is prevented by poor drainage and a shallow watertable or because of an inadequate throughput of poor quality (salty) irrigation water. The widespread occurrence of salt affected soils in the tropics ensures that salt tolerance in turfgrasses is a valuable attribute. There are two key criteria describing salt affected soils. The first is one of excessively high salt concentration in the soil solution. The 'salt' reflects contributions of relevant soluble salts of which $NaCl$ is usually the most important. The electrical conductivity of a soil extract in saturated $CaSO_4$ solution or that of irrigation water is a widely used criterion. Electrical conductivities in excess of $4000\,\mu S\,cm^{-1}$ require high salt tolerance. The other criterion relates to the possible deflocculation of soil clays due to a high proportion of exchangeable sodium (see Chapter 1). On even moderately clayey soils, deflocculation results in the production of an impermeable pan. Problems are likely when exchangeable Na comprizes more than 15% of the exchangeable bases (ESP > 15). Because sand-dominant soils are used for most of the intensively used turf surfaces, it is the salt concentration in the soil solution which is the more important.

The higher light intensities and temperatures in the tropics together provide more energy input for green plants and more favourable conditions for the utilization of that light energy than in temperate regions. A problem however is that in the high temperatures plant water stress is far more common. Plants respond to water stress by closing leaf stomata which in addition to reducing water loss also restricts gaseous exchange between air and leaf. This places a restriction on the entry of $CO_2$ which is chemically reduced to produce sugars in the so-called dark reactions of photosynthesis. The enzyme which catalyses the reaction responsible for the initial trapping of $CO_2$, ribulose bisphosphate carboxylase oxidase (RUBISCO), has a rather low affinity for $CO_2$ relative to $O_2$ and in consequence photosynthesis and therefore growth may be reduced. Systems have evolved naturally which enable some plants largely to overcome this potential problem. In essence they utilize an enzyme system which has a high affinity for $CO_2$ to synthesize compounds containing four carbon atoms to trap and store $CO_2$-carbon. This carbon is then used to form sugars by the same reaction sequences (involving RUBISCO) as in plants which do not have this mechanism. Tropical grasses which have

**Table 11.1.** Some of the characteristics which distinguish C-4 and C-3 plants. (From Noggle and Fritz, 1976.)

| Characteristic | C-4 plants | C-3 plants |
|---|---|---|
| $CO_2$ compensation point | $0-10 \, \mu L \, L^{-1}$ | $50-150 \, \mu L \, L^{-1}$ |
| Photorespiration | Slight to none | Present |
| Net rate of photosynthesis in full sunlight ($10-13 \times 10^4$ lux) | $40-80$ mg of $CO_2$ per $dm^2$ of leaf area per hour | $15-35$ mg of $CO_2$ per $dm^2$ of leaf area per hour |
| Response of net photosynthesis to increasing light intensity | Difficult to reach saturation even in full sunlight | Saturation intensity reached at $1-4 \times 10^4$ lux |
| Green mesophyll and bundle sheath cells | Present | Absent |
| Major pathways of photosynthetic $CO_2$ fixation | C-4-dicarboxylic acid and reductive pentose phosphate cycles | Only reductive pentose phosphate cycle |
| Optimum temperature for photosynthesis | $30-45\,°C$ | $10-25\,°C$ |

**Table 11.2.** Optimum temperatures for growth (°C) for warm and cool season turfgrasses. (From Beard, 1973.)

| Type | Roots (soil) | Shoots (air) |
|---|---|---|
| Cool season | $10-18$ | $16-24$ |
| Warm season | $24-29$ | $27-35$ |

this $CO_2$-trapping ability along with other species with similar ability are called C-4 plants. These are distinguished from temperate species which are called C-3 plants. Table 11.1 illustrates some of the important characteristics which distinguish C-4 from C-3 plants.

In addition to having a greater photosynthetic capacity, C-4 grasses have different temperature optima from cool season grasses (Table 11.2). Terminology originating in the USA which has become adopted in turfgrass science is to call tropical grasses 'warm season' and temperate grasses 'cool season' grasses.

# TURFGRASS MAINTENANCE PROBLEMS IN INTERMEDIATE CLIMATIC AREAS

Warm season grasses are found generally between about 36–38° north and south of the equator but occur locally in isolated areas as far as 53° north (Harlan *et al.*, 1970). As distance from the equator increases, seasonal changes in both daylength and temperature become more apparent. Latitudes are reached where warm season grasses become dormant in the cooler 'winter' season. They lose colour and become generally unattractive as playing surfaces. Cool season grasses grow satisfactorily under these conditions (around 5–20°C) and it is a widespread practice especially in golf green maintenance in these areas to overseed with cool season grasses into a warm season turf. The overseeded grasses produce an attractive surface in the cool season but they are outcompeted by the warm season grasses when temperatures rise once more (Meyers and Horn, 1969; Schmidt, 1969).

The problem in maintaining year-round turf surfaces are at an extreme in the 'transition zone'. This is defined as areas where cool season grasses have difficulty in surviving high summer temperatures and also where warm season grasses are often killed in the winter because of their sensitivity to low temperatures. In the USA this zone lies approximately between 38° and 40°N depending upon elevation. It extends approximately 50–200 km north and south of a line running from Washington DC through Cincinnati (Ohio), Louisville (Kentucky), St Louis and Kansas City (Missouri).

## CHARACTERISTICS AND USES OF WARM SEASON TURFGRASSES

### Sources and important species

Of the 5000 or so grass species which occur on earth only about 40 can be categorized as turfgrasses. Of these only about 14 species are warm season grasses (Musser, 1950; Hanson *et al.*, 1969; Beard, 1973). Whereas most cool season species are native to Europe and Eurasia, warm season species are of diverse geographical origin (Hartley, 1950; Hartley and Williams, 1956; Beard, 1973). For example *Cynodon* spp. are native to central and southern Africa and southern Asia, *Paspalum* to South America, *Buchloe* and *Bouteloua* to the Great Plains of North America and *Zoysia* to Korea and Japan. Table 11.3 lists the more widely used warm season turfgrasses. Warm season turfgrasses which have limited or specialized uses are given in Table 11.4.

**Table 11.3.** The most widely used warm season turfgrasses and a key for their identification.

| Scientific name | Common name |
|---|---|
| *Buchloe dactyloides* | Buffalograss |
| *Cynodon* spp.[a] | Bermudagrass, Daccagrass, devilgrass, quickgrass, wiregrass, doobgrass, Floridagrass, kweek, couch and others |
| *Eremochloa ophiuroides* | centipedegrass |
| *Paspalum notatum* | Bahiagrass |
| *Pennisetum clandestinum* | Kikuyugrass |
| *Stenotaphrum secundatum* | St Augustinegrass |
| *Zoysia* spp.[b] | Zoysia, Manila lawngrass, Koreangrass, Japanesegrass, mascarenegrass |

[a] Species are shown in Table 11.5 and discussed in Bermudagrass section.
[b] Species are shown in Table 11.6 and discussed in Zoysiagrass section.

**Table 11.4.** Warm season grasses for limited use in special conditions.

| Scientific name | Common name(s) |
|---|---|
| *Axonopus affinis* | Common carpetgrass |
| *A. compressus* | Tropical carpetgrass |
| *Bouteloua curtipendula* | Sideoats gramma |
| *B. gracilis* | Bluegramma |
| *Digitaria dactyloides* | Queenslands blue couch, blue couch, serangoon |
| *Eragrostis curvula* | Weeping lovegrass |
| *Hilaria belingeri* | Curley mesquite |
| *Paspalum vaginatum* | Seashore paspalum |
| *Sporobolus cryptandrus* | Sand dropseed |

## *Key*

1. Leaf folded in sheath; auricles absent, creeping stolons present .......... go to 2
   - Leaf rolled in sheath; auricles absent, sheaths round .......... go to 8

2. Blades petioled, sheaths greatly compressed .......... go to 3
   - Blades not petioled, sheaths compressed .......... go to 4

3. Ligule a fringe of very short hairs; sheaths with few hairs at margins and summit of keel; collar smooth; blades smooth, wavy, acute tip; culms branching ........ St Augustinegrass (*Stenataphrum secundatum*)
   - Ligule a ciliate membrane; collar hairy; blade margins cilliate; stolons thick, short-noded ............. Centipedegrass (*Eremochloa ophiuroides*)

4. Ligule a fringe of hairs ................................................. go to 5
   - Ligule a very short, blue membrane ............................... go to 7

5. Collar continuous, broad, hairy; leaves and sheaths very hairy; blades 4–5 mm wide, long acuminate tip; V-shaped; rhizomes and stolons present
   ....................................... Kikuyugrass (*Pennisetum clandestinum*)
   - Collar continuous, narrow, smooth, sparingly ciliate; sheaths and blades smooth or sparingly hairy ............................................. go to 6

6. Blades 1.5–3 mm wide, glat, acuminate tip; scaly rhizomes and flat stolons present .................................... Bermudagrass (*Cynodon dactylon*)
   - Blades 3–6 mm wide, acute tip; rhizomes absent; nodes hairy, especially on stolons; stolons elongate with short internodes .........................
   ................................................... Carpetgrass (*Axonopus affinis*)
   - Blades 5–10 mm wide, flat, hairy at base ....................................
   ................................................... Carpetgrass (*A. compressus*)

7. Blades 1 mm wide or less; margin of ligule irregular; stolons and rhizomes long, slender ............ African Bermudagrass (*Cynodon transvaalensis*)
   - Blades 4–8 mm wide; margin of ligule entire; stolons and rhizomes short, thick ......................................... Bahiagrass (*Paspalum notatum*)
   (Vernation variable, may also fall in group with leaves rolled.)

8. Collar hairy, at least at base or on margins ...................... go to 9
   - Collar not hairy; rhizomes and/or stolons present; perennial; ligule a fringe of hairs; collar broad, continuous ......................... go to 13

9. Sheaths not hairy ..................................................... go to 10
   - Sheaths and blades hairy ............................................. go to 12

10. Strongly stoloniferous ............................................... go to 11
    - Rhizomes short, stout, scaly, blades 1–2 mm wide; collar broad with hairy margins .................................... Blue grama (*Bouteloua gracilis*)

11. Blades 2–4 mm wide; collar broad, continuous; strongly stoloniferous; ligule a fringe of hairs; leaf margin smooth ...............................
    ........................................... Japanese lawngrass (*Zoysia japonica*)
    - Blades 2–3 mm wide; collar only sparingly hairy (otherwise same as *Z. japonica*) ............................. Manila lawngrass (*Zoysia matrella*)

12. Rhizomes present ............. Sideoats grama (*Bouteloua curtipendula*)
 – Rhizomes not present ........... Downy bromegrass (*Bromus tectorum*)

13. Rhizomes present; blades smooth ............................................
 ................. Mascarene grass or Korean velvetgrass (*Zoysia tenuifolia*)
 – Rhizomes absent; stolons well developed; blades sparsely pilose, 1–3 mm
 wide, curly, grey-green ............... Buffalograss (*Buchloe dactyloides*)

# Individual turfgrasses

## *Buchloe dactyloides* (Buffalograss)

Buffalograss is indigenous to the Great Plains of North America. It grows from southern Canada to northern Mexico. The common name derives from the American buffalo (bison) for which it served as primary grazing on short or low-growing grasses. Buffalograss survived close grazing to the exclusion of other herbage species. Like smooth bromegrass (*Bromus inermis*), ecotypes of Buffalograss appear to be adapted to either southern or northern areas. Thus the origin of improved cultivars needs to be considered when selecting Buffalograss for turf purposes.

Buffalograss is a dioecious sod-forming grass which spreads by stolons. It is very drought resistant. During periods of severe, extended drought, it becomes dormant, but it will initiate new growth following rain or irrigation.

Buffalograss has a grey greenish colour, soft fine leaf texture and produces a dense turf. Seed is contained in low positioned burrs that contain one or two caryopses. Turf can be established from seed, sod or plugs. Seed is difficult to harvest and the burrs need to be dehulled, chilled and sometimes scarified to improve germination.

'Sharp', 'Texoka', 'Mesa', 'Prairie' and 'Oasis' are named cultivars. 'Prairie', and 'Oasis', are the most recent cultivars. Each must be established vegetatively.

Improved seeded populations are being developed and release of these is anticipated by the mid 1990s.

Buffalograss is an excellent choice for domestic lawns in subhumid and semiarid regions. Newer cultivars selected for density and colour are suitable for parks, highways embankments, golf course roughs and fairways.

## *Cynodon* species (Bermudagrasses)

Bermudagrass is the most versatile and most widely used warm season turfgrass. Harlan *et al.* (1970) report that the genus *Cynodon* tribe *Chlorideae* comprises nine species and ten varieties. These are shown in Table 11.5.

*Chapter 11*

**Table 11.5.**  Species and varieties of the genus Cynodon. (After Harlan *et al.*, 1970.)

| Species | Chromosome number |
|---------|-------------------|
| 1.   *Cynodon aethiopicus* | $2n = 18, 36$ |
| 2.   *C. arcuatus* | $2n = 36$ |
| 3.   *C. barberi* | $2n = 36$ |
| 4a. *C. dactylon* | $2n = 36$ |
|    b.   var. *afghanicum* | $2n = 18, 36$ |
|    c.   var. *aridus* | $2n = 18$ |
|    d.   var. *coursii* | $2n = 36$ |
|    e.   var. *elegans* | $2n = 36$ |
|    f.   var. *polevansii* | $2n = 36$ |
| 5a. *C. incompletus* | $2n = 36$ |
|    b.   var. *incompletus* | Rarely 36 |
|    c.   var. *hirsutus* | $2n = 18$ |
| 6a. *C. nlemfuensis* | |
|    b.   var. *nlemfuensis* | $2n = 18$, rarely 36 |
|    c.   var. *robustus* | $2n = 18, 36$ |
| 7.   *C. plectostachyus* | $2n = 18$ |
| 8.   *C. transvaalensis* | $2n = 18$ |
| 9.   *C.* × *magennisii* | $2n = 27$ |

Of the nine species of *Cynodon* described by Harlan *et al.* (1970) only four are of current interest for turf.

**1.**  *C.* × *magennisii* is a natural hybrid (triploid) between *C. dactylon* and *C. transvaalensis*. The commercial cultivar 'Sunturf' was selected as a clone by W.W. Huffine (Huffine, 1957). 'Sunturf' is a vigorous, fine-leaved grass characterized by a dark green colour, and a low growth habit. It spreads rapidly by stolons and short, usually shallow, rhizomes. Under closely cut golf green conditions the grass has a tendency for stolons to 'loop' or grow on top of the sward. These are easily controlled by vertical mowing and are not a problem in lawn turf.

**2.**  *C. incompletus* var. *hirsutus* is recognized to include the lawn grass known as *C. bradleyi* (Harlan *et al.*, 1970). Bradleyigrass has medium textured leaves of greyish green colour. It forms a shallow rooted stoloniferous turf. There are no rhizomes (Youngner, 1956; Harlan *et al.*, 1970).

Both 'Sunturf' and Bradleyigrass are known to be winter hardy in Oklahoma (37° north latitude) (Harlan *et al.*, 1970).

In Australia 'Blue couch' is the name applied to *C. incompletus* (McMaugh, pers. comm.).

**3.** *C. dactylon* (L.) is generally referred to as 'common' Bermudagrasss. Common Bermudagrass is a tetraploid ($2n = 36$) except that in var. *aridus* it is a diploid ($2n = 18$) and both diploid and tetraploids are found in the var. *afghanicum* (Harlan *et al.*, 1970).

'Common' is distributed worldwide. It is propagated as sod (turf), sprigs, stolons and seed. It is the only Bermuda turfgrass generally established from seed. 'Common' is vigorous although not as much as the hybrids discussed later. It forms a rather dense turf at heights of cut ranging from 10 mm to 40 mm. It establishes rapidly, requires medium to high fertility and is considered to be heat and drought tolerant. Where overseeding for winter colour is practised, it serves as the base grass, or seedbed, for cool season grasses. Like all Bermudas, 'common' lacks shade and cold tolerance; although selected types including 'U-3', 'Tufcote', 'Vamont', 'Midway' and 'Guymon' are more tolerant of lower temperatures than is usual.

'Common' Bermudagrass begins to lose green colour when temperatures drop below 10°C and turns a light tan colour when frost occurs. It remains dormant throughout the winter months (Youngner, 1959).

Improved seeded cultivars of 'common' include 'U-3', 'Sahara', and 'Cheyenne'. 'Sahara' is a polycross of clones selected for colour, compactness, drought tolerance and resistance to Bermudagrass mite. The seed produces a uniform, medium to fine textured turf of a blue-green colour. Like all Bermuda grasses, it is not shade tolerant and 'Sahara' is less cold hardy than many of the cold-tolerant clones that must be planted from vegetative plant parts. 'Cheyenne' is a medium to coarse textured plant that forms a dense turf.

'Tiflawn', 'Texturf 10', 'Santa Ana', 'U-3', 'Vamont', 'Midway', 'Tufcote', 'Wintergreen' and 'Windsorgreen' are some of the improved types of 'common'. All must be established vegetatively.

A brief description of selected cultivars or strains of *C. dactylon* is presented below:

a) Until the late 1960s, 'U-3' was the most cold hardy of the Bermuda grasses. It survives winters as far north as Cleveland (Ohio) and central Pennsylvania (40°N). 'U-3' is used for sports fields, golf fairways and lawns.

b) 'Texturf 10 (T47)' selected from *C. dactylon* came from the tenth tee at the Corsicana Country Club, Texas. It was collected by A.W. Crain in 1949, selected by Watson in 1951 and released by Holt from Texas A & M University in 1959. 'Texturf 10' has a dark blue-green colour, a medium leaf texture, a low growth habit and produces dense heavy rhizomes. It has good low temperature colour retention in the cool season and begins growth earlier than 'common' in the spring. It is used on sportsfields and recreational lawns.

c) 'Santa Ana' has a bluish green colour with medium leaf texture. It has excellent salt and smog tolerance, good wear tolerance and colour retention, but has poor winter hardiness. It is used for sportsfields and lawns.

**d)** 'Wintergreen', selected by McMaugh in Australia is a semi-dwarf, fine-leaved, mid-green plant with an upright habit of growth. 'Wintergreen' is highly stoloniferous but produces few rhizomes. It has a medium rate of spread, produces a dense turf that retains green colour longer than 'common', but is not tolerant of very low temperatures. The plant has a high fibre content (27%) and is exceptionally wear tolerant.

**e)** 'Windsorgreen' is another McMaugh selection which is an induced mutation from 'Wintergreen'. It grows slightly taller, produces fewer seedheads and grows more slowly than 'Wintergreen'. 'Windsorgreen' has a bright emerald colour, produces a dense turf and grows better at temperatures slightly higher than those best for 'Wintergreen'. The plant produces an excellent playing surface for golf courses and is sought for landscaping purposes.

**4.** *C. transvaalensis* (African Bermudagrass) is a low-growing, fine-leaved vigorous grass of good density. It is a diploid ($2n = 18$) that readily hybridizes with *C. dactylon* ($2n = 36$). There are several cultivars of *C. transvaalensis* with good turf properties. Among them are 'Elliot' from the Frankenwald Turf Research Station, Johannesburg, South Africa; 'Germinston'; 'Skaapplas Fine'; 'Florida'; 'Royal Cape;' 'Waverly'; and 'Uganda'. Some of the above selections may be natural hybrids with *C. dactylon*. They all possess the vigour, potential for thatch build-up and in some cases the characteristic yellowish green colour of *C. transvaalensis*. Few are used widely for turfgrass purposes. In general they require intensive management.

**5.** *C. dactylon* × *C. transvaalensis* hybrids. Natural and man-made hybrids ($2n = 27$) of these two species are often referred to as 'improved' Bermudagrasses. Next to 'common', they are the most widely used of the Bermudagrasses for sportsfields, playgrounds, golf greens, tees and fairways, racetracks and improved lawns.

Naturally occurring hybrids have been produced in east-central and South Africa. Also, during the Second World War when golf greens which had been established using *C. transvaalensis* were not maintained (especially in Florida, USA), hybridization occurred between native and introduced grasses. In the late 1940s and early 1950s, a number of selections were made by various investigators (Beard, 1973). Among the named selections were 'Bayshore', 'Everglade' and 'Pee Dee'.

In the early 1950s Burton, of the Georgia Coastal Plains Experiment Station, Tifton, Georgia, initiated a breeding and selection programme involving *C. dactylon* × *C. transvaalensis* hybrids (Burton, 1947). From this programme came 'Tiffine' in 1953, 'Tifgreen' in 1956, 'Tifway' in 1960, and 'Tifdwarf' in 1956. These are widely used special-purpose warm season turfgrasses. They are found worldwide on golf courses and sportsfields in tropical and subtropical areas. They are vigorous, spread rapidly, form a dense turf and are exceptionally wear tolerant. These hybrids require high soil fertility and are very heat and drought tolerant. They perform better than 'common' in shade, but are not considered to be shade tolerant. They have good cold tolerance but do not survive in marginal zones of adaptation as well as the more cold-tolerant Bermudagrass selections.

a) 'Tifgreen' has a medium to light green colour. It is used on golf greens, fairways and tees. Ultraviolet rays (from sunlight) appear to produce mutations which often give 'Tifgreen' greens a mottled colour as the mutants spread.

b) 'Tifway I' (the original release) is a dark green, medium textured plant that produces a very dense turf. The plant spreads peripherally producing a dense circle with spreading stolons and rhizomes.

c) 'Tifway II' is a selection from the original plant. It is resistant to nematodes and has a lighter green colour than 'Tifway I'.

d) 'Tifdwarf' is a fine-leaved, low-growing, medium to dark green plant. It grows slowly and is not as tolerant of traffic or of overseeding as is 'Tifgreen'.

e) 'Tiffine' is a medium to fine-leaved grass that tends to swirl and produces numerous seed heads. It is found in India and Indonesia. 'Tiffine' has been superseded for the most part following the release of 'Tifgreen'.


## *Eremochloa ophiuroides* (centipedegrass)

Centipedegrass originated in southern China and is sometimes called Chinese lawngrass (Beard, 1973). Seed of centipedegrass was found in the baggage of Frank N. Meyer, the USDA plant explorer who disappeared on his fourth trip to China in 1916 (Hanson, 1965).

Centipedegrass is a medium to coarse textured plant that spreads by short compact stolons; the stems may be red, yellow or green. It produces a dense, vigorous turf but is not wear tolerant. It is adapted to a wide range of soil types but is best suited to acid, infertile, sandy soils. It grows poorly on alkaline soils. Centipedegrass exhibits iron deficiency symptoms (chlorosis) at moderate pH levels (6.5–6.8) and must be treated periodically with an iron supplement. Light applications of nitrogen ($0.5 \, \mathrm{g \, m^{-2}}$) once or twice per year are generally considered adequate. This grass has a strong tendency to produce thatch if over-watered and over-fertilized.

Centipedegrass is adapted to warm humid areas. It has moderate shade tolerance and, because of its poor wear resistance, is used primarily for lawns. Because of its relatively slow, low-growing habit, centipedegrass requires less frequent mowing than Bermudagrass to maintain a well-groomed appearance. It has limited cold tolerance, similar to St Augustinegrass. The grass is propagated by seed and sod. According to Jensen, pioneer grower in Tifton, Georgia, USA, as much sod as seed is sold (Jensen, 1988, pers. comm.).

## *Paspalum notatum* (Bahiagrass)

Bahiagrass originated in warm temperate and subtropical regions of eastern South America. Bahiagrass is light green to yellowish green in colour. It spreads slowly from short stolons and rhizomes. It tends to form tufts unless planted thickly because of its habit of growth. The root system is deep and profusely branched, contributing to its excellent drought tolerance. Bahiagrass is easily identified by its inflorescence which consists of a rapidly growing stem (axis) bearing *two* racemes; as opposed to three to five racemes found on dallisgrass (*P. dilatatum*) or the seven to nine racemes found on vaseygrass (*P. urvillei*).

Both diploid ($2n = 20$) and tetraploid ($2n = 40$) plants of Bahiagrass occur. Bahiagrass is mostly apomictic but cross pollination does occur (Burton, 1948).

Bahiagrass is adapted to a wide range of soil conditions and is found growing in poorly drained heavy soils as well as in sandy, infertile, droughty soils. Low temperature tolerance and colour retention are somewhat better than St Augustine and centipedegrass (Beard, 1973). It is best suited to warm subhumid, subtropical zones and coastal areas with neutral or mildly acidic soils. The grass succeeds best with a low cultural regime but responds to fertilization.

Bahiagrass forms an open turf of coarse texture. It is extremely tough and wear resistant. It is used for lawns, roadsides and highway embankments. A major use is for airfields as a landing area and to control dust. When used for lawn purposes, it should be mown at a height of 40 to 60 mm. Higher heights of cut and infrequent mowing are standard practice for most other uses. The rapidity of seedhead growth requires the use of rotary rather than cylinder mowers.

Bahiagrass is established from seed. Seed is plentiful and readily available commercially.

'Argentine', 'Paraguay' and 'Pensacola' are named turf-type cultivars of bahiagrass. Of these, 'Argentine' is considered the best selection for turfgrass purposes (Beard, 1973).

## *Pennisetum clandestinum* (Kikuyugrass)

Kikuyugrass was first described as occurring in Kenya and the common name derives from a native tribe.

Kikuyu is a dioecious grass; it is also a very vigorous and rapidly spreading grass that forms a dense, tough sod under close mowing. It produces a light green, almost yellowish green turf. Kikuyugrass is heat and drought tolerant but has poor low temperature tolerance. Hence, it is restricted to high and low altitude tropical and low altitude subtropical regions.

Kikuyugrass is used for turf worldwide in the regions of its adaptation, such as Australia, New Zealand, Colombia, Mexico, Spain and the United States. It was introduced to Australia in 1919 and during subsequent years it was disseminated throughout the subtropical regions. It was introduced to most areas as a pasture grass and for erosion control but it soon invaded areas adjacent to golf courses, roadsides, lawns and general purpose turf areas (Ribo, 1964). In southern California, it is designated as a noxious weed.

Youngner (1961) citing Edwards (1937) and Weintraub (1953), stated Kikuyugrass was introduced into the USA in 1920. Ribo (1964) stated that the first stolons of Kikuyu were brought to the USA from New Zealand in 1927, and to Mexico in the late 1940s. Ribo reported that the grass was introduced into Mexico as an ornamental grass for parks and gardens. By the mid to late 1970s, it was found in almost 90% of the green areas in Mexico City as well as on wasteland, in ditches, on dumps and in cultivated fields.

Youngner (1961), quoting Edwards (1937), described three ecotypes of *P. clandestinum*. Two, which he named 'Kabete' and 'Molo', are fertile and produce viable seed. A third, 'Rongai' is male sterile. Youngner (1961) reported that sterile strains exhibit increased vegetative vigour over the fertile types. The sterile types grow taller and produce a thicker mat. However, in mixed swards, the fertile types appear to completely resist invasion of the sterile. Youngner (1961), quoting Narayan (1955), reported that *P. clandestinum* is a facultative apomict and that male sterility is recessive to male fertility.

Kikuyugrass spreads by stolons, rhizomes and seed. Seed is retained in the leaf sheath which surrounds the inflorescence. When dry, the short lateral shoots of the inflorescence break off and are easily disseminated by wind and water. Seed has a long dormancy period, and germination is uneven. Kikuyugrass is quite persistent even under adverse conditions (Youngner, 1961).

Flowering is increased by close mowing and by grazing. Stamen emergence is reduced by low temperatures but stigma emergence is only partially reduced (Youngner, 1961).

Hathaway (1979) reported three varieties of seed: 'common', 'Breakwell' and 'Whittset'. Only 'Whittset' is certified. Seeds are a shiny, dark brown colour about 3 mm in length. They have a thick, hard, seed coat. Germination is unaffected by freezing and thawing, or drying. Scarified seed germinates approximately 30 days sooner than untreated seed (Youngner, 1961). Youngner (1961) and Youngner and Goodin (1961) described observations on the ecology, morphology and control of Kikuyu. Ribo (1964) also describes control techniques.

Hathaway (1979) reported that established Kikuyugrass has a small fertilizer requirement and that soluble nitrogen applied before growth slows, or when growth is reduced by low temperatures, will enhance colour in late autumn. He did not recommend fertilizing during long periods of growth, or during prolonged periods of rainfall. Unless mown closely, fertilized lightly and watered sparingly, Kikuyugrass becomes severely thatched.

### *Stenotaphrum secundatum* (St Augustinegrass)

St Augustinegrass is native to the West Indies and is used for lawn purposes in the warm humid regions of the world. It is the most shade tolerant of the perennial warm season grasses but is among the least tolerant of low temperatures.

St Augustinegrass is a medium to coarse textured plant that spreads by stolons. It roots readily at each node and produces a soft spongy turf unless periodically verticut and topdressed. It has fair drought resistance but is inferior to Bermudagrass and Zoysiagrass in this respect (Beard, 1973).

St Augustinegrass grows exceptionally well on organic soils and is adapted to any well-drained moist, fertile and slightly acid soil. Chlorosis is a major problem when St Augustinegrass is planted on alkaline soils. Sources of iron must be applied for successful turfgrass growth and quality on these soils. St Augustinegrass tolerates limited traffic and when mown at 40 to 50 mm, irrigated, and maintained at a medium fertility level, produces a very satisfactory turf especially for domestic lawns.

Although intolerant of low temperature, the grass will retain a green colour at tempertures well below those that cause Bermudagrass and *Zoysia* to lose colour.

'Floratam', 'Raleigh', 'Bitterblue' and 'Floratine' are selections of St Augustinegrass. They are finer textured, more tolerant of low temperatures and have a darker green colour than the common type. They also carry more resistance to insects, especially chinch bugs, and disease, especially *Rhizoctonia solani* (brown patch).

St Augustinegrass is established by sod, sod plugs and by stolons. Seed is not available commercially.

### *Zoysia* species (Zoysiagrass)

Zoysiagrass is indigenous to southern China, Korea and Japan. Three species of *Zoysia* are used for turfgrass purposes: *Z. japonica*, *Z. matrella* and *Z. tenuifolia*. Two other species, *Z. sinica* and *Z. macrostachya*, have been found growing along sea coast areas of Japan and around salt-beds in South Korea (Engelke and Murray, 1989). Lee (1991) cites Yeam. (1987) and Yoo (1990) as describing a sixth species, *Z. koreana*, which is reported to be the most salt-tolerant species (Table 11.6).

**Table 11.6.** Zoysiagrass species.

| Species | Common name |
| --- | --- |
| *Zoysia japonica* | Japanese lawngrass, shiba, Korean lawngrass |
| *Zoysia matrella* | Manila lawngrass, koshun-shiba |
| *Zoysia tenuifolia* | Mascarenegrass, korai, Korean velvetgrass |
| *Zoysia macrostachya* | Oni-shiba |
| *Zoysia sinica* | — |
| *Zoysia koreana* | — |

All *Zoysia* species are sexually compatible and a number of interspecific hybrids have been produced. The chromosome number, $2n = 20$, is stable across species, and all species cross pollinate and are self fertile (Engelke and Murray, 1989).

Zoysiagrass is a major warm season turfgrass in the warm temperate and subtropical areas of the world. In the USA it is frequently planted in the 'transition zones'. Zoysiagrass is the most widely grown turfgrass in Japan. *Z. japonica* has the widest distribution. It grows as far north as 40° 10′ along coastal areas and in the inland mountainous areas (Kitamura, 1989).

Zoysiagrass is used for all types of lawns. It is also used on golf course tees, fairways and greens in Japan and in the transition zones of the USA. A number of attributes make it a desirable, low maintenance turfgrass. It tolerates drought, salinity and the extremes of temperature. It has a slow growth rate which reduces mowing frequency. Under low fertility conditions and with enough supplemental water to sustain growth it competes favourably with weeds. However, its growth rate makes it slow to establish and to recover from damage.

*Zoysia japonica*, although quite cold hardy, loses colour with slight frost and remains dormant until mid to late spring. Beard (1973) reports a loss of green colour at 10°C to 12°C. This characteristic detracts from its usefulness as a lawn grass in marginal areas of adaptation.

Zoysiagrasses are inferior to Bermudagrass in terms of economy in water use, depth of rooting, drought resistance and dehydration, and wear tolerance (Watson, 1989). Kim (1987) and Beard (1973) report excellent heat tolerance and good shade and salt tolerance.

*Zoysia japonica* (common name Japanese or Korean lawngrass) is the most cold hardy of the species used for turfgrass purposes. It is coarser textured and spreads more rapidly than the other two species. 'Meyer', 'El Toro', 'Midway' and 'Belaire' are named cultivars in the USA.

*Zoysia matrella* (common name Manila lawngrass) has a narrower leaf than *Z. japonica*. In fact, it is the intermediate of the three species, possessing intermediate leaf texture, low temperature hardiness and density. It grows best in the warmer regions of its adaptation. At present there are no named cultivars of this species.

*Zoysia tenuifolia* (common names mascarenegrass, Korean velvetgrass and korai) is the least cold tolerant, has the finest leaf texture and is the most dense of the three species. Left unmown, the grass clumps into small attractive hummocks.

'Emerald' is a hybrid cultivar resulting from a cross between *Z. japonica* and *Z. tenuifolia*. It was released in 1955. It possesses the non-fluffy growth habit and faster rate of spread of the *Z. japonica* parent with the finer leaf texture, higher density and dark green colour of the *Z. tenuifolia* parent (Hanson, 1965).

'Meyer', named after the famous USDA plant explorer, Frank N. Meyer, was originally selected by Forbes and Ferguson in 1940, and was released in 1951 (Forbes and Ferguson, 1947; Hanson, 1965). 'Meyer' is intermediate in leaf texture between *Z. japonica* and *Z. matrella*. It is the most widely planted zoysiagrass in the United States. 'Meyer', like all improved cultivars of zoysiagrass, must be established from vegetative plant parts. When strip sodded or plugged, weeds must be controlled to ensure good coverage.

## Special-purpose Warm Season Turfgrasses for Limited Use

There are some species which are not used extensively but which have adaptations (e.g. salt tolerance) which make them valuable for specific uses and situations (Table 11.4).

### *Axonopus affinis* and *Axonopus compressus* (carpetgrasses)

Carpetgrasses are indigenous to Central America and the West Indies (Hanson, 1965). They are coarse textured grasses with a light green colour. They are low-growing perennial grasses that form a medium dense turf. Carpetgrasses spread by stolons that root at each node and they have poor wear tolerance. They are well adapted to wet, infertile soils especially those of a sandy nature.

*A. affinis* is generally called 'common' carpetgrass, while *A. compressus* is known as 'tropical' carpetgrass. 'Common' is more widely distributed and is used to a limited extent as a turfgrass. 'Tropical' is more tolerant of droughty conditions (Beard, 1973).

The quality is similarly poor for the two species. One of the major detractions of carpetgrass in lawns is the production of long seedhead stems. These are best mown with a rotary mower at 25–50 mm. Because of its adaptation to infertile, acid soils it is suitable for use on highway embankments and for erosion control.

## *Bouteloua gracilis* (bluegramma)

Bluegramma is native to the Great Plains of North America. It is a densely tufted plant with distinct grey-bluish green leaves. Bluegramma has an extensive, but relatively short root system.

It spreads by short rhizomes and forms a medium dense turf. Bluegramma is of limited use as a turfgrass. It is sometimes planted as a lawn grass in semiarid areas. Its primary use is for erosion control along highways, and for land reclamation purposes.

Bluegramma is a good forage plant and is adapted to a wide range of soil conditions in the warm subhumid and semiarid zones. It is cold hardy and drought tolerant.

Bluegramma is established from seed and has a relatively slow rate of growth.

Improved turfgrass selections are under development at the University of Nebraska and at Colorado State University. There are a number of named forage-type cultivars but most of these are not suitable for turfgrass.

## *Digitaria dactyloides* (Queensland blue couch, blue couch or serangoon)

Blue couch is a fine-leaved (2–3 mm wide) blue-green grass used for lawns and putting greens in tropical regions of Australia. It is adapted as far as north as southern Malaysia and Asia, where it is called serangoon. Blue couch produces a dense, soft turf subject to thatching. Few attempts have been made to select improved strains. It is preferentially avoided by army worms. Queensland blue couch does not have widespread use on golf courses and is being rapidly replaced by 'Tifdwarf' Bermudagrass on golf greens (McMaugh, pers. comm.).

## *Hilaria belangeri* (curley mesquite)

Curley mesquite is native to the southern Great Plains of North America. It is found growing in heavy, deep clays and droughty sands in subhumid and semiarid regions of the southwestern United States, Mexico and in Central America.

Curley mesquite is stoloniferious, drought resistant and tolerates a 25 to 50 mm height of cut. Currently it has limited value for turfgrass purposes.

It is primarily a forage plant. However, Mancino and Kneebone at the University of Arizona are selecting turf types from an extensive collection.

## *Paspalum vaginatum* (seashore paspalum)

Seashore paspalum is native to the subtropical and tropical regions of north and south America. It grows along seashores (hence its name), and in brackish areas along coastal marshes. In the USA it grows in the coastal areas of North Carolina, South Carolina, Georgia, Florida and Texas. It is also found in Mexico and Argentina.

*Paspalum vaginatum* is highly salt tolerant. Henry *et al.* (1979) reported that 'Adalayd' and 'Futurf' were growing well in soil with an electrical conductivity (EC) of 2500 $\mu$S cm$^{-1}$, a level at which 'Santa Ana' Bermudagrass had weakened or died out. At a second site, a golf course fairway in San Clemente, California, 'Futurf' was growing in soils with EC readings ranging from 4000 to 5000 $\mu$S cm$^{-1}$. Earlier tests showed that 'Futurf' did not tolerate conductivity in excess of 5000 $\mu$S cm$^{-1}$.

In addition to its salt tolerance, Harivandi and Gibeault (1983) reported that, based on five years of research at the University of California, South Coast Field Station, *Paspalum vaginatum* had good tolerance to drought and high temperatures. It tolerates light shade and mowing heights as low as 7 mm. It is not as tolerant of low temperatures as is Bermudagrass (Ibitaya *et al.*, 1981). Ability to tolerate salinity and produce an acceptable turf are its most valuable characteristics.

## Miscellaneous Warm Season Grasses

*Bouteloua curtipendula* (sideoats gramma) and *Sporobolus cryptandrus* (sand dropseed) are native to the North American Great Plains. They are used for re-grassing rangelands, for erosion control along highway rights-of-way and in roughs and non-use golf course sites. They spread by seed. There are no improved turf types of either species. Their use for turf is limited.

Both grasses are drought tolerant and are adapted to warm, low rainfall areas where irrigation is impracticable. They have a tufted habit of growth, thin foliage and are established from seed.

*Eragrostis curvula* (weeping lovegrass) is indigenous to Africa. Its use is limited to roadsides, roughs and bunker facings on golf courses and on other untrodden areas for erosion control. It is a vigorous, tall-growing, perennial, bunch-type plant. It is adapted to warm subhumid and humid regions. Weeping lovegrass is established from seed, grows rapidly and is highly drought resistant. Weeping lovegrass has an extensive fibrous root system which quickly binds soil on slopes and other erodible areas.

None of these three bunch-type plants tolerates close mowing. Also in most cases, they should be mown only once or twice per year.

# REFERENCES

Beard, J.B. (1973) Warm season turfgrasses. In: *Turfgrass Science and Culture*. Prentice Hall, Englewood Cliffs, New Jersey, Chapter 4.

Burton, G.W. (1947) Breeding bermudagrass for the southeastern United States. *Journal of the American Society of Agronomy* 39, 551–559.

Burton, G.W. (1948) The method of reproduction in common bahiagrass, *Paspalum notatum*. *Journal of the American Society of Agronomy* 40, 443–352.

Burton, G.W. and Deal, E.E. (1962) Shade studies on southern grasses. *Golf Course Reporter* 30, 26–27.

Burton, G.W. and Elsner, J.E. (1965) Tifdwarf – bermudagrass for golf greens. *USGA Green Section Record* 2, 8–9.

Engelke, M.C. and Murray, J.J. (1989) Zoysiagrass breeding and cultivar development. In: Takatoh, H. (ed.) *Proceedings of the 6th International Turfgrass Conference*. Tokyo, Japan, pp. 423–426.

Forbes, I. and Ferguson, M.H. (1947) Observations on Zoysiagrasses. *The Greenkeeper Reporter*, GCSAA, Lawrence, Kansas.

Halfacie, R.G. and Barden, J.A. (1979). Chapter 6 of *Plant Metabolism in Horticulture*. McGraw-Hill, New York.

Hanson, A.A. (1965) *Grass Varieties in the United States*. Agricultural Research Service, United States Department of Agriculture. Agriculture Handbook No. 170. Superintendent of Documents, Washington, DC.

Hanson, A.A. and Juska, F.V. and Burton, G.W. (1969) Species and varieties. In: Hanson, A.A. and Juska, F.V. (eds) *Turfgrass Science*. American Society of Agronomy, Monograph 14. Madison, Wisconsin, pp. 370–409.

Harivandi, M. Ali and Gibeault, V.A. (1983) Fertilizing seashore paspalum. *California Turfgrass Culture* 33, 8.

Harlan, J.R., De Wet, J.M.J., Huffine, W.W. and Deakin, J.R. (1970) *A Guide to the Species of* Cynodon *(Gramineae)*. Oklahoma Agricultural Experiment Station Bulletin B-673.

Hartley, W. (1950) The global distribution of tribes of the Gramineae in relation to historical and environmental factors. *Australian Journal of Agricultural Research* 1, 335–373.

Hartley, W. and Williams, R.J. (1956) Centres of distribution of cultivated pasture grasses and their significance for plant introduction. In: *Proceedings of the 7th International Grassland Conference*, pp. 190–210.

Hathaway, P.M. (1979) Kikuyugrass – like it or not it's here to stay. *Golf Course Management*, January 1979, pp. 44–46.

Henry, J.M. and Gibeault, V.A. (1985) *Paspalum vaginatum* winter colour management study. *California Turfgrass Culture* 35, 4–7.

Henry, J.M., Gibeault, V.A., Youngner, V.B. and Spaulding, S. (1979) *Paspalum vaginatum* 'Adalayd' and 'Futurf', *California Turfgrass Culture* 29(2), Spring.

Holt, E.C. (1969) Turfgrass under warm humid conditions. In: Hanson, A.A. and Juska, F.V. (eds) *Turfgrass Science*. American Society of Agronomy Monograph 14. Madison, Wisconsin, pp. 513–528.

Huffine, W.W. (1957) Sunturf bermuda, a new grass for Oklahoma lawns. *Oklahoma Agricultural Experiment Station Bulletin* B-494.

Huffine, W.W. (1966) Bermudagrass around the globe. *Turfgrass Times* 1, 18–24.

Ibitaya, O.O., Butler, J.B. and Burke, M.J. (1981) Cold hardiness of bermudagrass and *Paspalum vaginatum*. *Southwest Horticultural Science* 16(5), 638–684.

Juska, F.V., Cornman, J.F. and Hovin, A.W. (1969) Turfgrasses under cool, humid conditions In: Hanson, A.A. and Juska, F.V. (eds) *Turfgrass Science*. American Society of Agronomy Monograph 14. Madison, Wisconsin, pp. 491–508.

Kim, Ki S. (1987) Comparative drought resistance mechanisms of eleven major warm season turfgrasses. Unpublished PhD Thesis. Texas A & M University and Seoul National University.

Kim, Ki S., Beard, J.B. and Seifers, S.I. (1988) Drought resistance comparisons among major warm season turfgrasses. *Green Section Record* 17(3), pp. 1–4.

Kitamura, F. (1989) The climate of Japan and its surrounding areas and the distribution and classification of zoysiagrasses. In: Takatoh, H. (ed.) *Proceedings of the 6th International Turfgrass Research Conference*. Tokyo, Japan, pp. 423–426.

Lee, Geung Joo (1991) Comparative salt tolerances of eight zoysiagrasses (*Zoysia* spp.). Unpublished Masters Thesis, Seoul National University, Seoul, Korea.

Meyers, H.J. and Horn, G.C. (1969) The two-grass system in Florida. In: *Proceedings of the 1st International Turfgrass Research Conference*. Sports Turf Research Institute, Bingley, pp. 110–117.

Musser, H.B. (1950) Special purpose turfgrasses. In: *Turf Management*. McGraw-Hill, New York, pp. 84–114.

Noggle, G.R. and Fritz, G.J. (1976) Photorespiration. In: *Introductory Plant Physiology*. Prentice Hall, Englewood Cliffs, New Jersey, p. 214.

Ribo, C.A. (1964) Kikuyugrass. Article based on Masters Thesis from Chipingo Agricultural College, Mexico.

Schmidt, R.E. (1969) Overseeding cool-season turfgrasses on dormant bermudagrass turf. In: *Proceedings of the 1st International Turfgrass Research Conference*. Sports Turf Research Institute, Bingley, pp. 124–126.

Watson, J.R. (1989) The USGA Research Committee activities with particular reference to the utilization, problems and perspective of Zoysiagrass in the USA. In: Takatoh, H. (ed.) *Proceedings of the 6th International Turfgrass Research Conference*. Tokyo, Japan, pp. 437–440.

Wilkins, M.B. and Hipkins, M.F. (1984) Photosynthesis. In: *Advanced Plant Physiology* Pitman, London.

Youngner, V.B. (1956) Evaluation of new bermudagrass species and strains. *Southern California Turfgrass Culture* 6, 13–14.

Youngner, V.B. (1959) Growth of U-3 bermudagrass under various day and night temperatures and light intensities. *Agronomy Journal* 51, 557–559.

Youngner, V.B. (1961) Observations on the ecology and morphology of *Pennisetum clandestinum*. *Orton* 16(III), 77–84.

Youngner, V.B. (1980) Zoysiagrasses in California. *California Turfgrass Culture* 30(1).

Youngner, V.B. and Goodin, J.R. (1961) Control of *Pennisetum clandestinum* (kikuyugrass). *Weeds* 9(2) 238–242.

# INDEX